Yale University
Mrs. Hepsa Ely Silliman
Memorial Lectures

Ross Granville Harrison
January 13, 1870—September 30, 1959

Organization and Development of the Embryo

by Ross Granville Harrison
edited by Sally Wilens

New Haven and London, Yale University Press, 1969

Library of Congress catalog card number: 69–15465
Designed by Helen Frisk Buzyna,
set in Garamond #3 type,
and printed in the United States of America by Connecticut Printers, Inc., Hartford, Conn.
Distributed in Great Britain, Europe, Asia, and Africa by Yale University Press Ltd., London; in Canada by McGill University Press, Montreal; and in Latin America by Centro Interamericano de Libros Académicos, Mexico City.

The Silliman Foundation

On the foundation established in memory of Mrs. Hepsa Ely Silliman, the President and Fellows of Yale University present an annual course of lectures designed to illustrate the presence and providence of God as manifested in the natural and moral world. It was the belief of the testator that any orderly presentation of the facts of nature or history contributed to this end more effectively than dogmatic or polemical theology, which should therefore be excluded from the scope of the lectures. The subjects are selected rather from the domains of natural science and history, giving special prominence to astronomy, chemistry, geology, and anatomy. The present work constitutes the forty-second volume published on this foundation.

Foreword

In the spring of 1949 Professor Ross G. Harrison delivered the series of Silliman Lectures at Yale University. It is a great loss that illness in his last years did not permit preparation of the lectures for publication. Fortunately he had written numerous drafts of preparatory notes as well as the final notes that he used during the lectures; these give the scope of the material he planned to incorporate.

The five Harrison papers that are reprinted in this volume are the major source of his Silliman Lectures, which were presented under the title "Organization and Development of the Embryo." As was his custom when preparing a manuscript or lecture, he had within reach for reference a vast number of reprints and books (mainly from his personal library).

It was my privilege to work with Professor Harrison as his research assistant and to assist him in the preparation of the lectures. This has facilitated the task of assembling the reprinted papers and certain excerpts from his Silliman Lecture notes which, in many cases, bridge the interval between date of publication of his papers and delivery of the lectures. All additions to the papers are either enclosed in brackets or indented; some notations labeled R.G.H. were inserted by him after publication.

A number of figures taken from the lists of projected lantern slides have been reproduced and added to the text, particularly to two papers ("Cellular Differentiation and Internal Environment" and "On the Status and Significance of Tissue Culture"), which had been published without illustrations. These figures are labeled according to the number of the lecture of which they were an integral part; for example, Fig. I–1 designates Fig. 1 taken from Silliman Lecture I; Fig. II–1 designates Fig. 1 from

Silliman Lecture II, etc. Figures used in Lectures I and II have been added to the "Cellular Differentiation" paper, which is relevant to both lectures.

The synopsis of the Silliman Lectures that precedes the reprinted papers is taken from Professor Harrison's program notes. It gives a cohesive plan for the presentation of his thoughts concerning embryonic organization and development based mainly on this reprinted combination of his writings and augmented by the work of other investigators. The papers which appear in this book as Chapters 1, 2, and 5 were used for preparing Lectures 1 and 2; Chapter 3 for Lecture 3; Chapter 4 for Lecture 4; Chapter 5 for Lecture 5; and Chapter 6 for Lecture 6. Professor Harrison's summary of the lectures, entitled "Recapitulation and Conclusion" in his synopsis, has been added as a final chapter. In this section it is apparent that, although the fundamental concepts he envisioned are beginning to receive experimental substantiation, the primary questions he asked concerning life and the living processes remain unanswered and continue to challenge the investigator.

Professor Harrison concluded his first lecture with seven lantern slides that depict his series of 46 stages of normal embryonic development of the spotted salamander *Amblystoma punctatum*. The reproductions of these figures (in Chapter 2) assist in the visualization of a developing embryo. Further assistance in understanding the stages of development of embryos mentioned in various experiments is provided by Professor Harrison's description of the stages; according to notes in his files, this account would have been rearranged and consolidated for publication. The short outline of the descriptions was prepared by me under his supervision—a few notations are recent additions.

I take responsibility for preparation of the supplementary illustrations and for compilation of the Silliman Lectures bibliography. I wish to thank the publishers who extended permission to reprint the papers and the added illustrations and the authors who allowed reproduction of their figures. The publishers include the following: Accademia Nazionale dei Lincei, Rome; American Association for the Advancement of Science, Washington, D.C.; The American Physiological Society, Washington, D.C.; A. & C. Black Ltd., London; Cambridge Philosophical Society, Cambridge; The Connecticut Academy of Arts and Sciences, New Haven; Gustav Fischer, Stuttgart; *Growth*, Philadelphia; The Rockefeller Institute Press, New York; The Royal Society of London; Springer-Verlag, Heidelberg; Stazione Zoologica Napoli; The Williams & Wilkins Company, Baltimore; The Wistar Institute Press, Philadelphia.

For valuable counsel and encouragement that made possible the preparation of this book for publication I am deeply grateful to the following members of the laboratory staff: Professors Edgar J. Boell, G. Evelyn Hutchinson, Clement L. Markert, Donald

F. Poulson, and J. P. Trinkaus, all of whom have a special regard for Professor Harrison and his work.

In conclusion I should like to express my deep appreciation of the interest and assistance of the Yale University Press, and in particular, my gratitude to Miss Jane Olson for her wise and skillful contribution to the production of this volume.

New Haven, Connecticut S. W.
June 1968

Preface

A complete bibliography of Professor Harrison's published papers is part of this volume. It will be noticed that between the years 1893 and 1948 the publication list totals 83 titles. His rate of production, 1.5 papers per annum, might arouse wonder in the minds of many researchers today, but this was in the era of the long monograph, when frequent publication of short papers was suspect. The five papers selected for reprinting give some insight into the past and the current significance of his work. They also reveal the beautiful precision of Professor Harrison's carefully planned and executed experiments and his genius for penetrating analysis of the findings. The fundamental nature of his interpretations of the "organized complexity" of the developing embryo has special value for the modern investigator who uses new techniques to attack old and unsolved problems. The central theme of his Silliman Lectures is "molecular configuration and orientation in the organization of living matter"; he looks to the chemist and physicist as well as the biologist for experimental substantiation of this concept.

Six notebooks written by Professor Harrison in Bonn, 1892–1894, at the Anatomisches Institut and in the last year also at the Johns Hopkins University give clear indication of the lines of research that were to lead to a life of brilliant and creative productivity. His curiosity concerning the process of embryonic development of a wide variety of vertebrate forms and his initial interest in the intricate ramifications of the embryonic nervous system are apparent. He obviously was stimulated by his fortunate association with Professor Moritz Nussbaum, under whose tutelage he worked on his Ph.D. problem. In the notebooks are records of his academic activities: the collecting of young embryos (*Rana, Triton,* and *Bufo*) in the environs of Bonn, from the aquarium in the Institut library and from a sink in the garden where certain frogs "had spent their entire lives"; and the observing and staging of normal embryos, which were preserved and sectioned for histological study. Many hours must have been spent on the carefully recorded and intensive survey of the literature that was related to the problems that held his interest.

After returning to Johns Hopkins he resumed working under William K. Brooks and was awarded his Ph.D. degree in 1894. Between the years 1893 and 1895 five papers on the development of the fins of the teleost fish were published; a short account of his doctoral thesis in English (1894) was followed by the definitive paper in German (1895). During this period the influence of Brooks and Franklin P. Mall was very great. Mall was a member of the prestigious Johns Hopkins medical staff with the "four doctors," Welch, Osler, Kelly, and Halsted, and was head of the Department of Anatomy in which Harrison was appointed instructor in 1896. A large photograph of Brooks with the inscription, "March 25, 1898, Yours truly, W. K. Brooks," and another of Mall had prominent places on the walls of his room in the Osborn Zoological Laboratory.

Following discovery and publication of the technique of tissue grafting by Born (1894–1897), Professor Harrison in 1897 reported his first grafting experiments at the 18th Annual Meeting of the American Morphological Society at Cornell University. The short account in *Science* (1898) shows his enthusiasm for this new technique. In a longer paper published in the same year, he writes:

> In the method of grafting we have a means of experimentation for which no substitute is offered. Born's discovery that certain amphibian embryos lend themselves with readiness to such operations is of especial importance in that it renders the method applicable to the study of developmental problems. How important this form of experiment is may be well appreciated from a careful study of the paper in which Born records the results of his work in this field.

In his paper "The Growth and Regeneration of the Tail of the Frog Larva," Professor Harrison describes the combining of two differently pigmented embryos of different species which resulted in a chimaeric frog that was carried through metamorphosis. In his Silliman Lecture VI notes he speaks of his interest in this experiment as follows:

> Chimaeras were studied in the first experiments in embryonic transplantation made by Born. . . . This was in the heyday of the Weismann theory of the continuity of

> the germ plasm. Thinking that such combined organisms might throw light on the question of the influence of soma on germ, I began a series of experiments to that end, which demonstrated that their actual execution would involve much detailed labor and require facilities for rearing the animals not at that time available. But the original experiments opened up other problems which could advantageously be studied by combining parts derived from two different species which differ from one another in pigmentation.

In the early years the method was applied to the study of the growth and regeneration of the tail, the relationship of developing muscle and the nervous system, and the outgrowth of the lateral line system of sense organs and the factors determining the path.

From 1900–1910 Professor Harrison's main interest was in the question of how the nervous connections become established during the development of the embryo. In the 1893 notebooks written during his first visit to Bonn, many pages give evidence of critical reading of the literature; also there are records of the study of his sectioned material, in which he traces in great detail, through successive stages, the origin and development of the teleost fins and muscle innervation. In the summer of 1899, on one of his return visits to Bonn, he finished the requirements and received his M.D. degree. At the Anatomisches Institut, and later at Johns Hopkins, he worked on his first long paper on the histogenesis of the nervous system in the salmon. In this paper, published in German (1901), he incorporated his early notes and further observations from the sections loaned to him by Professor Nussbaum. Numerous publications that follow this paper show the degree to which Professor Harrison became involved in the lively controversy for and against the neurone theory of nerve outgrowth.

In 1904 at the Eighteenth Meeting of the Anatomische Gesellschaft in Jena, he participated in the discussion of the papers by von Kölliker (con) and Schultze (pro) and reported his view against spindle cell origin of the nerve fiber based on his own observations of the naked developing fiber (devoid of spindle cells) in young frogs at the tadpole stage. Professor Harrison's photograph of the members who attended this meeting (he joined the Society in 1898) makes one aware of the impressive occasion and the fact that the young associate professor from Baltimore was in the presence of the most eminent and venerable German scientists of the day. These men and the few foreign members were anatomists, histologists, and pathologists with divergent interpretations of nerve fiber origin. Ramón y Cajal, who supported the neurone theory with his beautiful histological preparations, gives in his autobiography a dramatic account of the ordeal of receiving the 1906 Nobel Prize in Physiology and Medicine with the co-sharer, the eminent C. Golgi, who had opposing views.[1] Golgi emphasized this difference in his acceptance speech; according to Cajal this lack of tact caused great suffering to himself and enormous embarrassment to their Swedish hosts and the audience.

The years soon after the turn of the century were active and noteworthy ones for Professor Harrison. There were many publications resulting from his engrossing experimental program, as well as numerous lectures (his first Harvey Lecture in 1908), reports at meetings, the demands of the teaching schedule, and the assumption of new responsibilities. He still found time to assist with the founding of the *Journal of Experimental Zoology*. With E. B. Wilson, Morgan, and Conklin the arrangements were carried out, and the *Journal* started functioning in 1904. He was prevailed upon by these men and Mall to become the managing editor. Some of the initial clerical and business duties elicited this remark in volume 100, published in 1945: "It was an uneconomic use of the time of a scientific man and did at times interfere with research, but those were still pioneering days and this was the only feasible way of carrying on the enterprise" (*Retrospect 1903–1945*, p. xxi). No doubt this was counterbalanced by the pleasure of association with his colleagues who constituted the editorial board and the satisfaction of establishing with them the high standards of the *Journal.*

Perhaps the stamina which was always available to meet the unlimited demands of his burgeoning program is explained by Professor Harrison's early undergraduate experience as president of the Hopkins Tramp Club. John C. French, librarian-historian and author of *A History of the University Founded by Johns Hopkins* (1946, p. 246), wrote in his description of student societies:

> The members of the Tramp Club, who were at pains to explain that they were not what a later generation would have called hobos, were hardy and resolute walkers. They made weekly expeditions into Maryland and long interstate foot-journeys during vacations. Candidates for membership had to qualify by a thirty-mile hike in the company of two or more members of the Club. Their record for a day was a jaunt of sixty-nine miles accomplished within nineteen hours.

In a letter dated June 1949, after seeing Professor Harrison in Baltimore, French asked, "Do you possibly owe your present vigorous health to the strenuous exercise prescribed by the group?" The long walks and elaborately planned camping trips

1. Santiago Ramón y Cajal, *Recollections of My Life,* translated by E. Horne Craigie, Memoirs Amer. Phil. Soc., Vol. VIII, Part II (1937), pp. 544–55.

with many distinguished colleagues, and especially mountain climbing were often the source of his fondest reminiscences.

Prior to 1904 Professor Harrison's study of the peripheral nerves was based on observations of their development in successive stages of normal embryos. This was followed by an extensive program of experimentation in which surgical methods were used in operations upon embryos (*in vivo*) to investigate the neural crest as the source of Schwann sheath cells (1904, 1906); the bearing of transplanted limbs on development of nerves (1907); the behavior of spinal nerves in "altered surroundings" (1910). In a 1912 paper[2] the purpose of these experiments is given in this manner:

> By a rigorous process of elimination, done by means of operations upon frog embryos, it was possible to show that only the nerve cells, and not the sheath cells, had anything to do with the formation of the axone of the nerve fiber. It was to find a definite solution of the first named problem that led me in 1907 to the experiment of completely isolating small pieces of living embryonic tissue in culture media, especially in clotted lymph, observing their behavior under these conditions outside the body.

Another undertaking in 1907 was Professor Harrison's move from Baltimore to New Haven. This review of the event was given by Professor William Lyon Phelps at the 1939 Commencement in his customary role of presenting candidates for honorary degrees:

> In 1907 a very good thing happened to Yale. Dr. Harrison came hither as Bronson Professor of Comparative Anatomy. The chair was well named, for he had been teaching first at Bryn Mawr, and then at Johns Hopkins. Equipped with a Johns Hopkins Ph.D. and also a charming wife and an M.D. degree acquired at Bonn on the Rhine, the Sheffield mansion became his workshop and the house afterward called Mory's his home. In a few years he created the first university department at Yale, with the perilous honor of serving on both the Sheff and Academic faculties as well as those of the Graduate School and the School of Medicine.

In his laboratory on Hillhouse Avenue he continued the tissue culture experiments. The first report of this work was read before the Society for Experimental Biology

2. "Biology and Medicine," in *Physician and Surgeon, 34* (1912), 49–65.

and Medicine in New York, May 22, 1907, and published in the *Anatomical Record,* June 1, 1907. The classic paper "The Outgrowth of the Nerve Fiber as a Mode of Protoplasmic Movement," published in 1910, gives a full account of the experiments. As has been emphasized many times, Professor Harrison was the first to grow animal tissues in culture outside the body, and he needed this technique for crucial evidence of the "outgrowth" of the nerve fiber from a single ganglion cell. What has not been stressed is that this was a miracle of achievement for one man working alone. The preparation of the tissues, the routine of sterilization, countless hours of careful observation of growth and movement of the structures within each culture, the records and the exquisite drawings of the live material must have meant a tremendously strenuous program.

Professor Harrison carried out another series of experiments in which "the three principal factors involved in the cultivation of tissues, viz. the tissue itself, the fluid medium, and the solid support" were varied to permit observation of cell movement and reaction. Published in 1914, this paper marked the end of his utilization of tissue culture. The method had now attracted many investigators who developed the technique, adding refinements of apparatus and media which led to its use as a standard method of research in histology, pathology, and serology—particularly in the study of tumors. Many recent significant and beneficial applications have come from its use by investigators in both biological and medical laboratories. Of the highest importance is the virus culture research of Dr. John F. Enders, which made possible the discovery of the poliomyelitis vaccine. For this work he shared the Nobel Prize in Medicine with Robbins and Weller in 1954.

It has been recorded that in 1917 the Nobel Prize Committee voted the award to Professor Harrison but it could not be given that year because of the World War.[3] It is paradoxical that the Committee later failed to recognize Professor Harrison as the originator of the tissue culture method or to recognize the importance of growing animal tissues, nervous and non-nervous, outside the body. This is apparent both in the 1933 review of their previous decision and in their official account of the award to Enders, in which credit is given exclusively to Alexis Carrel, then at the Rockefeller Institute. The fact is that in 1912 Carrel was awarded the first Nobel Prize in Medicine for the United States "in recognition of his works on vascular suture and the transplantation of blood vessels and organs."

In the Introduction of his Silliman Lectures, Professor Harrison tells of a pleasant interlude that occurred soon after his arrival in New Haven:

3. *Nobel, the Man and His Prizes,* "The Prize in Physiology and Medicine," p. 245, G. Lilijestrand, edited by The Nobel Foundation (Norman, University of Oklahoma Press, 1951).

> One of my first duties at Yale when I came here in 1907 was to look after the Silliman Lecturer, who that year was the late William Bateson of Cambridge University. Seven years before, after a diapause of 35 years, the science of what is now called Genetics, a name given to it by Bateson himself, had started on its modern development, one of the most remarkable movements in biology of our day.
>
> Bateson's lectures were notable and gave an account of a young and vigorous science which has since been built into an imposing edifice. My reason for referring to it here, aside from my friendship and regard for Bateson, is that it is closely related to the subject of the present lectures, embryology. The two fields deal with two different phases of the reproduction of organisms, each being complementary to the other.

Since these words were written, an even closer alliance between the two fields has taken place. Most significant is the formulation of the structural chemistry of the gene, which profoundly stimulates investigation of the patterns of selective protein synthesis. His deep interest in the expression of molecular configuration in living protoplasm made Professor Harrison a pioneer molecular model-builder of protein constituents—at first using brown cork balls (the beautifully colored atoms were not available) and wire bonds cut to suitable length from knitting needles. Aside from their scientific usefulness, the symmetrical relations of the crystalline structures delighted him. He also made and designed numerous models in wax and wood for study and demonstration of the experimental findings that formed the basis for his rules concerning the symmetry of developing organs such as the limb and ear.

For the *Yale Alumni Weekly* (*17*, 1907, 198–99) Professor Harrison wrote a report on *The Bateson Lectures. Review of the Year's Course on "Problems of Genetics,"* (Silliman Lectures Foundation). This was his first publication as Head of the Department. Another report of his appeared in the same journal in 1914 (*23*, pp. 667–72), in which the new Osborn Memorial Laboratories and staff were described with great pride. Records show his large correspondence with the architects and numerous conferences in their New York office; attention was paid to the smallest detail in construction and furnishings.

The move to the new laboratory marked the end of the first third of the years devoted to active research. In subsequent years Professor Harrison used his skillful adaptation of the transplantation technique to study embryo morphogenesis. The fledgling research on the origin of the fins of the salmon was followed by interest in a related structure, the amphibian extremity—the forelimb. While still a student in Bonn, Professor Harrison had read a paper by Julia B. Platt ("Ontogenetischen Differenzierung des Ectoderms Necturus," *Arch. f. mikr. Anat.*, *43*, 1894) and had listed the many findings in his 1894 notebook. The tenth finding had a check mark: "Lateral von der Vorniere ensteht die Anlage der vorderen Extremität (Lateral to the pronephros evolves the rudiment of the forelimb). This organ and the ear were to keep him close to his microscope through many operating seasons.

In some biographical notes Professor Harrison included this account of his work:

> Later became interested in the symmetry relations of organisms. Organs that occur on both sides of the body, such as the limbs and the ear, are in two forms, rights and lefts, which are the mirror image of each other. The cells composing these organs in early stages are essentially alike in that any part of the rudiment can give rise to any part of the organ. By grafting these organ rudiments in different stages of development and varying their orientation, it is found that they give evidence of polarization, first in a head-tail direction and later dorso-ventrally. This can be explained by the assumption that certain particles in the living protoplasm are definitely oriented and that they behave, in this respect, as do molecules in certain liquid crystals.

In the large-scale program of the final years, Professor Harrison continued and extended the ear symmetry experiments.[4] He gave equal time to the study of various organs such as the eye and the limb by interchanging component tissues between embryos of different species (his second Harvey Lecture, 1933). A wide range of developmental problems concerning these and other organs provided substantial doctoral theses for his students; in many instances their initial experiments were the inspiration for their future research. Professor Harrison cited many experiments, his own and those of his students, in his Silliman V Symmetry Lecture and summarized them as follows:

> The significance of all this detail, which may seem trivial, is that it gives crucial evidence that these two different systems, ear and forelimb, are made up of units arranged in a repeat pattern which gradually become more and more definitely oriented. These facts and many others that might be added necessitate the assumption that the processes of development leading up to the final form of the organism are but the expression of the forces involved in the molecular configuration of the egg protoplasm.

4. A lantern slide showing a tabulation of experiments and the evaluation of factors that influence embryonic ear development was projected in Silliman Lecture II. This table has been added to the reprinted 1945 paper, "Relations of Symmetry."

The excerpt that follows is taken from Professor Harrison's last paper—his part in the Memorial Meeting at Woods Hole (1948) for his friend, Frank R. Lillie:

> The phase of Dr. Lillie's life assigned to me for consideration has to do with his relations to the National Academy of Sciences and the National Research Council. Now I find that the records of these services are bound up in annual reports and minutes of meetings, material as dry as dust and as soulless as stone. . . . It is not possible, of course, to separate the character and personality of a man from his aims and achievements, and so we must look behind the published records to get the full story of his life, even as regards his formal relations to the organizations he served.

In these brief background notes I have attempted to tell of Professor Harrison's aims and achievements with the hope that they reflect his personality in some measure. His contribution to the world of science brought him many honors, national and international. These include membership in the Accademia Nazionale dei Lincei (1929), the Royal Society, London (1940), and the Academie des Sciences, Institut de France (1946). In 1956 he was awarded the Accademia Nazionale dei Lincei International Prize of the "Antonio Feltrinelli" Foundation (Biological Science and Application). He served on the Board of Scientific Advisers of the Rockefeller Institute for Medical Research from 1939 to 1953, and for most of these years he was Chairman of the Executive Committee; after 1953 he was a member of the Board of Trustees. The scientific stature achieved by his students is a monument to his influence as preceptor. The special quality of his gentleness, reserve, and distinction elicited from students, colleagues, and the many others with whom he came in contact through the years a warm response compounded of affection and respect. He was the "Chief" in every sense of the word. All of this is poignantly told by Victor C. Twitty, devoted student and friend, in the book written a short time before Twitty's death.[5]

His formal relations to the organizations he served were best conveyed by his friend President Charles Seymour when he conferred the honorary Sc.D. degree on Professor Harrison soon after he assumed the role of emeritus professor:

> Revered and beloved leader, eminent scholar, selfless and inspiring personality, for the services you have rendered to this university in its laboratories, its schools of the arts, science, medicine—undergraduate and graduate—for your investigations and discoveries, and for the example you have given to American universities

5. V. C. Twitty, *Of Scientists and Salamanders,* San Francisco, W. H. Freeman, 1966.

as the perfect scientist, Yale honors herself in conferring upon you the degree of Doctor of Science, admitting you to all its rights and privileges.

Later a second honor came with the invitation to present the Silliman Lectures. Their preparation was unavoidably postponed for several years because of his demanding schedule as Chairman of the National Research Council, which from 1938 to 1946 divided his days between Washington and New Haven. Professor Harrison enjoyed enormously the close liaison with Dr. Frank B. Jewett, President of the National Academy of Sciences, and the members of various committees who worked with him on vitally important projects during the Second World War.

Professor Harrison was a magnetic figure to scientists who came to his laboratory from all parts of the world. Many of them were house guests in the beautiful Harrison home and enjoyed the gracious Old World atmosphere created by Mrs. Harrison. In his first visit to Bonn, Professor Harrison was introduced by Professor Nussbaum to Ida Lange, who happened to be visiting her sister, whose husband was a prominent psychiatrist and member of the medical staff at the University. The young couple were married in Altona, Germany, on January 9, 1896. After some months in Naples at the Zoological Station, they moved to Baltimore. Mrs. Harrison's perfect dinner parties were, in some measure, due to Professor Harrison's avocation as wine-maker —his admirable red and white wines became internationally famous. These two people were unusually gifted in the art of making their guests comfortable, whether they were "homeless" students rescued during the holidays or friends entertained on more formal occasions.

To the laboratory he planned and its recent towering addition, Professor Harrison bequeathed a lasting heritage.

Sally Wilens

Synopsis of the Silliman Lectures

Lecture I. *Introduction*

Living matter, its properties and degrees of organization—cells and organisms; reproduction of living units; the principal events of development; descriptive and experimental methods of investigation

Lecture II. *The egg and early stages of development*

Polarity and symmetry of the egg; its finer structure; nucleus and cytoplasm; the whole from a part and a single individual from two; embryonic induction—whole organism, central nervous system, sense organs

Lecture III. *The autonomy and dependence of cells and tissues in development*

Tissue culture; tissue affinity; dissociation and reconstitution in sponges and hydroids; involution and reconstitution in nature; reproductive cycle in slime molds; pigment cells and color patterns

Lecture IV. *The nervous system*

Gross form and its relation to the embryonic axis; differentiation of nerve cells and fibers; conditions of outgrowth in tissue cultures and in the embryo; equilibrium between periphery and nerve centers; functional and form regulation in spinal cord and brain; regeneration

Lecture V. *The symmetry of organisms*

Fundamental asymmetry masked; asymmetry of proteins and protoplasm; rights and lefts; situs inversus; mirror symmetry of paired organs; experiments on the development of the ear and limb; crystal analogies; the paracrystalline state

Lecture VI. *Development and growth in complex systems*

Growth of parts in relation to the whole (heterauxesis); organisms composed of parts from two or more species (chimeras); mutual effects on growth rates and size—experiments on eye, limbs, and other organs; functional effects—experiments on heart and central nervous system. Recapitulation and conclusion

Contents

1

Cellular Differentiation and Internal Environment

It may not be inappropriate in beginning this discussion to refer to a passage from Psalm 139, which seems to be germane to the subject.

14. I will praise Thee; for I am fearfully and wonderfully made: marvelous are Thy works; and that my soul knoweth right well.
15. My substance was not hid from Thee, when I was made in secret, and curiously wrought in the lowest parts of the earth.
16. Thine eyes did see my substance, yet being unperfect; and in Thy book all my members were written, which in continuance were fashioned, when as yet there was none of them.

There are several remarkable things about this text. First, it refers to development as a personal experience; and, indeed, all of us have been through it. This in itself should arouse our curiosity, but unfortunately we have no recollection of the adventure. It is also to be noted that the process of development is described here as a sort of epigenesis. "And in Thy book all my members were written, which in continuance were fashioned, when as yet there was none of them." It does not picture development, in the naive manner, as a mere unfolding or "*evolutio*" of something already present.

This paper was presented in a Symposium at Stanford University, June 1939, to commemorate the advances made during the 100 years after formulation of the cell theory by Schleiden and Schwann. Reprinted from *The Cell and Protoplasm,* Publication No. 14 of The American Association for the Advancement of Science, Washington, D.C., The Science Press, 1940, pp. 77–97.

While this symposium deals with the cell in all of its aspects, the topic assigned to me, "Cellular Differentiation and Internal Environment," clearly concerns the development of the organism and implies that the treatment of the subject should be in terms of the cell as a unit. With this I am wholly in agreement, for in my opinion none of the attacks that have been made upon the cell theory have shaken it to any appreciable degree.

It seems to be a characteristic of the human mind to describe the phenomena of nature in terms of units or particles. Physics, chemistry, and crystallography have all profited immensely by this device, even though modern physics tells us that nature cannot be fully described in such terms alone. Such theories are oversimplifications and are sooner or later modified; witness the atomic theory of the 19th century. But it may be doubted if science can advance without the aid of over-simplified hypotheses.

The cell theory differs from other particulate theories in the circumstance that the units can actually be seen under the microscope and sometimes even by the naked eye. Each unit is composed of two main constituents—nucleus and cytoplasm. However, the fact of individual visibility does not mean that there are no difficulties or uncertainties connected with the theory.

One of the notions that has frequently aroused opposition is the concept of the organism as a cell state made up of quasi-independent or autonomous units, the cells. The inadequacy of this concept was emphasized by Conklin, who declared that there is something in the organization of the individual that makes it more than the sum of its parts. While this statement in itself may be unobjectionable, the mystical haze that often surrounds the word "organization" has no place in science.

I doubt if there is any astute and serious observer who thinks of particulate units at any level as wholly independent of one another. The *relations* of the particles are part of the system, and it is their behavior in relation to one another that constitutes "organization." These relations are amenable to experimental investigation and are usually expressed in some kind of a configuration. This occurs at all levels, and may concern electrons, atoms, molecules, cells, or individual organisms. No particle or unit can be wholly understood or its behavior predicted unless its reactions with others are taken into consideration. The cell theory of the organism concerns, then, not only the cells as independent units but also their structural and functional relations to each other.

Whitman's notable essay on "The Inadequacy of the Cell Theory of Development" (1893) is often cited as opposed to the cell concept of the organism. However, this paper is in essence a plea for the recognition of a microstructure as the basis of organization, and is in no sense the negation of the cell as a living unit, although "cellular interaction" as an explanation of formative processes, is summarily dismissed as a term of reference. "If the formative processes," says Whitman, "cannot be referred to cell-division, to what can they be referred? To cellular interaction? That would only be offering a misleading name for what we cannot explain; and such an answer is not simply worthless, but positively mischievous, if it put us on the wrong track. Loeb's experiments in heterogenesis [heteromorphosis?] furnish a refutation of the interactive theory. The answer to our question may be difficult to find, but we may be quite certain that when found it will recognize the regenerative and formative power as one and the same thing throughout the organic world. It will find, as Wiesner has so well insisted, a common basis for every grade of organization, and it will abolish those fictitious distinctions we are accustomed to make between the formative processes of the unicellular and multicellular organisms. It will find the secret of organization, growth, development, not in cell-formation, but in those ultimate elements of living matter, for which *idiosomes* seems to me an appropriate name."[1]

The cell theory may be subjected to various experimental tests of its adequacy, many of which have been devised since Whitman wrote his essay. In the sense of those who emphasize "organization" as the chief characteristic of living creatures, "inadequacy" may be measured to a limited extent by the differences in behavior between cells associated in the organism and those acting alone or in small groups. This has been done by means of tissue culture and other types of isolation, including separation of blastomeres. A cell from the neural tube of an embryo, isolated in a suitable medium, forms a nerve fiber, and a presumptive heart cell may develop striations and beat rhythmically by itself (Burrows, 1912). However, the nerve fiber does not function as such without connections, and the heart muscle cell does not form an organ that pumps blood. On the other hand isolated blastomeres may give rise to a whole organism. Thus a cell, when isolated, may do either more or less than it would presumably have done if left in normal relations to its associated elements.

That isolated cells often achieve so little is no indication of limitation in what they can do, since the conditions under which they are isolated are too restricted. Until such cells are tested under a great variety of conditions, or better still, under all such as are possible, their full capacity will not be known.

One of the more concrete matters of disagreement over the cell theory concerns the mode of connection between cells in the organism. Are they completely separated from each other by membranes or is their internal substance or cytoplasm continuous? Much importance has been attached to alleged protoplasmic bridges between

1. Whitman (1893), p. 657.

cells and to the supposed syncytial nature of the tissues. Adam Sedgwick (1894) once described the vertebrate embryo as an organism with nuclei lying in a continuous reticulum, using this misconception to bolster up an erroneous theory of the histogenesis of the nerve fiber. That this question is not of such fundamental importance as some would attribute to it is shown by many considerations.

While the eggs of insects and many other arthropods proceed through their early development as syncytia without cell membranes, the cells in other embryos are from the first sharply divided from one another. There may be regional cytoplasmic differentiations in single cells that are as sharply defined as cells are from one another, but as division proceeds the various regions become separated from one another by the cell walls.

Certain tissue elements, such as striated muscle fibers in vertebrates and arthropods, are multinucleate and surrounded by a common membrane. Connective tissue cells are often described as forming a syncytium, but their ability to move about independently and the difficulty of resolving optically the delicate connections between fine cell processes invite caution in accepting this interpretation. Cells growing in culture media outside the body have thrown light upon this question. While fibroblasts, muscle cells, neuroblasts have often been described as forming syncytia, it has been shown that supposed nerve nets may often be resolved later by separation of the apparently fused processes (Harrison 1910; Levi 1917).[2] Similar observations have been made on mesenchyme (Lewis 1922) and heart muscle (Levi 1925).

Viewed pragmatically, the experimental analysis of the processes of development is greatly facilitated by considering the embryo as made up of discrete units associated with each other in various degrees of intimacy.

In relation to development, the most important fact in connection with the cell theory is that each organism begins its independent existence as a single cell, which, barring parthenogenesis, is the product of fusion of two unlike cells, one derived from the body of each parent. This is the narrow bottle-neck through which we have all passed, the book in which all our members were written, when as yet there was none of them.

Since every organism develops in a certain environment within rather narrow physical limits, its developmental processes can be properly considered only in relation to this environment, as Child has already emphasized. In arranging this symposium Taylor has distinguished between external and internal environment, and has left to me the task of interpreting the meaning of the latter. I shall therefore define

2. For the contrary view see Bauer (1932).

the internal environment of a cell as the sum of all those factors with which it is in relation that lie within the organism but outside of the cell in question. These factors affecting the differentiation of the cell change with the successive phases of the development of the organism.

Strictly speaking, a single cell has only an external environment. However, the germ cells that unite to form the fertilized egg do have in a certain sense an internal environment inside the body of the parent organism. When they are cast loose they cease for a time to have such an environment, and it is only after cleavage begins that the resulting cells acquire a new internal environment in their sister cells. For a short period this is the only internal factor that counts, there being as yet no general *milieu intérieur.* However, during this period, which in the vertebrate embryo includes segmentation, gastrulation, and neurulation, the most important differentiations occur. Later, after the establishment of a circulating medium and the primitive nerve paths, an internal environment of the ordinary sort is again present.

The ovum, as it grows, acquires a large amount of cytoplasm which enters into the composition of the embryo, though much of the material may be mere inclusions of a purely nutrient nature. So long as the egg is part of the maternal body, its exact relations to the contiguous cells of the parent organism, and, secondarily, to the circulating medium are of prime importance in setting up its organization or configuration. At this time the egg is either attached to the maternal tissues at one pole or is tightly enclosed in a capsule lined by follicle cells. In many, if not in all cases, but especially in those of the first group, the material is laid down in the egg around a polar axis but with marked gradients along this axis, by which a visible polarity is established in the egg. The nucleus lies on this axis but usually eccentrically placed with reference to the egg as a whole. Substances emanating from the nucleus may therefore act vectorially on material entering the egg from without.

At maturation, when a large amount of fluid is discharged from the nucleus into the cytoplasm, and accompanying or following the entry of the sperm, a major commotion occurs, during which there is a lively streaming of cytoplasmic materials, resulting in their partial rearrangement. After this many eggs become visibly bilaterally symmetrical, while others show a clear asymmetry. It is at this period that the egg loses its original internal environment and is discharged either to the exterior or to the oviduct of the maternal parent, where a greater or lesser part of its development may take place.

The polar differentiation acquired while the egg was part of the maternal organism may last in part through maturation and fertilization in spite of the streaming and rearrangement that occurs then. After fertilization the new arrangement of ma-

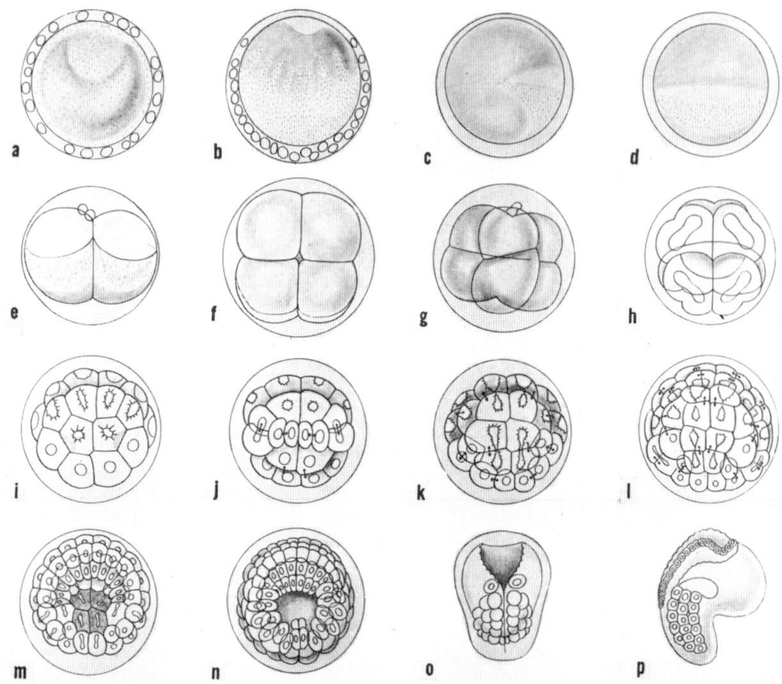

FIG. I–1. Rearrangement of substances in the living ascidian egg, *Cynthia partita,* following fertilization. (a) unfertilized egg: "test cells" within the chorion (outer membrane), upper clear cytoplasm derived from germinal vesicle, inner gray cytoplasm surrounded by peripheral layer of yellow cytoplasm containing yellow granules (mitochondria). (b) egg five minutes after sperm entry at the lower pole: clear cytoplasm visible at animal pole, streaming of yellow cytoplasm toward vegetal pole with exposure of the gray yolky cytoplasm. (c) side view of egg before first cleavage: on the right, subequatorial yellow crescent with clear crescent above and gray crescent on the left; between areas of yellow and gray crescents, the area of gray yolk. Later in development cells in the yellow crescent give rise to mesoderm, the gray crescent to neural plate and notochord, gray yolk to endoderm, upper region to ectoderm. (d) posterior view of the same stage as (c). (e–l) various views of cleavage stages showing distribution of varicolored cytoplasm. (m, n) cell pattern in two early gastrulation stages. (o) dorsal view of early tadpole with neural plate, neural tube and tiers of mesenchyme and muscle-forming cells. (p) side view of tadpole, later stage: toward the left the neural tube, mesenchyme, and muscle rudiments; gastral endoderm in bulge on the right. After Conklin, 1905.

terials follows a certain pattern peculiar to the species, as is well exemplified in the eggs of ctenophores, molluscs, annelids, and ascidians [Fig. I–1], though less definitely marked in many other forms. When cleavage occurs the resulting cells are thus already to a certain extent differentiated from one another, as measured by what they do in forming the primitive organs of the embryo.

Little is known with regard to how these differentiations come about, but it may be assumed that they are related in some way to the physico-chemical properties of the material entering into the composition of the egg and to the direction from which they enter it. Of first importance in this connection are the protein molecules, which to the best of our knowledge are polarized and asymmetric. They would thus tend to orient themselves within the cell as they increase. Other substances, either combined with the proteins or separate, would tend to concentrate near one or the other pole of the egg according to their electrical charge.

While the asymmetry of protein molecules does not preclude them from combining to form crystals of a high degree of symmetry, as in the case of pepsin, it may be assumed that, in some way related to this molecular asymmetry, there arises sooner or later in the development of the embryo a macroscopic asymmetry, which characterizes most if not all organisms and is merely masked by a superficial bilateral or radial symmetry.

Superposed upon the polar arrangement there are also differences between cortex and deeper portions of the egg cytoplasm. This, too, may be due primarily to molecular orientation at the surface. In this cortical or ectoplasmic layer, which is usually free from inclusions, we probably have the characteristic protoplasm of the species in its purest form. The behavior of this material in early stages of development, i.e. during maturation and segmentation, has been followed by many observers in a great variety of eggs, since G. F. Andrews (1897) first described the amoeboid spinning movements in the cortical layer of starfish and sea urchin eggs. In the eggs of *Beroë* a very peculiar streaming of this layer takes place during cleavage (Spek 1926) [Fig. I–2]. Everything points to the importance of this cytoplasmic constituent in the early processes of development (Just 1939).

The use of the term protein molecule in the foregoing requires some further explanation. As stated by Chambers, the proteins extracted from cells or those in dead cells are different from living protoplasm and are doubtless much less complex in their configuration. Such arrangements as occur in the living are upset at death or by the chemical processes used in separating the constituents known as pure proteins. These changes are, so far as we know at present, irreversible, so that study of the

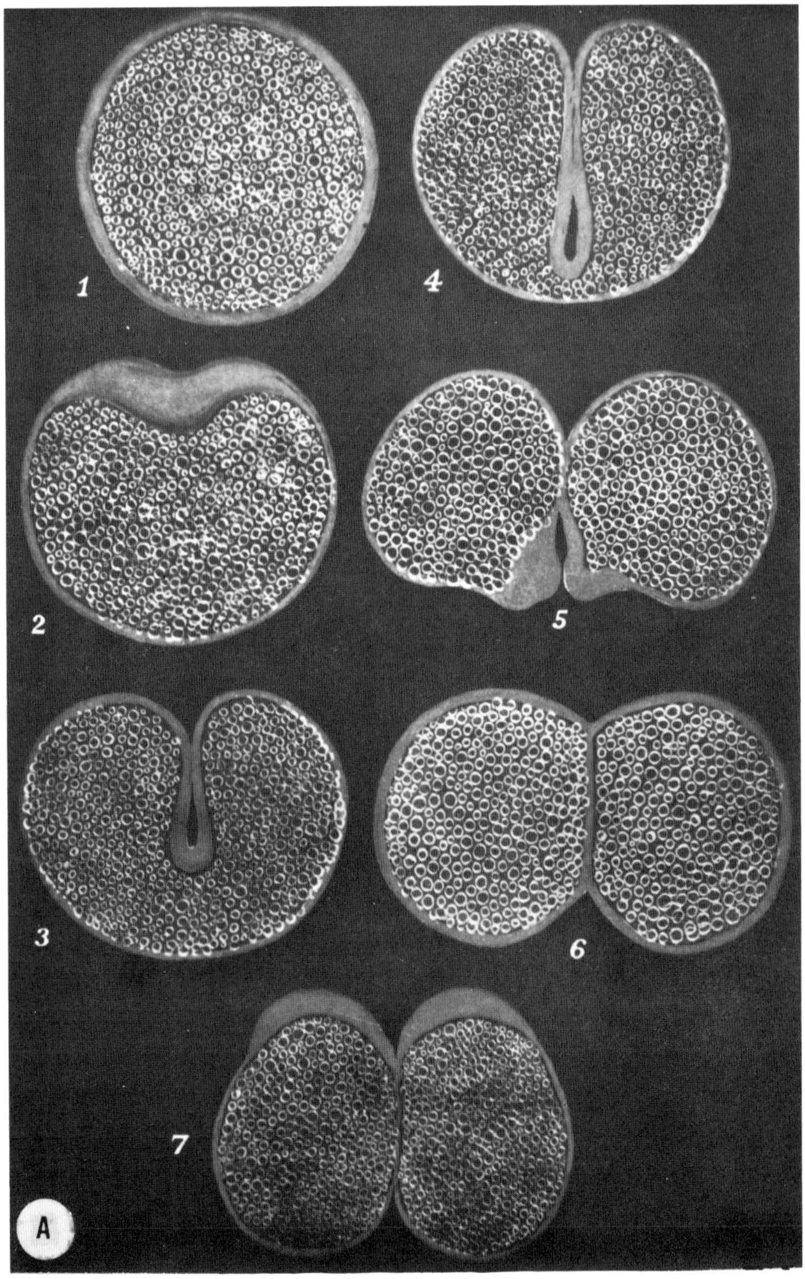
1
4
2
5
3
6
7
A

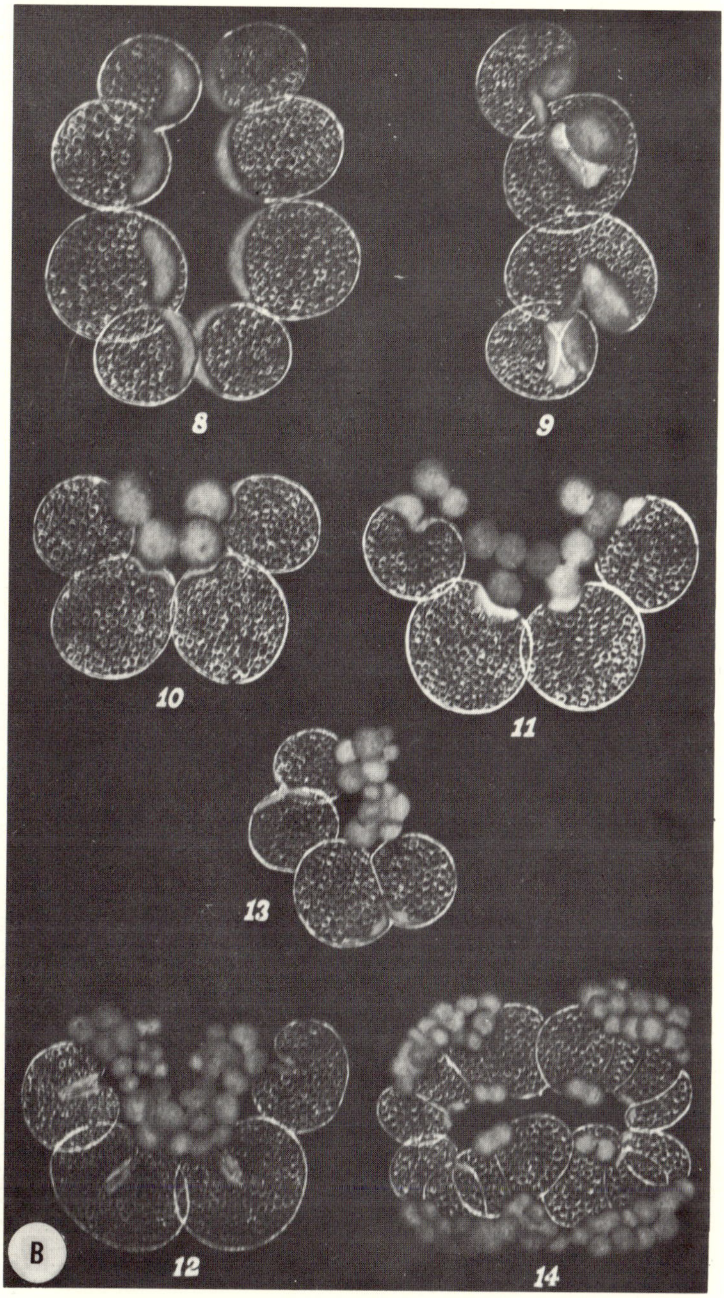

FIG. I–2. A. The development of the Ctenophore egg, *Beroë ovata,* as seen with dark-field illumination; the clear cytoplasmic cortex appears green. B. After a series of changes in the egg due to active streaming, the green cortical cytoplasm, at the end of cleavage stages, becomes isolated and forms the micromeres. Later in development the micromeres give rise to the ectoderm including the swimming plates or costae. After Spek, 1926.

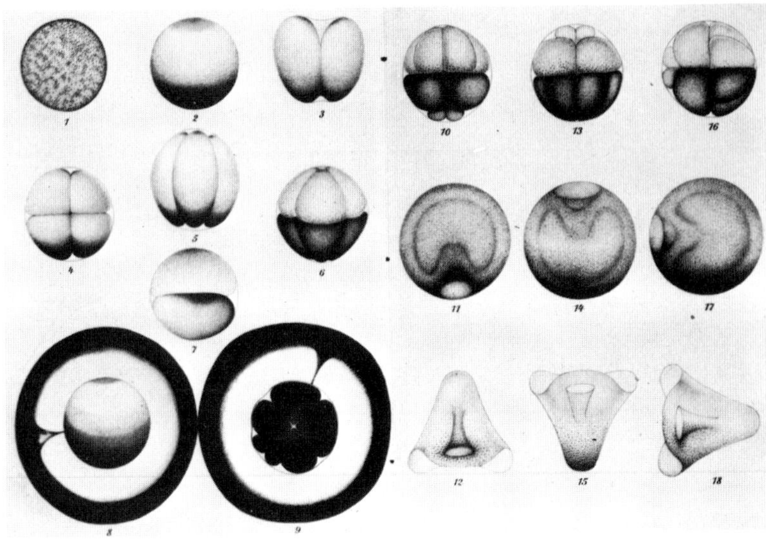

FIG. I–3. Maintenance of the primary egg axis and normal embryonic development of sea urchin (*Arbacia*) eggs after centrifugation: (8) uncleaved, (9) 16-cell centrifuged eggs, the india ink-stained funnel or micropyle at the animal pole with micromeres at the opposite vegetal pole as in non-centrifuged eggs. Variations in the plane of stratification of visible substances; centripetal, centrifugal, or the "side" shown by the position of the red pigment layer (dark area). Micromere formation at the vegetal pole which is the site of gastrula invagination, and later embryonic differentiation as pluteus are independent of disarrangement of visible cell substances following centrifugation. After Morgan and Spooner, 1909.

properties of pure proteins alone can never give an adequate picture of living protoplasm, though knowledge of these substances is essential to an understanding of it.

Both the arrangement of materials in the ovarian egg and the later arrangement assumed after maturation and fertilization show regional differences, which must be the result of a sorting out process. This requires energy and, according to some, even intelligence. It may be related in some way to the bioelectric potentials of both the ovum itself and the surrounding maternal tissues.

In the orientation of the protein molecules of the egg cell, due to their dipole character, a paracrystalline configuration is set up, and large numbers of molecules of water and electrolytes enter into definite relation with it. There are also associated carbohydrates and lipoids. Some of the material is built up into the living protoplasm, and the composition of this protoplasm shows regional differences corresponding to the main axis of the egg. The ground cytoplasm of the egg varies along this axis, retaining throughout, however, the configuration characteristic of the species. In addition to the material that becomes a part of the ground cytoplasm, there is also other material which forms aggregates that do not enter into the composition of the cytoplasm so intimately but take on granular or vesicular form and constitute visible inclusions, such as pigment granules, oil droplets, mitochondria, and yolk.

The inclusions are also redistributed in a definite way after fertilization and since they are parceled out to particular cells which give rise to particular organs, they have often been termed organ-forming substances. Experiments in centrifuging eggs show, however, that they usually play no such role, for they may be distributed very abnormally by this treatment without producing abnormalities in the resulting embryo [Fig. I–3].

In accordance with the foregoing, the polarity of the egg may be based upon three factors: the dipole character of the protein molecules and their resultant orientation, regional gradational differences in the composition of these molecules, and regional differences due to cytoplasmic inclusions. The first is a polarity of orientation (*Richtungspolarität*), the third a polarity of stratification (*Schichtungspolarität*) (Boveri 1901; Driesch 1908; Przibram[3] 1913), while the second is a polarity in which probably both orientation and stratification are concerned [see pp. 195–98]. The position of the nucleus in the cytoplasm also has a definite relation to the polarity of the egg.

According to the hypothesis here adopted, the polarity of the egg, and hence that of the resulting organism, is referable primarily to the orientation of large molecules, the arrangement of which is determined in relation to their internal environment within the maternal body as well as by their mutual relations. The asymmetry of the organism may likewise be assumed to follow ultimately from the asymmetric configuration of the molecules [see p. 171]. While these relations are normally fixed, they may, however, be inverted by external factors, as in the production of *situs inversus* (Spemann 1906; Pressler 1911) or in the inversion of the polarity of the *Fucus* egg by means of directed light and other physical factors (Whitaker 1931, 1936).

While the correlatives of the form and structure of the adult organism are to be sought in the molecular characteristics of the substances composing the egg—and these must be considered as persisting throughout all stages of development—a more definite macroscopic topographic relation between egg and organism than existed before is entered into after the streaming that follows maturation and fertilization comes to repose. To what extent this secondary arrangement is correlated with the

3. Op. cit., p. 35.

differentiation of the embryo is a question to which many have addressed themselves. There is much variation in the answer according to the species of egg studied.

Rapid changes take place in the egg cytoplasm during maturation and fertilization and immediately thereafter. If fragments are removed from the egg in the early part of this period, the embryo may develop without defect except in size, especially when the egg is cut vertically, i.e. parallel to its axis. As the time interval between fertilization and removal of the fragment increases, the defects become more marked and more specific. These statements are based upon experiments with eggs of nemerteans (Wilson 1903; Zeleny 1904), *Dentalium* (Wilson 1904), and ascidians (Dalcq 1932). One of the most striking cases of defect after removal of portions of the cytoplasm of an unsegmented egg is found in the ctenophore, *Beroë ovata* (Driesch and Morgan 1895; Fischel 1903; Yatsu 1912). In this form there is a definite relation between the position and the amount of the cytoplasm removed and the number of rows of swimming plates present in the embryo which develops from the remainder.

Experiments of this kind show that there are regional differences in a given egg and also differences between species of eggs with respect to their germinal localization. There is likewise variation in the extent to which defects in the early stages of development may be made up by the material that is left, and this holds both for the unsegmented egg and the egg which has undergone cleavage. Driesch thought all cells of the sea urchin blastula to be alike, as measured by what they can do in the course of development (prospective potency), and that what they actually do is a function of their position in the whole. On the other hand, gastropod, annelid, ctenophore, and ascidian eggs have been characterized as mosaics, in which the role of the constituent cells is fixed from the beginning and cannot be changed. Cleavage has correspondingly been described as indeterminate or determinate and the course of development in the respective classes as regulatory or mosaic.

It is obvious that eggs do differ in this respect, but no egg is wholly in either class. If mosaic development were the rule my theme would end here, for subsequent changes in the embryo would then take place according to the initial structure of the germ, and could not be modified by the internal environment.

In the egg of the molluscs *Ilyanassa* (Crampton 1896), *Dentalium* and *Patella* (Wilson 1904, 1904a), each cell when separated from the rest gives rise to just what it would, were it left in place in the embryo. Here, then, the internal environment of blastomeres may be assumed to have no effect on differentiation. Wilson found, however, that each fragment of an unfertilized egg, when divided vertically and fertilized, develops into a whole larva, thus showing regulation within the single cell. On the other hand, removal of the polar lobe results in a deficiency of the apical organ and the post-trochal region that is not made up. The case of *Tubifex,* an oligochaete, which like the gastropods and scaphopods has a spiral cleavage and a mosaic development, is nevertheless interesting in showing a considerable amount of regulation under certain conditions (Penners 1924, 1925). When the polplasm is divided equally at the first cleavage, which takes place frequently as a result of low temperature, both cells contain teloblast material and the result is a double embryo (*duplicitas cruciata*). When the AB cell is killed by ultraviolet radiation, the CD half, and probably also the D-quarter when all three remaining cells are destroyed, give rise to whole embryos. The above facts imply considerable powers of regulation in these eggs with highly determinate cleavage and also that in such eggs the cells constituting a mutual internal environment do normally restrict the activity of neighboring elements.

The ascidian egg, which according to Conklin (1905, 1905a, 1911, 1931) holds to a very rigid cleavage mosaic, shows considerable lability before fertilization. It may then be deprived of large amounts of cytoplasm and still develop normally (Reverberi 1931; Dalcq 1932). Even two complete embryos, one of which is merogonic, may sometimes be obtained from a single egg. There is, however, indication of bilateral symmetry at this early stage. Evidences of regulation in the fertilized egg have been found by von Ubisch (1938), who describes the development of a single giant larva from two fused eggs. In embryos derived from one of the first two blastomeres there may also be a restricted regulation, as shown by the movement of myotome cells around the tip of the notochord to the defective side (Cohen and Berrill 1936). Evidence of induction of cerebral vesicles with sensory cells by means of certain underlying cells has also been obtained (Rose 1939; Vanderbroek, unpublished[4]). Von Ubisch (1939), however, failed to find such induction in *Ascidiella.*

Again, in *Clavelina,* which probably also develops through a strictly "determinate" cleavage, there are certain totipotent reserve cells which enter into the winter buds in unorganized clumps and later give rise to new ascidiozooids (Spek 1927; Brien-Gavage 1927). In regeneration also, the several types of cells and tissues show much lability in their powers of differentiation. These facts indicate that the strictly mosaic cleavage pattern is only of temporary significance as far as germinal localization is concerned. There is a possibility that it is due to nothing more fundamental than the peculiar aggregate state of the cytoplasm at the time [p. 102].

Likewise in the ctenophore egg, which also undergoes determinate cleavage, there are evidences of regulatory development in the formation of the aboral sense organ

4. Cf. Dalcq (1938), p. 104.

and the gastral pouches (Driesch and Morgan 1895). Chun's (1892) observations on the later stages of development indicate that the partial larvae which develop out of separated blastomeres are transformed by a mode of "postgeneration" into whole ctenophores.

While the three best known types of "mosaic" eggs thus show that "determinate" cleavage may lead to an embryo having some lability with respect to differentiation, an examination of the so-called "regulation" eggs shows, on the other hand, a considerable degree of localization within the cytoplasm. In the Hydromedusae (*Clytia, Laodice*) single blastomeres of the eight-cell stage may give rise to a whole organism (Zoja 1895–6) [cf. Teissier—R.G.H.]. However, in many hydrozoan eggs there is a concentric stratification with definite ectoplasmic and endoplasmic layers, so arranged that each of the early blastomeres receives both substances, and it is only later that cells of pure ectoplasm or pure endoplasm arise by cleavage parallel to the surface.

It was with reference to the sea-urchin egg that the [notion of the—*R.G.H.*] complete totipotence of individual blastomeres was first maintained. This interpretation goes back to Driesch (1891, 1892), who found that blastomeres isolated in the two- or four-cell stage, while segmenting as fractions, close over in the blastula, gastrulate normally, and give rise to normal plutei of reduced size. Later Driesch claimed that even single blastomeres of the 32-cell stage would gastrulate and he showed conclusively that two eggs or blastulae could be fused together so as to give a single normal embryo. However, the sweeping conclusion drawn by Driesch (1894, 1929[5]) from the above facts, that "the prospective value of any blastula cell is a function of its position in the whole" does not hold. The experiments on which it was based were not sufficiently controlled to warrant the conclusions, nor were the methods available at that time adequate to carry out the requisite experiments.

Driesch's generalization implies that cells in their differentiation must be subjected to the action of surrounding cells, their internal environment. That is, the term "position" means their location with reference to other cells. However, the gastrular invagination begins at a certain point on the spherical blastula, and surrounding cells are carried in with it; this must mean either that there is something peculiar to the cells at that point, which is contrary to Driesch's hypothesis, or that the point is determined by environmental factors of an accidental nature. The latter alternative is highly improbable.

It remained for von Ubisch (1925, 1925a, 1931, 1933, 1933a, 1936, 1939) and

5. Driesch (1929), p. 55.

Hörstadius (1928, 1935, 1937, 1938, 1939), by means of ingenious devices, to explore the ways in which blastomeres act upon one another in gastrulation and in various processes of differentiation. The methods used are similar to those first successfully employed with amphibian eggs, i.e. transplantation and vital staining, besides methods of separating individual blastomeres and parts of older embryos with due regard to the exact origin of each cell or cell group used in the experiments.

The sea-urchin egg shows graded differences along its main axis, about which the cytoplasmic substances appear superficially to be radially symmetrical (Boveri 1901a). The polar axis is not shifted by centrifuging but its degree of fixity is a matter of dispute. Tennent, Taylor, and Whitaker (1929) claim that in *Lytechinus* it may be shifted by cutting the egg, while Hörstadius (1937) finds that in the egg of *Arbacia* it is fixed. Development of egg fragments shows, too, that even in the unfertilized egg there is a considerable degree of cytoplasmic localization. While the eggs of but few echinoderms have actually been shown to be bilaterally symmetrical, with dorso-ventral polarization, this is probably true for the group generally, though the plane of symmetry may be shifted and the polarity of the dorso-ventral axis fixed by such external factors as stretching (Boveri 1901; Lindahl 1932), centrifugation (Runnström 1926; Lindahl 1932a; Pease 1939), or chemicals (Lindahl 1932). However, the details are not all in agreement. When the egg of *Dendraster,* the Pacific Coast sand dollar, is centrifuged, the centripetal pole is found to lie on the ventral side though frequently not in the median plane. Pease concludes from his experiments on this form that two factors are concerned in fixing the dorso-ventral axis, one, located in the cortical layer, which is not movable, and the other, located in the endoplasm, which may be shifted by centrifuging. The direction of the axis, and with it the median plane, is the resultant of the interaction of the two factors, the ventral side being in the region of maximum reactivity. When eggs are cut in the frontal plane during cleavage stages, the dorso-ventral polarity is reversed in the dorsal half (Hörstadius and Wolsky 1936). The bilateral symmetry of the embryo usually first becomes visible when the primary mesenchyme cells group themselves on the two sides, prior to laying down the larval skeleton, in response to stimuli emanating from the ectoderm (Driesch 1896).

By means of isolation and grafting experiments Hörstadius (1935) has made an exhaustive analysis of the action of the material in the successive layers along the primary axis of the egg [Figs. II–1, II–2, II–3]. Of great interest are the cases in which the segmenting egg of from 16–24 cells is divided "horizontally" by tiers, which are allowed to develop by themselves or in various combinations. The results show that the combination must contain material from both the animal and vegetative regions

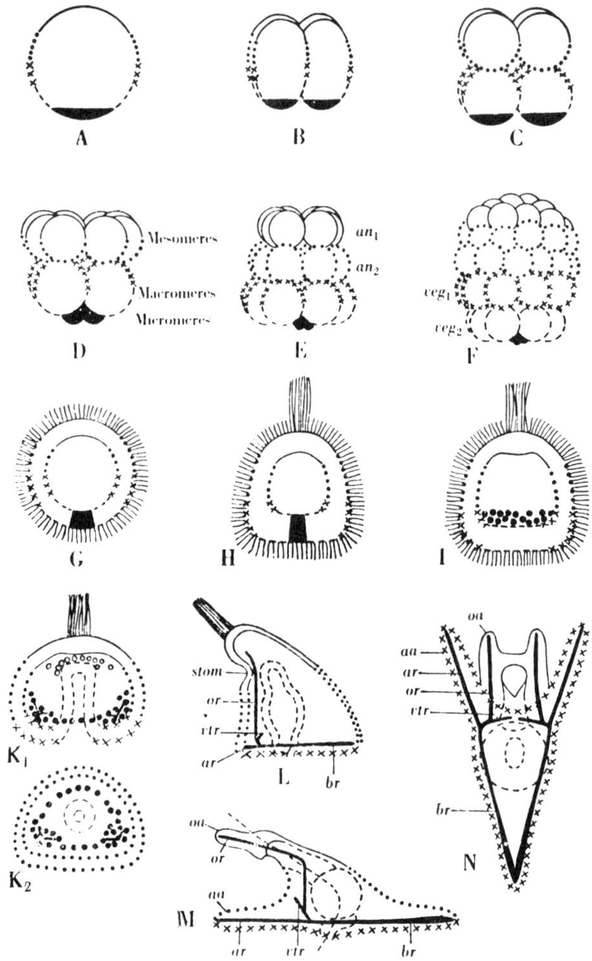

FIG. II–1. Diagram of the normal development of *Paracentrotus;* an, mesomeres; veg, macromeres. A, uncleaved egg. B, 4-celled stage. C, 8-celled stage. D, 16-celled stage. E, 32-celled stage. F, 64-celled stage. G, young blastula. H, later blastula. K_1 gastrula with: secondary mesenchyme and two triradiate spicules formed. K_2: transverse optical section of the same gastrula; bilateral symmetry established. L, stomodaeum invaginating; the prism stage. M, pluteus larva from the left side; the broken line indicates the position of the egg axis. N, pluteus from the anal side, aa, anal arm; ar, anal rod; br, body rod; oa, oral arm; or, oral rod; stom., stomodaeum; vtr, ventral transverse rod. After Hörstadius, 1935.

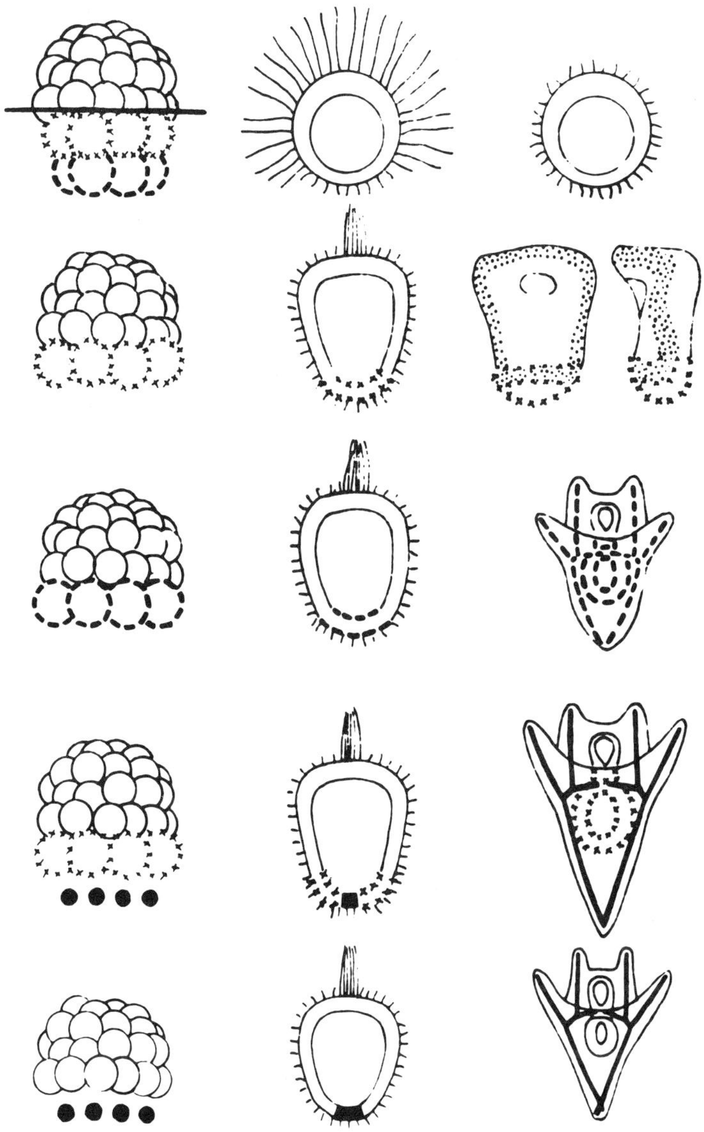

FIG. II–2. Diagram of the influence of the vegetative region consisting of macromeres; veg_1 (crosses) and veg_2 (broken lines) with micromeres, black and the animal region mesomeres; an_1 and an_2 (continuous lines) in eggs with equatorial cleavage, i.e. no veg material included in the animal region. After Hörstadius, 1939.

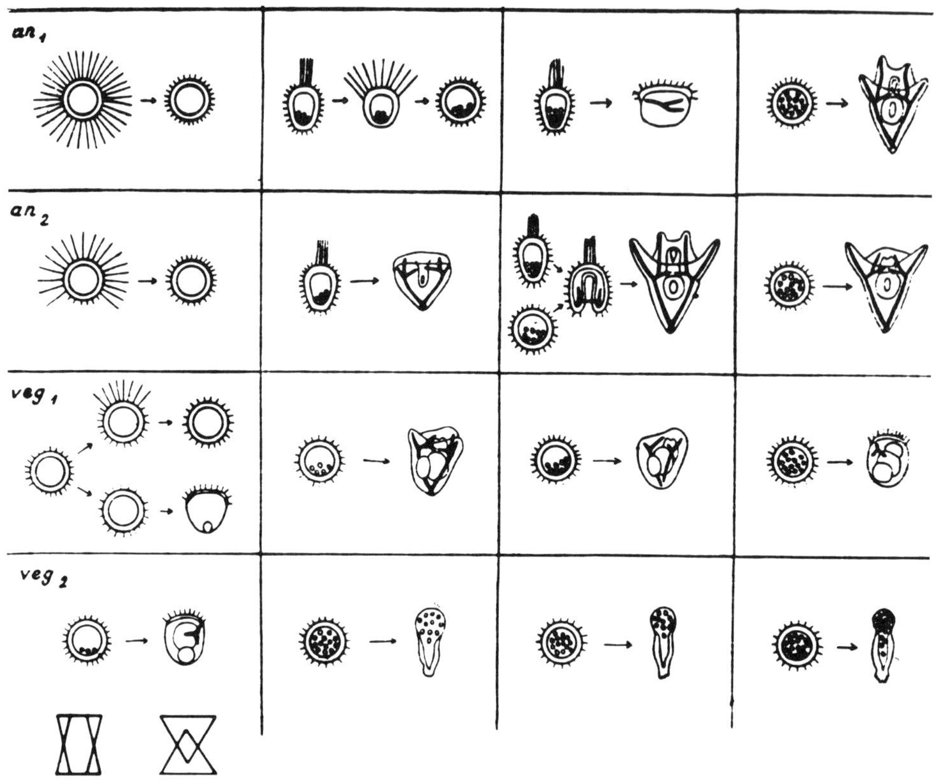

FIG. II–3. Diagram of the development of the layers an_1, an_2, veg_1 and veg_2, isolated (first column) and with 1, 2, and 4 implanted micromeres (second to fourth columns). After Hörstadius, 1935.

of the egg in proportions that do not deviate too far from the normal. An animal half of the egg or the uppermost tier of cells in the animal hemisphere gives rise to a blastula in which the apical tuft of stiff cilia is much enlarged. If micromere cells from the opposite pole are added, the tuft is reduced to proper proportions and a normal small pluteus may develop. In fact, two normal plutei may be obtained from a single egg after horizontal division, provided the middle part is left intact and the micromeres are joined to the "most animal" cells. This involves a considerable amount of regulation, in which particular cells give rise to structures quite different from those they would normally form. Animal (ectodermal) cells may be endodermized by the presence of micromeres, and a certain amount of vegetative material is necessary to inhibit the spread of the apical tuft. On the other hand some animal material is necessary for gastrulation to take place.

In such processes of differentiation there must be a lively reaction between the changing cell and its internal environment of other cells. This must either take place through direct contact, as when the micromeres induce presumptive ectoderm to invaginate (gastrulation), or be mediated by substances diffusing through the blastocoele from cells at a greater distance, e.g. in the inhibition of the spread of the apical tuft by the presence of micromeres at the opposite pole. In contrast with the mosaic eggs referred to above, the sea-urchin egg gives much more evidence of interaction between parts after development has started, but there is no such complete isotropy [equipotence—*R.G.H.*] in the sea urchin egg as Driesch supposed. The material of the vegetative portion of the egg is different from that of the animal portion, and some of each is necessary for normal development.

According to the Runnström-Hörstadius theory, there are in the egg two material gradients oriented in opposite directions; one diminishes as the other increases from one pole to the other. At the animal pole, the "most animal" region of the egg, the power to form structures characteristic of that region is most intense; at the vegetative pole, the "most vegetative" region, the power to invaginate and form structures normally arising there (archenteron, mesenchyme) is at a maximum. When any portion of the egg is cut out there is a tendency toward a rearrangement of these materials so that the gradients are restored approximately to their original scale. Before the third (equatorial) cleavage membrane is established this might take place by rearrangement of coarsely particulate material, but the fact that the phenomenon occurs even after the egg is divided into many cells indicates that the reestablishment of the gradients after disturbance is brought about by subtler changes in the egg materials.

While there are many processes in the development of the sea urchin that are either of a self-regulatory character or are dependent upon the internal environment for realization, the marked regional differences in the several layers of the egg as regards their powers of differentiation show that the mosaic quality of this egg is not negligible.

The amphibian egg, thought by Roux (1888) to be the archetype of mosaic development [This was based on experiments in which one of the two first blastomeres was killed by a hot needle thrust into it; the remaining blastomere then developed into a half-embryo, either lateral or anterior (Lecture II notes).], is now known to be about as regulable as the sea-urchin egg. However, all amphibian eggs that have been studied show a certain amount of germinal localization at the time cleavage begins—a cortical layer, the gray crescent, and peculiarities of yolk and pigmentation in the two hemispheres. These eggs are enclosed in a tight membrane and are very fluid, so that removal of parts of the unsegmented egg by pricking to test their function in development has not yielded results that can be clearly interpreted (Brachet 1923) [see Pasteels 1932a,b—*R.G.H.*]. On the other hand, the effects of separation and rearrangement of cell constituents by means of the centrifuge and by gravitation (inversion of eggs) have given results of interest.

When the egg of a frog is turned there is a marked rearrangement of constituents. The cortical layer, which holds most of the pigment, and the main part of the gray crescent remain in position, and the main mass of heavy white yolk rotates as a unit. The position of the dorsal lip of the blastopore in such eggs indicates that it is formed in definite relation to the gray crescent (cortical field) and the yolk mass (Weigmann 1927; Dalcq and Pasteels 1937) [p. 198]. In other words, turning the egg does not result in indiscriminate mixing of the elements but in an orderly rearrangement, which may be such that a normal embryo may develop from it. When the egg is inverted in the two-cell stage the rearrangement takes place independently in the two cells. This may result in the formation of two separate dorsal blastoporic lips followed by the development of a twin embryo (Schultze 1894). All of this goes to show that even in the earliest stages, development is affected by the interaction of cell constituents according to an orderly topographic arrangement. The internal environment of the cell constituents within the cell thus plays an important role. The nature of the reactions is, however, in total obscurity.

One of the first experiments, carried out by Pflüger (1883), was to fix the just fertilized amphibian egg (*Bombinator*) in abnormal orientation either upside down or in some intermediate position. Many such turned eggs develop normally but the pigment, which is more or less stuck in the cortical layer, becomes abnormally distributed; there is less in the outer layer.

Pflüger drew the sweeping conclusion that the egg was isotropic and that the axes were determined by gravitation, the animal-vegetative directly so and the median plane by the plane passing through the original egg axis and the newly imposed one. It was shown by Born (1885), however, that the contents of the egg rearrange themselves and that the main yolk mass rolls over to the bottom of the egg and the animal cytoplasm and inclusions rise to the upper regions.

The yolk mass may be divided and may land in different positions with regard to the gray crescent, which becomes blurred. The cortical layer, which is thin, apparently remains fixed, and is called by Dalcq and Pasteels the "cortical field" centering at the middle of the gray crescent.

Gastrulation in the normal embryo begins at the nearest point of the yolk mass to the center of the gray crescent. This may occur in inverted eggs, but in other cases where the yolk mass is divided or turned toward another part of the cortex, the site of gastrular invagination corresponds to the rearrangement of the yolk mass so that there may be double embryos or lopsided ones. In fact, it is stated by Dalcq and Pasteels (1937, 1938) that these invaginations may take place at any point on the surface of the egg. This being the case, it is difficult to see that the gray crescent has any essential role in determining the median plane, and yet these same authors concede its importance, as do many others who have investigated the question. The subject is thus left in an unsatisfactory state requiring further investigation, particularly the extent of the cortical field on the surface, as well as into the depths. [Lecture II notes.]

The amphibian egg shows perhaps better than any other the combination of mosaic and regulative qualities. The normal scheme of germinal localization has been mapped in great detail by vital staining, first applied in a wholly satisfactory and exhaustive way by Vogt (1925, 1929). The areas determined by this means are, however, not organ-forming areas in the strict sense, for practically any piece of the egg of sufficient size may, if isolated or placed in new position, give rise to more than it would have—or at least to something different—had it been left in place. This is shown by many types of isolation and recombination experiments (Bautzmann 1929; Holtfreter 1929, 1931, 1938).

It has been shown beyond doubt that the egg after fertilization has a bilaterally symmetrical structure, which may, in fact, even be antecedent to fertilization. When the first cleavage furrow divides the egg approximately in the plane of symmetry, which is the presumptive median plane of the embryo, separation of the blastomeres is followed by the development of each into a complete embryo [Fig. II–4] (Ruud 1925) [Herlitska 1895, 1897; McClendon 1909; Spemann 1919]. The coblastomere of the two-cell stage therefore constitutes an internal environment which, if present, restricts its mate to forming half an embryo. Here, however, the regulative process may be little more than a flow of materials within the single remaining cell, for a dead cell, if not removed, may have the same effect as the living cell (Roux 1888). The amphibian egg contents are very fluid and in them there is obviously not the same hindrance to flow as in ascidian eggs. When the first cleavage is at right angles to the plane of symmetry, it divides the germinal material into presumptive dorsal and ventral halves. The former, when separated from the rest of the egg, has sufficient power to readjust its internal arrangement to form a normal whole embryo. The latter, which contains no gray crescent material, lacks this power and forms a very defective *"Bauchstück"* or ventral piece, which is without notochord and nerv-

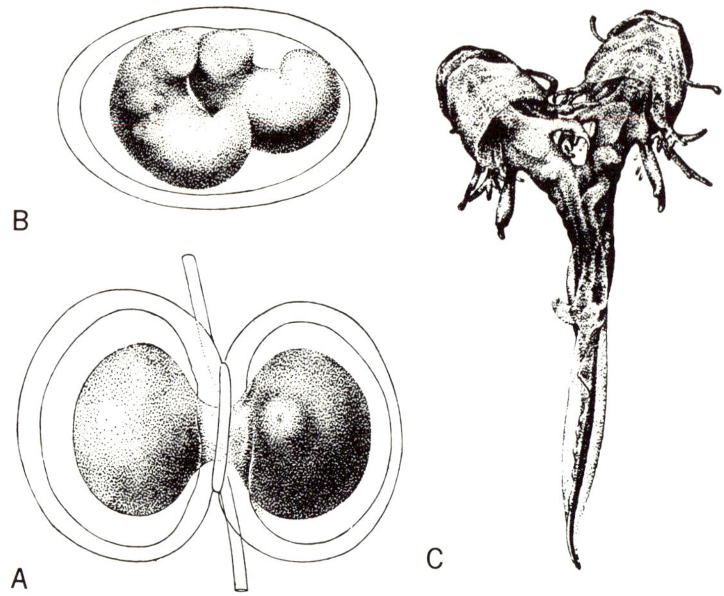

FIG. II–4. A. Constriction of the *Triton taeniatus* egg with a hair loop, a short time after fertilization.

B. Complete median constriction and the formation of 2 twin embryos.

C. Partial constriction at the 2-cell stage with the production of anterior doubling and inversion of situs viscerum in the right-hand member. After Spemann, 1919.

ous system. Some readjustment of material takes place in this piece too, but it is insufficient to produce a normal embryo. An essential constituent is wanting. Two eggs fused together in the two-cell stage may form either a single normal embryo of large size or double or multiple monsters, according to the relative position of the gray crescents in the combination. This, likewise, shows the power of adjustment of the egg substance and the limitations imposed upon it by cytoplasmic differentiation (Mangold and Seidel 1927) [Figs. II–5 and II–6].

Even through the blastula stage the two lateral halves of the newt embryo retain the ability to readjust and form a complete embryo after separation (Spemann and Falkenberg 1919; Ruud and Spemann 1922), except that the side originally turned toward the other cell may be defective. The right-hand partners in pairs of twins produced in this way may have their asymmetry (*situs viscerum*) reversed. Regional areas along other axes of the blastula become more and more specialized as development proceeds, but at the same time show much dependence upon surrounding parts (internal environment) for their proper differentiation.

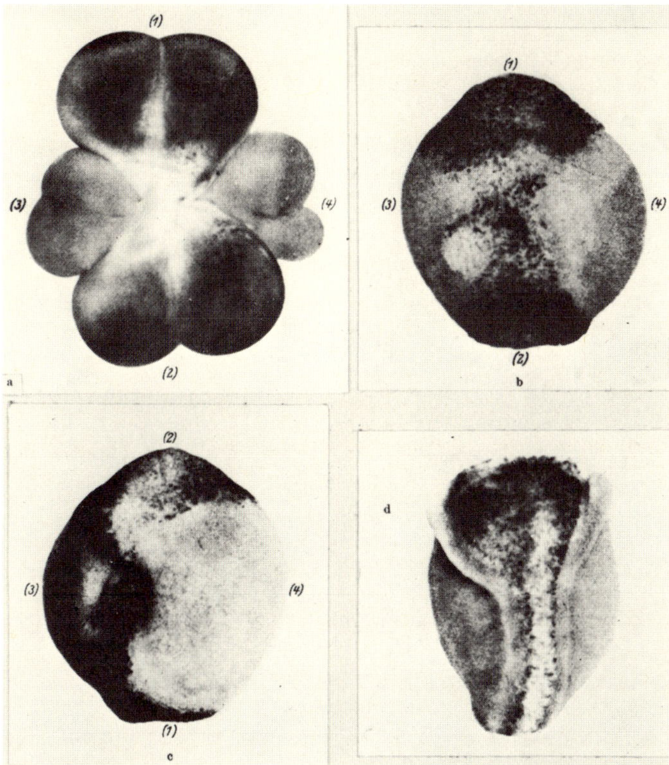

FIG. II–5. Fusion of two eggs of different species and pigmentation (*T. taeniatus* and *T. alpestris*) which were laid across each other in the 2-cell stage; (b, c) intermediate stages; (d) resulting giant neurula. After Mangold and Seidel, 1927.

The most striking case of dependent differentiation is that of the central nervous system, which is formed in the outer layer of the embryo by reaction with the material underlying it, which is turned in during gastrulation. This material, because of its marked general effect upon the development of the embryo, has been termed the organizer by Spemann (1924, 1927), its discoverer. The material which is invaginated and which acts upon the overlying tissues forms the pharyngeal roof, head mesenchyme, axial musculature, and notochord.

Several of the earlier experiments leading to the conception of the organizer are the following: (1) When two newt's eggs are divided down the midline in the blastula stage and the two right halves or the two left halves are grafted together, so that the two half dorsal lips are not exactly opposite one another and point in opposite directions, each one forms or organizes a whole embryonic axis with the incorpora-

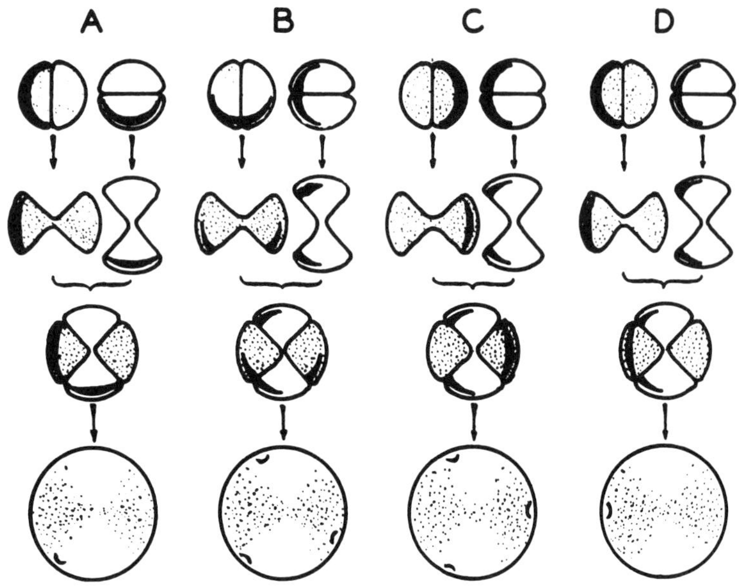

FIG. II–6. Diagram showing the principal combinations resulting after fusion of two eggs in the 2-cell stage.

Top row: Eggs before fusion showing position of gray crescent.

Second row: Eggs with changed contour preparatory to fusion.

Third row: Eggs after fusion.

Fourth row: Gastrulae after fusion, showing position of blastopores (one for each contiguous gray crescent). A and D produce single giant embryos. B and C produce embryos with triplet components. Twins and quadruplets result from intermediate combinations. After Mangold and Seidel, 1927.

tion of adjacent material from the other half, and double embryos result. (2) When half the dorsal lip and adjacent material is grafted into another less specialized region of the embryo, it invaginates and reacts upon its surroundings so as to produce a whole new embryonic axis in the host organism [Figs. II–7 and II–8]. (3) Pieces of presumptive ectoderm and presumptive neural plate, when interchanged in the blastula or early gastrula, develop in harmony with their new surroundings (Spemann 1918). (4) Pieces of presumptive ectoderm, when grafted into the dorsal lip of the early gastrula, acquire the properties of the latter and are able to act as organizer material (Spemann and Geinitz 1927).

Finally a simple method was devised in Spemann's laboratory whereby inductive power of any tissue or substance alive or dead could be tested. This is the so-called

"Einsteckmethode" whereby the tissue to be tested is placed in the blastocoele and so after gastrulation comes to lie between ectoderm and mesoderm. The ease with which the method could be applied has led to a flood of tests which has left us with a large mass of factual information that is difficult to interpret, and leaves the subject of organizer action, or embryonic induction, in a very unsatisfactory condition.

Experiments with the natural organizer showed that the material of the dorsal blastoporal lip which is first turned in and becomes pharyngeal roof and prechordal mesoderm, i.e. which normally underlies the brain, induces head structures when it is lodged either in the trunk (tail) region or in the head region; while trunk (tail) organizer, which is invaginated later, tends to induce trunk or tail structures —though in the head region it may induce head structures. [Lecture II notes.]

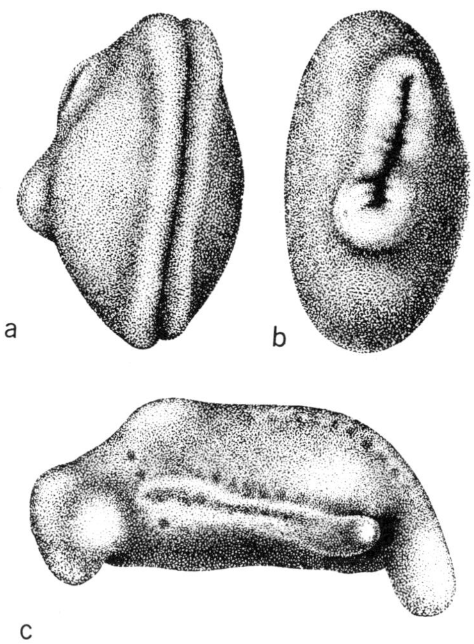

FIG. II–7. *Triton taeniatus* embryo with secondary neural folds; a, dorsal view; b, lateral view; c, same embryo at a later stage—secondary embryo on left side with ear vesicles in anterior part of induced medullary tube, somites and a small tail. After Spemann and Mangold, 1924.

What it is that gives the organizer cells these properties, which are so effective as an internal environment to the overlying cells, and what the nature of the reaction in the surrounding tissues is has not been ascertained. Spemann (1931a) himself exper-

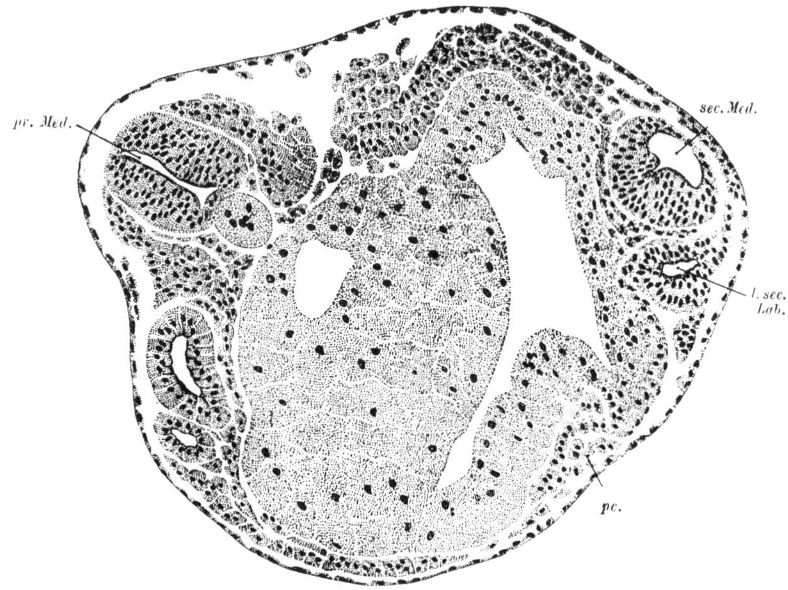

FIG. II–8. Cross section of embryo shown in Fig. 7c. Primary nervous system on the left, secondary neural tube and ear vesicle on the right. After Spemann and Mangold, 1924.

imented with organizer material the structure of which had been destroyed. Later Holtfreter (1933, 1933a, 1934) discovered that a great variety of animal tissues both adult and embryonic and from almost any phylum could, when placed in the blastocoel so that it ultimately came to lie under the ectoderm, stimulate the formation of a thickened area (neural plate) in the overlying tissue, which might fold over and close in as the neural tube does. Even pure chemical substances held in an agar matrix may have a similar effect. There is a difference, however, between the response to living normal organizing material and that to dead material, the former stimulating the formation of a complete neural plate and tube with normal regional differences, while the latter produces simply a thickened plate which may close to a tube without, however, showing such regional differentiation. [This is not correct. Dead material may induce a great variety of structures but in a great many cases the induction is very irregular and sometimes chaotic. *R.G.H.*]

Much effort has been spent to find out the nature of the chemical substance that produces these effects, and several different conclusions have been tentatively reached. The fact that certain embryonic tissues which have no organizing effect when living do so act when dead has led to the view that the substance responsible for the effect is

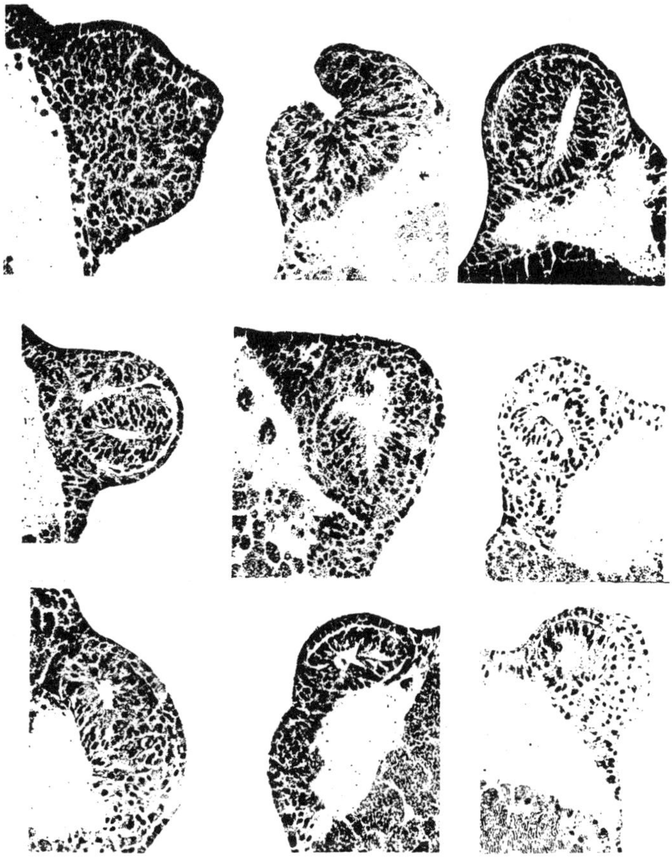

FIG. II–9. Induction of secondary neural structures in the chick. After Waddington, 1938.

free only in the active regions but exists in a bound condition in other parts of the embryo. Needham, Waddington, and co-workers (1935, 1938) have isolated sterol derivatives from the organizer and have found that synthetic sterols when applied in an agar matrix do induce the formation of neural-plate-like structures [Fig. II–9]. Another fact that may be of significance is the occurrence of an intense carbohydrate metabolism in the organizer (Woerdeman 1933; Heatley and Lindahl 1937; Boell et al. 1939; Brachet 1939).

Waddington's distinction between evocation and individuation. Evocation: thickening and palisade formation and folding over to form neural tube sometimes with accessory structures. Individuation: brought about only by living organizer, characterized by formation of a more or less complete secondary embryo. Normally the organizer has field properties with a gradient as has the material on which it works. These are in normal development, very precisely adjusted to one another.

Toivonen's experiments, however, do not tally with the above. Toivonen has found that certain adult tissues preserved in alcohol have very specific inductive capacity. Liver of perch, adder, or guinea pig induces forebrain structures, i.e. brain, eye, nose, balancer. Kidney of perch and guinea pig is a mesoderm inductor, giving rise primarily to trunk mesoderm which in turn may induce hindbrain and spinal cord. Kidney of adder and pig is an inductor of the posterior part of the head: mesoderm and rhombencephalon and secondarily, ear vesicle. Guinea pig thymus is a typical lens inductor.

These results, however, are not confirmed by the somewhat similar work of Chuang.

All of the above evidence indicates that the properties of the "embryonic field" extend throughout the embryo even after many features, according to some tests, appear to be determined or fixed—localized. This indicates that one must not take the specific localization pattern as anything macroscopically stereotyped. [Lecture II notes.]

Differences between the action of the natural organizer in its normal position and that of dead material and chemical substances may be due to the circumstance that the former is distributed in a gradient having different local concentrations. Another factor that may have some influence is the possible variation of the intensity of action with respect to time. A normal induction might be achieved only through the action of a living system, because this might require the action to start with a low intensity, increase to a maximum, and then taper off; or it might require several cycles of variation in intensity, which dead material could not afford. Regional differences in the reacting material must also be taken into consideration, as shown by the different respective reactions of presumptive head and trunk regions to similar implanted organizer material (Spemann 1931).

What happens in the ectodermal cells that become transformed into neural plate is of much importance in any theory of organizer action, but it is almost totally unknown. The neural plate contains relatively more water than the embryo as a whole (Glaser 1914), and it seems to be the inner portions of the neural plate cells that are the most swollen. We might assume, then, that the organizer produces a chemical change, which leads first to an orientation of particles that makes the cells columnar, followed by a greater hydration of the protein lattice localized in the basal halves of the cells. This would lead to the folding of the plate to form a groove and its ultimate closing into a tube. The changes underlying the process would thus be of a chemical and paracrystalline nature.

It must also be borne in mind that the agencies effecting the differential changes in the central nervous system are heterogeneous in composition. Underlying the fore brain is the pharyngeal roof and some mesenchyme, and under the rest of the brain and spinal cord the notochord occupies the midline with the myotomes on each side. All of these factors are known to have a specific effect on the overlying neural plate (Lehmann 1926, 1928, 1938).

Since the word "organizer" connotes a master regulator which creates the organization, and since there are in the course of development many actions of the same general character that could hardly be accorded such a role, it is perhaps more appropriate to use the word "induction" to denote processes of this kind. Outside the area of the central nervous system there are many other ectodermal organs—balancer, gills, hypophysis, etc.—that are dependent upon other structures for their differentiation. Whether the ectodermal structures just referred to arise *in situ* through some localizing factor in the ectoderm itself, or whether they are produced by reaction with the underlying tissue is still uncertain in some cases.

So much attention has been concentrated on the study of the induction of the neural plate and the embryonic axis that the lesser organs have been neglected and yet there are problems concerned with them that are of great interest. The last topic that will be considered is an analysis of the factors concerned in the development of the labyrinth or internal ear. [Lecture II notes.]

An experimental study of the development of the ear shows that, beginning with the late gastrula or early neurula stage, at least four factors are concerned in its differentiation (Harrison 1935a, 1938a) [see pp.174–79]. Three of these lie within its internal environment. The cellular material is, as is well known, supplied by the ectoderm. While an ear vesicle may arise from any ectoderm of the gastrula, in the early neurula stage the cells of the auditory placode, out of which it normally develops, are already predisposed toward ear formation (otogenesis). The underlying mesoderm is still involved at this time and it is probably this layer, acting at a still earlier stage, that has induced the just mentioned condition in the ear placode. A second effective factor in the internal environment of the developing ear is a certain spot in the hind brain region of the neural tube with which the ear placode is brought into contact as the neural folds rise and close over. A fourth factor is the position in the embryo in which the ear vesicle develops. By means of grafting experiments all of these factors have been varied, but the results do not indicate that they singly have any markedly specific role to play, except that the hind brain seems to be significantly associated with the development of the endolymphatic duct and sac. The development of a normal ear vesicle seems rather to be dependent on a series of factors acting in turn, and furthermore the orientation within the organism of at least two of them —the ear placode and the myelencephalon—has at the appropriate time an effect upon the asymmetry of the ear that develops.

Again nothing is known with regard to the actual processes involved in the development of this organ and the nature of the action of one part upon the other. There is evidence, however, that an orientation of some structural element of the earplate cells takes place about the time of closure of the neural folds (Harrison 1936).

There is a certain similarity between the formation of the neural groove and that of the auditory pit in that cells elongate in a direction perpendicular to the surface and a vesicle is ultimately folded off from a plate. There are differences in degree of elongation, in shape, and in the fact that the neural folds rise above the surface, while the auditory pit sinks in and the folds remain even with the surface. One might again assume a hydration in the protein lattice of the auditory cells and a consequent change in the lattice spacing, with a resulting change in the shape of the cells which brings about a modification of the shape of the rudiment from a plate to a vesicle. Subsequent changes, which produce the internal ear or labyrinth out of a simple vesicle, involve complicated folding and constriction of its walls. Certain conditions in the internal surroundings, possibly the failure of the endolymphatic duct and sac to develop, often bring about osmotic disturbances in the vesicle, which then swells considerably and becomes cystic. Normal folding fails to take place and the sensory areas develop incompletely.

The induction of the crystalline lens by the optic vesicle is another notable example of dependent differentiation. This case is complicated, however, by the fact that in some species of amphibia and fish the lens is capable of self-differentiation without the influence emanating from the optic cup. The natural conclusion to draw from these facts is that the lens rudiment itself includes the factors necessary to form a lens and that the optic cup is needed in some cases to set off the chain of reactions that produce it out of its ectodermal rudiment, while in other cases this specific action is not required.

The amphibian limb is an organ that is localized at least as far back as the gastrula or perhaps even earlier, as shown by both transplantation and defect experiments (Detwiler 1929; Suzuki 1928). However, it was found by Balinsky (1925, 1931, 1933), and later confirmed by others, that by implanting in the flank of an embryo such organ rudiments as ear placodes, nasal placodes, pieces of brain, or, in some cases, even small pieces of celloidin, more or less perfect limbs could be induced. Here, cells that would ordinarily form axial musculature and mesenchyme on the flank of the organism are now activated to form an appendage with all of the complex arrangements in muscle, cartilage, bone, etc., that constitute a limb and are quite foreign to that region. No satisfactory explanation of this has been given.

All the reactions described in the foregoing, with the possible exception of the inhibition of the spread of the apical tuft in the sea urchin, apparently take place between cells that are in contact. When neither the nervous system nor the circulation is developed, it is indeed difficult to imagine a reaction at a distance. Protoplasmic transmission is a slow process, and there is in fact almost no evidence of reaction between cells that are far removed from one another in the early stages of embryonic development. The clearest case is a phenomenon observed by Twitty (1937) in experiments in which small grafts (eyes) from *Triturus torosus* were implanted in amphibian embryos of other species. Such grafts completely paralyze the host embryo for a time, and this action takes effect before the circulation begins. It can only be explained by the diffusion of the toxic substance through the tissues of the host.

In growth, as well as in differentiation, cells and cell groups are dependent upon one another. The proportions of an organism are acquired through the maintenance of the normal relative growth of its parts. Delicate adjustments are here necessary, but they are little understood, though known to be governed by both genetic and environmental factors. Rates of growth are specific, and each organ or part has its characteristic rate relative to that of the whole. Relative growth rates are susceptible to external and internal environmental influences, though in a much less degree than the general growth of the whole organism.

The specificity of growth rates is strikingly shown by grafting organs, such as limbs and eyes, between different species (e.g. *Amblystoma tigrinum* and *A. punctatum*) which grow at different rates. Grafted organs retain their characteristic rates and the result is an organism, often grotesque in appearance, with a limb or an eye that is either much too large or too small for it. Such cases show that although subjected to the circulating medium of the host, congenital specific factors control the relative growth rate of the graft (Harrison 1924, 1935).

Germ layer chimaeras in the case of limbs, i.e. individuals in which either the ectoderm or the mesoderm is transplanted alone, show that the relative growth rate of the appendage as a whole is governed almost entirely by that of the mesoderm, the ectoderm having very little effect upon it. In the eye, however, both optic cup and lens are concerned in maintaining the relative growth rate of the whole organ. Each of these components stimulates or retards the growth of the other, according to relative growth rates in the two species, so that an eye of approximately correct internal proportions, though still out of scale with the animal as a whole, develops out of the heteroplastic combination. This effect must be attributed to the mutual regulation of growth and division rate of cells by their associated cells constituting the internal environment (Harrison 1929).

Within the nervous system there is much interdependence in respect to growth and differentiation between functionally associated nerve centers and between center and end organ. Both hypoplasia and hyperplasia of centers have been observed, according as connecting tracts are diminished or increased in volume (Detwiler 1920, 1936; May 1933; Hamburger 1934, 1939). A large eye, for instance, sends out a correspondingly large optic nerve which stimulates the mid-brain of the opposite side to hyperplasia (Harrison 1929; Twitty 1932). There are also influences emanating from nerve centers that affect growth and development of other organs, but this subject is still largely controversial. One of the most striking effects of this kind is the tendency to atrophy or stoppage of growth in muscles after severance of the nerve supply. However, the initial differentiation of muscle is not dependent upon its connection with the nervous system (Harrison 1904; Hamburger 1928).

As stated above, there is almost no evidence that regions which are far apart affect one another during early stages of development, but as soon as the circulation and nervous connections develop, factors influencing differentiation become effective through these media. An exhaustive treatment of this subject would include a large part of the field of endocrinology, with all of the complications involved in the origin and control of secondary sex characters. Since this would require space greatly in excess of that available, I shall limit myself here to a single topic, the development and distribution of pigment.

Pigmentation in the lower vertebrates, as well as in certain invertebrate forms, is due to cells, known as chromatophores, which are scattered in the mesenchyme or concentrated on membranes surrounding other organs such as blood vessels, fascia, peritoneum and meninges. In amphibian larvae many lie embedded between the epidermal cells. The pigment of the early embryonic stages is already developed in the ovarian egg and is merely distributed, mainly to the ectoderm cells, during segmentation.

The pigmentary system of vertebrates, both in its developmental stages and later, is subject to a variety of influences—intercellular, nervous, humoral and outer environmental (temperature and light). Moreover, the several actions may be mediated in different ways, either directly or reflexly. The processes involved are therefore complicated, and much ingenuity has been exercised in working out experimentally the interplay of factors [pp. 111, 242].

We are concerned here not with transitory changes in pigment cells, which are largely adaptive to external conditions, but with quasi-permanent or irreversible transformations involved in the development of the pigmentary system. In the embryo the prospective chromatophores are wandering cells and hence their place of origin must be ascertained before their distribution can be understood.

It has now been established beyond doubt that the neural crest is the principal, if not the sole, source of pigment cells (exclusive of the retinal tapetum) in at least two classes of vertebrates, amphibians (DuShane 1935) and birds (Dorris 1939). The neural crest is a strip of cells lying between the neural plate and the ectoderm, which becomes separated from both of these structures either during or shortly after the closure of the neural folds. The pigment cells thus start from a position on the dorsal side of the spinal cord and wander from there ventrally under the ectoderm and in the interstitial tissue. Associated with presumptive pigment cells in the neural crest are other cells destined to give rise to a variety of tissues—visceral cartilages, mesenchyme, spinal ganglion cells, sheath cells, sympathetic ganglia, chromaffine tissue.

This origin of pigment cells has been demonstrated by means of tissue culture, transplantation and defect experiments with both amphibian and chick material. While the recorded observations deal mainly with melanophores, the above probably applies also to the xanthophores, which contain a diffuse yellow lipochrome and to the guanophores containing granular pigment with metallic luster.

When the cells of the neural crest start out from their dorsal position, the potential chromatophores are indistinguishable from their accompanying cells. They are at first without pigment and reach remote positions in the embryonic body before showing their differentiation. There is evidence that many of them are not self-differentiating but require activation by something derived from other cells in order to produce pigment (DuShane[6] 1935; Harrison 1938).

The above facts are shown by simple experiments. When the neural crest is removed bilaterally from the trunk region of an amphibian embryo, the whole flank remains without pigment. When the neural crest is transplanted to the abdominal region, much pigment develops and spreads from the site of the graft. Neural crest explanted *in vitro* develops many pigment cells (Dorris 1938; Twitty and Bodenstein 1939).

6. Op. cit., p. 18.

Pigmentation is frequently not diffuse but is distributed according to patterns that have definite relations to organs and regions. By means of transplantation experiments with embryos of three closely related species of *Triturus* and their hybrids, Twitty (1936) has shown that such markings are the result of interplay of a multiplicity of factors, some inherent and some environmental.[7] The latter may be either of the intimate cell-to-cell type or humoral.

The stripe along the dorsal margin of the myotomes characteristic of larvae of *Tr. torosus* is not formed when neural crest of *Tr. rivularis,* which does not have stripes but a much more extensive diffuse pigmentation, is grafted in place of the neural crest of the *torosus* embryo. Reciprocally, neural crest of *torosus* forms stripes on *rivularis*. However, the stripes do not form in the absence of the myotome ridge in either species. Both grafting and tissue culture experiments show that the *rivularis* neural crest cells tend to spread much farther from their seat of origin than do those of *torosus*. This seems to be correlated with the slowness of *rivularis* cells to differentiate pigment as compared with cells from *torosus*. Movement and differentiation are in inverse relation to one another.

In the amphibian embryo the pigmentary system is the first tissue that can be observed to respond to hormonal action through the medium of the blood. When the hypophysis rudiment is removed in early embryonic stages, the melanophores, which begin to differentiate about the time the circulation of the blood starts, do not expand. The larvae acquire a silvery or yellow appearance, according to species, owing to the preponderance of other pigment-bearing elements. This effect is produced primarily through the distribution of the pigment within the individual melanophores, and in this respect is like the adaptive color changes in organisms in response to temperature, light, and color of the background. However, it has been shown that continued absence of the hypophysis results in the disappearance of some of the non-melanophore pigment in the epidermis, just as the presence of more than the normal amount (by grafting supernumerary hypophyses) produces an organism with more than the normal amount of pigment, due partly to "expansion" of the melanophores and partly to increase in the non-melanophore pigment in the epidermis (Blount 1932). The pigmentary system thus may participate in more rapid changes of a physiological nature and in slower differentiating changes that may be more or less permanent, both taking place as a result of hormonal action.

In many organisms the pigment pattern is very sharply defined and highly specific,

7. See also Twitty and Bodenstein (1939).

often in great detail, and in certain fishes, amphibians, and birds it is known to be associated with sex and sexual activity through mediation of hormones. Experiments have also shown that other endocrine secretions and related substances, e.g. thryo-globulin and thyroxin, have a marked effect on the plumage of birds.

The feather is a complicated structure and no complete understanding of the effect of the internal medium upon its differentiation is possible without detailed knowledge of its development and growth. For the present purpose, however, it will suffice to say that the epithelial feather germ grows from its base, the tip of the feather being pushed out from within (Lillie and Juhn 1932). The more distal barbs are therefore the older. The individual barbs grow likewise from the base, and their growth rate is higher at first, i.e. when the tips of the barbs are formed, and it gradually tapers off. The barbs are attached obliquely to the shaft so that a transverse bar on a feather cuts successive barbs at different levels. When a feather is plucked, it regenerates from the remaining follicle in essentially the same way as the original feather develops. Thus it is possible to carry out on adult birds experiments which show the effect on the feather pattern of substances circulating in the internal medium.

The response to injections of thyroxin is expressed by a dark pigment spot or stripe on the brown or yellow feathers of the brown Leghorn chicken. The same injection, if the amount is appropriately gauged, may affect individual feathers differently according to their position on the body of the bird. This is owing to certain factors, which vary according to position. The threshold of reaction and the rate or amount of reaction are directly proportional to the growth rate, and the latent period is inversely proportional (Lillie 1932). Since these factors are different in the individual barbs, the effect produced on the feathers may vary all the way from a narrow spindle-shaped black spot centered on the shaft to a transverse bar, running clear across the feather. As many as five or six successive injections, if properly spaced, may be recorded in this manner on a single feather. The female sex hormone has similar effects when injected into capons. According to dosage and position it produces brown spots or bars on feathers that without treatment would have become black, or it may produce whole brown feathers. These hormones also affect the form and texture of the feathers.

The natural barring of feathers, such as occurs in the Plymouth Rock breed, is apparently brought about in a different way (Montalenti 1934). It is to be referred rather to a rhythmic activity of the genes in each feather follicle, the exact nature of which is unknown. The fact that the same phenotypic effect may be produced in such different ways is of great interest.

Conclusion

The foregoing sketch, which makes no pretense to completeness, calls attention to some of the facts regarding the action of the internal environment upon the differentiation of cells, i.e. upon the process of dependent differentiation in contradistinction to self-differentiation (Roux 1885). With the growing mass of evidence for the former and, hence, for the theory of epigenesis, an exhaustive treatment of the subject would extend far beyond the scope of a single paper. So, bearing in mind the examples given above, we may now review the salient points in the theory of development here presented.

The egg cell from which each organism is derived is composed of protoplasm peculiar to the species and different from that of any other species. This substance is laid down with collaboration of the nucleus and in continuity with the tissues of the maternal parent, perpetuating its specific characteristics thereby. The growth period of the ovum is thus an important one in the development of the individual—one that has been relatively neglected by the embryologist.

The material is deposited in the egg according to a certain configuration which is oriented in relation to the surrounding parental tissues. This configuration must be conceived as related to the molecular structure of the protoplasm, more particularly to that of the proteins. Although the gross pattern of the egg is much changed at the time of maturation and fertilization, the new arrangement also is a function of its molecular constitution and has a definite topographic relation to the form of the future organism.

Since cellular differentiation takes place in the cytoplasm, we are concerned here more directly with this constituent of the cell, bearing in mind, however, that the cytoplasm is accompanied and ultimately controlled by the genic complex in the chromosomes. Since the latter is presumably the same for all cells of the organism, differences between cells must arise through interaction between the constant genom and the locally variable cytoplasm, in which they ultimately become visible.

In the sense that the organism is made up of the same specific protoplasm as that occurring in the egg, together with its products or differentiations, the characters of the organism are predetermined in the germ. The actual realization of these characters, however, comes about during development in a variety of ways. In the so-called mosaic eggs, i.e. those with "precocious segregation," to use Lankester's (1877) term, local differentiation arises before cleavage and a minimum of modification in response to internal environmental factors is involved in subsequent development.

In other eggs the protoplasm long remains labile with respect to its prospective differentiation, and in these many organs of the embryo develop only through the interaction between the cells producing them and other elements in the internal environment, i.e. other cells and the internal medium. In such eggs local predetermination is reduced to a minimum. Their members in continuance are fashioned, when as yet there is none of them.

References

Andrews, G. F. 1897. Some spinning activities of protoplasm in starfish and sea-urchin eggs. *J. Morph., 12:* 367–89.

Balinsky, B. I. 1925. Transplantation des Ohrbläschens bei *Triton. Roux' Arch., 105:* 718–31.

———. 1931. Zur Dynamik der Extremitätenknospenbildung. *Roux' Arch., 123:* 565–648.

———. 1933. Das Extremitätenseitenfeld, seine Ausdehnung und Beschaffenheit. *Roux' Arch., 130:* 704–46.

Bauer, Karl. 1932. Beobachtungen über das Wachstum von Nervengewebe in vitro. *Zeit. f. mikr. anat. Forsch., 28:* 47–80.

Bautzmann, H. 1929. Ueber bedeutungsfremde Selbstdifferenzierung aus Teilstücken des Amphibienkeimes. *Naturwiss., 17:* 818–27.

Blount, R. F. 1932. Transplantation and extirpation of the pituitary rudiment and the effects upon pigmentation in the urodele embryo. *J. Exp. Zool., 63:* 113–41.

Boell, E. J., H. Koch, and J. Needham. 1939. IV. Respiratory quotient of the regions of the amphibian gastrula. *Proc. Roy. Soc.* (B), *127:* 374–87.

Boell, E. J., J. Needham, and V. Rogers. 1939. Morphogenesis and metabolism. I. Anaerobic glycolysis of the regions of the amphibian gastrula. *Proc. Roy. Soc.* (B), *127:* 322–56.

Boveri, Th. 1901. Ueber die Polarität des Seeigeleies. *Verh. Phys.-Med. Gesell. Würzburg., 34:* 145–76.

———. 1901a. Die Polarität von Ovocyte, Ei und Larve des *Strongylocentrotus lividus. Zool. Jahrb.,* Abt. Anat. u. Ont., *14:* 630–53.

Brachet, A. 1923. Recherches sur les localisations germinales et leurs propriétés ontogénétiques dans l'oeuf de *Rana fusca. Arch. Biol., 23:* 343–430.

Brachet, J. 1939. Étude du métabolisme de l'oeuf de Grenouille (*Rana fusca*) au cours du developpement. V. Le métabolisme protéique et hydrocarbone de l'oeuf en relation avec le problème de l'organisateur. *Arch. Biol., 50:* 233, 267.

Brien-Gavage, P. and E. 1927. Contribution à l'étude de la Blastogénèse des Tuniciers. III. Bourgeonnement de *Clavelina lepadiformis* Müller. *Rec. Inst. Zool. Torley-Rousseau, 1:* 31–81.

Burrows, M. T. 1912. Rhythmische Kontraktionen der isolierten Herzmuskelzelle ausserhalb des Organismus. *Münch. med. Woch.,* pp. 1–10 (pagination of reprint).

Chun, C. 1892. Die Dissogenie, eine neue Form der geschlechtlichen Zeugung. Festschrift für Leuckart, Leipzig. (Cited from Schleip, p. 50.)

Cohen, A., and N. J. Berrill. 1936. The development of isolated blastomeres of the ascidian egg. *J. Exp. Zool., 74:* 91–117.

Conklin, E. G. 1905. The organization and cell-lineage of the ascidian egg. *J. Acad. Nat. Sci. Philadelphia,* 2nd ser., *13:* 5–119.

———. 1905a. Mosaic development in ascidian eggs. *J. Exp. Zool., 2:* 145–223.

———. 1911. The organization of the egg and the development of single blastomeres of *Phallusia mamillata. J. Exp. Zool., 10:* 393–407.

———. 1931. The development of centrifuged eggs of ascidians. *J. Exp. Zool., 60:* 1–119.

Crampton, H. E. 1896. Experimental studies on gasteropod development. *Roux' Arch., 3:* 1–19.

Dalcq, A. 1932. Étude des localisations germinales dans l'oeuf vierge d'ascidie par des expériences de mérogonie. *Arch. d'Anat. Microsc., 28:* 223–333.

———. 1938. *Form and Causality in Development.* Cambridge, Univ. Press, 197 pp.

Dalcq, A., and J. Pasteels. 1937. Une conception nouvelle des bases physiologiques de la morphogénèse. *Arch. Biol., 48:* 669–710.

Detwiler, S. R. 1920. On the hyperplasia of nerve centers resulting from excessive peripheral loading. *Proc. Nat. Acad. Sci., 6:* 96–101.

———. 1929. Transplantation of anterior-limb mesoderm from Amblystoma embryos in the slit-blastopore stage. *J. Exp. Zool., 52:* 315–24.

———. 1936. *Neuroembryology.* New York, Macmillan, 218 pp.

Dorris, F. S. 1938. The production of pigment in vitro by chick neural crest. *Roux' Arch., 138:* 323–33.

———. 1939. The production of pigment by chick neural crest in grafts to the 3-day limb bud. *J. Exp. Zool., 80:* 315–45.

Driesch, H. 1891, Entwicklungsmechanische Studien. I. Der Werth der beiden ersten Furchungszellen in der Echinodermenentwicklung. Experimentelle Erzeugung von Teil- und Doppelbildungen. *Zeit. wiss. Zool., 53:* 160–78.

———. 1892. Entwicklungsmechanische Studien. III. Die Verminderung des Furchungsmaterials und ihre Folgen (Weiteres über Teilbildungen). *Zeit. wiss. Zool., 55:* 1–10.

———. 1894. *Analytische Theorie der organischen Entwicklung.* Leipzig, Engelmann. 184 pp.

———. 1896. Die taktische Reizbarkeit der Mesenchymzellen von *Echinus microtuberculatus. Roux' Arch., 3:* 362–80.

———. 1908. Zur Theorie der organischen Symmetrie. *Roux' Arch., 26:* 130–45.

———. 1929. *The Science and Philosophy of the Organism,* 2d ed. London, Black, 344 pp.

Driesch, H., and T. H. Morgan. 1895. Zur Analysis der ersten Entwickelungsstadien des Ctenophoreneies. I. Von der Entwickelung einzelner Ctenophorenblastomeren. II. Von der Entwickelung ungefurchter Eier mit Protoplasmadefekten. *Roux' Arch., 2:* 204–26.

DuShane, G. P. 1935. An experimental study of the origin of pigment cells in Amphibia. *J. Exp. Zool., 72:* 1–30.

Fischel, A. 1903. Entwickelung und Organ-Differenzirung. *Roux' Arch., 15:* 679–750.

Glaser, O. C. 1914. On the mechanism of morphological differentiation in the nervous system. I. The transformation of a neural plate into a neural tube. *Anat. Rec., 8:* 525–51.

Hamburger, Viktor. 1928. Die Entwicklung experimentell erzeugter nervenloser und schwach innervierter Extremitäten von Anuren. *Roux' Arch., 114:* 272–363.

———. 1934. The effects of wing bud extirpation on the development of the central nervous system in chick embryos. *J. Exp. Zool., 68:* 449–94.

———. 1939. The development and innervation of transplanted limb primordia of chick embryos. *J. Exp. Zool., 80:* 347–89.

Harrison, R. G. 1904. An experimental study of the relation of the nervous system to the developing musculature in the embryo of the frog. *Am. J. Anat., 3:* 197–220.
———. 1910. The outgrowth of the nerve fiber as a mode of protoplasmic movement. *J. Exp. Zool., 9:* 787–846.
———. 1924. Some unexpected results of the heteroplastic transplantation of limbs. *Proc. Nat. Acad. Sci., 10:* 69–74.
———. 1929. Correlation in the development and growth of the eye studied by means of heteroplastic transplantation. *Roux' Arch., 120:* 1–55.
———. 1935. Heteroplastic grafting in embryology. Harvey Lectures, 1933–34. *29:* 116–57.
———. 1935a. Factors concerned in the development of the ear in *Amblystoma punctatum. Anat. Rec., 64:* 38–39.
———. 1936. Relations of symmetry in the developing ear of *Amblystoma punctatum. Proc. Nat. Acad. Sci., 22:* 238–47.
———. 1938. Die Neuralleiste. *Verh. Anat. Ges., Erg. Heft Anat. Anz., 85:* 3–30.
———. 1938a. Further investigation of the factors concerned in the development of the ear. *Anat. Rec., 70:* 35.
Heatley, N. G., and P. E. Lindahl. 1937. Studies on the nature of the amphibian organization centre. V. The distribution and nature of glycogen in the amphibian embryo. *Proc. Roy. Soc.* (B), *122:* 395–402.
Holtfreter, J. 1929. Ueber die Aufzucht isolierter Teile des Amphibienkeimes. I. Methode einer Gewebezüchtung in vivo. *Roux' Arch., 117:* 421–510.
———. 1931. Ueber die Aufzucht isolierter Teile des Amphibienkeimes. II. Züchtung von Keimen und Keimteilen in Salzlösung. *Roux' Arch., 124:* 404–66.
———. 1933. Nachweis der Induktionsfähigkeit abgetöteter Keimteile. Isolations- und Transplantationsversuche. *Roux' Arch., 128:* 584–633.
———. 1933a. Eigenschaften und Verbreitung induzierender Stoffe. *Naturwiss., 21:* 766–70.
———. 1934. Der Einfluss thermischer, mechanischer und chemischer Eingriffe auf die Induzierfähigkeit von *Triton*-Keim Teilen. *Roux' Arch., 132:* 225–306.
———. 1938. Differenzierungspotenzen isolierter Teile der Urodelengastrula. *Roux' Arch., 138:* 522–656.
———. 1938a. Differenzierungspotenzen isolierter Teile der Anurengastrula. *Roux' Arch., 138:* 657–738.
Hörstadius, Sven. 1928. Ueber die Determination des Keimes bei Echinodermen. *Act. Zool., 9:* 1–191.
———. 1935. Ueber die Determination im Verlaufe der Eiachse bei Seeigeln. *Pub. Staz. Zool. Napoli, 14:* 251–479.
———. 1937. Investigations as to the localization of the micromere-, the skeleton-, and the entoderm-forming material in the unfertilized egg of *Arbacia punctulata. Biol. Bull., 73:* 295–316.
———. 1938. Schnürungsversuche an Seeigelkeimen. *Roux' Arch., 138:* 197–258.
———. 1939. The mechanics of sea-urchin development, studied by operative methods. *Biol. Rev., 14:* 132–79.
Hörstadius, Sven, and Alexander Wolsky. 1936. Studien über die Determination der Bilateralsymmetrie des jungen Seeigelkeimes. *Roux' Arch., 135:* 69–113.
Just, E. E. 1939. *The Biology of the Cell Surface.* Philadelphia, Blakiston, 383 pp.
Lankester, E. R. 1877. Notes on embryology and classification. *Quart. J. Micr. Sci., 17:* 399–454.

Lehmann, F. E. 1926. Entwicklungsstörungen in der Medullaranlage von *Triton,* erzeugt durch Unterlagerungsdefekte. *Roux' Arch., 108:* 243–82.

———. 1928. Die Bedeutung der Unterlagerung für die Entwicklung der Medullarplatte von *Triton. Roux' Arch., 113:* 123–71.

———. 1938. Regionale Verschiedenheiten des Organisators von *Triton,* insbesondere in der vorderen und hinteren Kopfregion, nachgewiesen durch phasenspezifische Erzeugung von lithiumbedingten und operativ bewirkten Regionaldefekten. *Roux' Arch., 138:* 106–59.

Levi, G. 1917. Connessioni e struttura degli elementi nervosi sviluppati fuori dell' organismo. *Mem. R. Accad. Lincei,* ser. 5, *12:* 142–82.

———. 1925. Conservazione e perdita dell' indipendenza delle cellule dei tessuti. *Arch. Exp. Zellforsch., 1:* 1–57.

Lewis, W. H. 1922. Is mesenchyme a syncytium? *Anat. Rec., 23:* 177–84.

Lillie, F. R. 1932. Growth rate and hormone threshold with reference to physiology of development. *Am. Nat., 66:* 171–79.

Lillie, F. R., and Mary Juhn. 1932. The physiology of development of feathers. I. Growth-rate and pattern in the individual feather. *Physiol. Zool., 5:* 124–84.

Lindahl, P. E. 1932. Zur experimentellen Analyse der Determination der Dorsoventralachse beim Seeigelkeim. I. Versuche mit gestreckten Eiern. *Roux' Arch., 127:* 300–22.

———. 1932a. Zur experimentellen Analyse der Determination der Dorsoventralachse beim Seeigelkeim. II. Versuche mit zentrifugierten Eiern. *Roux' Arch., 127:* 323–38.

Mangold, O., and F. Seidel. 1927. Homoplastiche und heteroplastiche Verschmelzung ganzer Tritonkeime. *Roux' Arch., 111:* 593–665.

May, R. M. 1933. Réactions neurogéniques de la moelle à la greffe en surnombre, ou à l'ablation d'une ébauche de l'anoure, *Discoglossus pictus,* Otth. *Bull. Biol. France et Belg., 67:* 327–49.

Montalenti, Giuseppe. 1934. A physiological analysis of the barred pattern in plymouth rock feathers. *J. Exp. Zool., 69:* 269–345.

Pasteels, J. 1932a. Expériences de piqûre localisée sur l'oeuf fécondé et non segmenté des Anoures. *C. R. Assoc. Anat., 27 Reúnion, Nancy, 27:* 440–49.

———. 1932b. Étude des localizations germinales de l'oeuf insegmenté des Anoures. *Arch. Biol., 43:* 522–74.

Pease, D. C. 1939. An analysis of the factors of bilateral determination in centrifuged echinoderm embryos. *J. Exp. Zool., 80:* 225–47.

Penners, A. 1924. Experimentelle Untersuchungen zum Determinationsproblem am Keim von *Tubifex rivulorum* Lam. I. Die Duplicitas cruciata und organbildende Keimbezirke. *Arch. mikr. Anat. u. Entw.-Mech., 102:* 51–100.

———. 1925. Experimentelle Untersuchungen zum Determinationsproblem am Keim von *Tubifex rivulorum* Lam. II. Die Entwicklung teilweisse abgetöteter Keime. *Zeit. wiss. Zool., 127:* 1–140.

Pressler, Kurt. 1911. Beobachtungen und Versuche über den normalen und inversen Situs viscerum et cordis bei Anurenlarven. *Roux' Arch., 32:* 1–35.

Przibram, H. 1913. *Experimentalzoologie,* 4. *Vitalität.* Wien, Fr. Deuticke, 179 pp.

Reverberi, Giuseppe. 1931. Studî sperimentale sull'uovo di Ascidie. (Il comportamento e lo sviluppo dei frammenti d'uovo di *Ciona* nei diversi momenti compresi fra la deposizione e la segmentazione) *Pub. Staz. Zool. Napoli, 11:* 108–93.

Rose, S. Meryl. 1939. Embryonic induction in the ascidia. *Biol. Bull., 77:* 216–32.

Roux, Wilhelm. 1885. Einleitung zu den Beiträgen zur Entwickelungsmechanik des Embryo. *Zeitsch. f. Biol., 21:* 411–28. Also in *Gesam. Abh., 1895, 2:* 1–23.

———. 1888. Beiträge zur Entwickelungsmechanik des Embryo. Nr. V. Ueber die künstliche Hervorbringung "halber" Embryonen durch Zerstörung einer der beiden ersten Furchungszellen, sowie über die Nachentwickelung (Postgeneration) der fehlenden Körperhälfte. *Virch. Arch., 114:* 113–53, 246–91. Also in *Gesam. Abh.,* 1895, *2:* 419–521.

Runnström, J. 1926. Experimentelle Bestimmung der Dorso-Ventralachse bei dem Seeigelkeim. *Ark. f. Zool., 18A:* 1–6.

Ruud, Gudrun. 1925. Die Entwicklung isolierter Keimfragmente frühester Stadien von *Triton taeniatus. Roux' Arch., 105:* 209–93.

Ruud, Gudrun, and Hans Spemann. 1922. Die Entwicklung isolierter dorsaler und lateraler Gastrulahälften von *Triton taeniatus* und *alpestris,* ihre Regulation und Postgeneration. *Roux' Arch., 52:* 95–166.

Schleip, W. 1929. *Die Determination der primitiven Entwicklung.* Leipzig, Akad. Verlags, 914 pp.

Schultze, O. 1894. Die künstliche Erzeugung von Doppelbildungen bei Froschlarven mit Hilfe abnormer Gravitationswirkung. *Roux' Arch., 1:* 269–306.

Sedgwick, A. 1894. On the inadequacy of the cell theory of development, and on the early development of nerves, particularly of the third nerve and of the sympathetic in elasmobranchii. *Quart. J. Mic. Sci., 37:* 87–101.

Spek, Josef. 1926. Ueber gesetzmässige Substanzverteilungen bei der Furchung des Ctenophoreneies und ihre Beziehungen zu den Determinationsproblemen. *Roux' Arch., 107:* 54–73.

———. 1927. Ueber die Winterknospenentwicklung, Regeneration und Reduktion bei *Clavellina lepadiformis* und die Bedeutung besonderer "omnipotenter" Zellelemente für diese Vorgänge. *Roux' Arch., 111:* 119–72.

Spemann, H. 1906. Ueber embryonale Transplantation. *Naturwiss. Rund., 21:* 1–7 (pagination of reprint).

———. 1918. Ueber die Determination der ersten Organanlagen des Amphibienembryo I–VI. *Roux' Arch., 43:* 448–555.

———. 1924. Ueber Organisatoren in der tierischen Entwicklung. *Naturwiss., 12:* 1092–94.

———. 1927. Organizers in animal development. *Proc. Roy. Soc.* (B), *102:* 177–87.

———. 1931. Ueber den Anteil von Implantat und Wirtskeim an der Orientierung und Beschaffenheit der induzierten Embryonalanlage. *Roux' Arch., 123:* 389–517.

———. 1931a. Das Verhalten von Organisatoren nach Zerstörung ihre Struktur. *Verh. Deutsch. Zool. Ges.,* 129–32.

Spemann, H., and Hermann Falkenberg. 1919. Ueber asymmetrische Entwicklung und Situs inversus viscerum bei Zwillingen und Doppelbildungen. *Roux' Arch., 45:* 371–422.

Spemann, H., and B. Geinitz. 1927. Ueber Weckung organisatorischer Fähigkeiten durch Verpflanzung in organisatorische Umgebung. *Roux' Arch., 109:* 129–75.

Suzuki, S. 1928. Defektversuche an ventralen und lateralen Bezirken der Randzone von Pleurodeleskeimen. *Roux' Arch., 114:* 371–457.

Teissier, G. 1931. Étude experimentale du développement de quelques Hydraires. *Ann. Sci. Nat. Zool. 10e sér. 14:* 5–60.

Tennent, D. H., C. V. Taylor, and D. M. Whitaker. 1929. An investigation on organization in a sea-urchin egg. *Carnegie Inst. Wash. Pub. 391:* 1–104.

Twitty, V. C. 1932. Influence of the eye on the growth of its associated structures, studied by means of heteroplastic transplantation. *J. Exp. Zool., 61:* 333–74.

———. 1936. Correlated genetic and embryological experiments on *Triturus.* I. and II. *J. Exp. Zool., 74:* 239–302.

———. 1937. Experiments on the phenomenon of paralysis produced by a toxin occurring in *Triturus* embryos. *J. Exp. Zool., 76:* 67–104.

Twitty, V. C., and D. Bodenstein. 1939. Correlated genetic and embryological experiments on *Triturus.* III. and IV. *J. Exp. Zool., 81:* 357–98.

von Ubisch, L. 1925. Entwicklungsphysiologische Studien an Seeigelkeimen. I. Ueber die Beziehungen der ersten Furchungsebene zur Larvensymmetrie, und die prospektive Bedeutung der Eibezirke. *Zeit. wiss. Zool., 124:* 361–80.

———. 1925a. Entwicklungsphysiologische Studien an Seeigelkeimen. II. Die Entstehung von Einheitslarven aus verschmolzenen Keimen. *Zeit. wiss. Zool., 124:* 457–86.

———. 1931. Untersuchungen über Formbildung mit Hilfe experimentell erzeugter Keimblattchimären von Echinodermenlarven. *Roux' Arch., 124:* 181–240.

———. 1933. Formbildungsanalyse an Seeigellarven. *Naturwiss., 21:* 183–86.

———. 1933a. Keimblattchimären. *Naturwiss., 21:* 325–29.

———. 1936. Ueber die Organisation des Seeigelkeims. *Roux' Arch., 134:* 599–643.

———. 1938. Ueber Keimverschmelzungen an *Ascidiella aspersa. Roux' Arch., 138:* 18–36.

———. 1939. Keimblattchimärenforschung an Seeigellarven. *Biol. Rev., 14:* 88–103.

———. 1939a. Ueber die Entwicklung von Ascidienlarven nach frühzeitiger Entfernung der einzelnen organbildenden Keimberzirke. *Roux' Arch., 139:* 438–92.

Vogt, Walther. 1925. Gestaltungsanalyse am Amphibienkeim mit örtlicher Vitalfärbung. Vorwort über Wege und Ziele. I. Teil: Methodik und Wirkungsweise der örtlichen Vitalfärbung mit Agar als Farbträger. *Roux' Arch., 106:* 542–610.

———. 1929. Gestaltungsanalyse am Amphibienkeim mit örtlicher Vitalfärbung. II. Gastrulation und Mesodermbildung bei Urodelen und Anuren. *Roux' Arch., 120:* 384–706.

Waddington, C. H. 1938. Studies on the nature of the amphibian organisation centre. VII. Evocation by some further chemical compounds. *Proc. Roy. Soc.* (B), *125:* 365–72.

Waddington, C. H., J. Needham, W. W. Nowinski, and R. L. Lemberg. 1935. Studies on the nature of the amphibian organisation centre. I. Chemical properties of the evocator. *Proc. Roy. Soc.* (B), *117:* 289–310.

Weigmann, Rudolph. 1927. Ueber die Bestimmung der Medianebene im Froschei. *Zeit. wiss. Zool., 129:* 48–102.

Whitaker, D. M. 1931. Some observations on the eggs of *Fucus* and upon their mutual influence in the determination of the developmental axis. *Biol. Bull., 61:* 294–308.

Whitaker, D. M., and E. W. Lowrance. 1936. On the period of susceptibility in the egg of *Fucus furcatus* when polarity is induced by brief exposure to directed white light. *J. Cell. Comp. Physiol., 7:* 417–424.

Whitman, C. O. 1893. The inadequacy of the cell-theory of development. *J. Morph., 8:* 639–58.

Wilson, E. B. 1903. Experiments on cleavage and localization in the nemertine-egg. *Roux' Arch., 16:* 411–60.

———. 1904. Experimental studies on germinal localization. I. The germ-regions in the egg of *Dentalium. J. Exp. Zool., 1:* 1–72.

———. 1904a. Experimental studies in germinal localization. II. Experiments on the cleavage-mosaic in *Patella* and *Dentalium. J. Exp. Zool., 1:* 197–268.

Woerdeman, N. W. 1933. Ueber den Glykogenstoffwechsel des Organisations-zentrums in der Amphibiengastrula. *Proc. Acad. Sci. Amsterdam, 36:* 189–93.

Yatsu, N. 1912. Observations and experiments on the ctenophore egg. III. Experiments on germinal localization of the egg of *Beroë ovata. Annot. Zool. Jap., 8:* 5–13.

Zeleny, C. 1904. Experiments on the localization of developmental factors in the nemertine egg. *J. Exp. Zool., 1:* 293–329.

Zoja, R. 1895. Sullo sviluppo dei blastomeri isolati dalle uova di alcune meduse (e di altri organismi). *Roux' Arch., 1:* 578–95; (1896) *2:* 1–37.

2

Harrison Stages and Description of the Normal Development of the Spotted Salamander, *Amblystoma punctatum* (Linn.)

The need for standardized stages in the embryonic development of various organisms for the purpose of accurate description of normal development and for utilization in experimental work has long been recognized and has led to practical application in the hands of many investigators. By far the most comprehensive of such efforts is the series of "Normal Plates," which were begun under the editorship of the late Professor Keibel and now have reached sixteen volumes.

Several series of normal stages of amphibian development have been brought out, including *Necturus, Molge vulgaris* (*Triton*), and *Megalobetrachus* (*Cryptobranchus*) *japonicus,* in the Keibel series; *Diemictylus* (*Triturus*), by Oyama (1930); *Rana sylvatica,* by Pollister and Moore (1937); and *Rana pipiens,* by Shumway (1940). Unfortunately, the numbering of the stages has been done independently in each series and is not the same in any two, except in the case of *R. sylvatica* and *R. pipiens.*

In connection with experimental work in which embryos of the spotted salamander (*Amblystoma punctatum*) were mainly used, a series of normal plates of this species was begun under my direction many years ago and brought to near completion in 1927 in the form of wash drawings by Miss Lisbeth Krause, and are reproduced here at about half scale.

The spotted salamander lays its eggs in small pools in the woods or, less frequently,

Prepared by Ross G. Harrison. Previously unpublished.

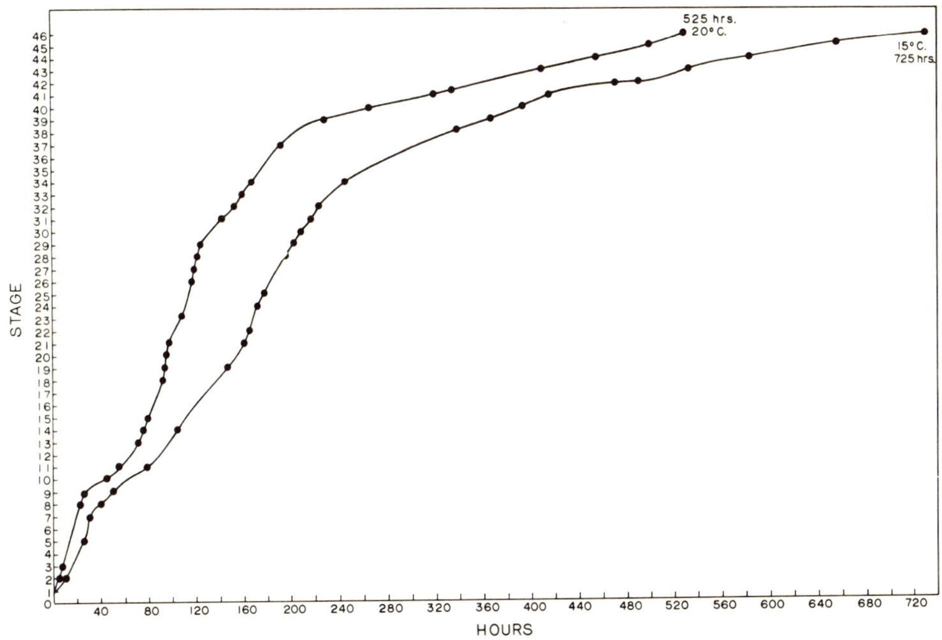

GRAPH 1. Normal embryonic development of *Amblystoma punctatum* between fertilization and yolk resorption at 15° and 20° C. Black dots represent averages of several determinations. After Harrison, unpublished.

in open country in early spring. In the vicinity of New Haven freshly laid eggs are found in the latter part of March and the first part of April. In the South they may be obtained in January and February, and in northern New England early in May. Freshly laid eggs have been collected on Mount Mansfield, Vermont, late in May. This species does not respond well to pituitary treatment, and it does not seem feasible to extend the spawning season by this means.

Development is a continuous process with an indefinite number of stages. Out of these, forty-six have been chosen to designate those between fertilization and complete resorption of the yolk. They are like a few frames taken from a motion picture film; they are chosen so as to be readily recognizable and are based mainly on external form. The time interval between stages is not constant, being short during the process of segmentation and neurulation and longer in gastrulation. After closure of the neural folds the intervals gradually lengthen. These features are shown in Graph 1, in which the numbered stages are plotted against time. The whole course of development between fertilization, which immediately precedes laying, and yolk resorption requires about 725 hours at 15°C and 525 hours at 20°C.

Each stage number marks the midpoint between the preceding and the succeeding stages. Since experiments and observations often have to be made upon embryos which either have not reached or have passed this point, the designation minus or plus has to be added to the stage number on occasion. In the gastrulation stages, unfortunately, the intervals are too far apart, and succeeding stages are designated 9+, 10–, etc.

Maturation and fertilization have not been studied cytologically in *Amblystoma punctatum*, but in the axolotl (*A. mexicanum*), according to Fick (1893), maturation begins, as in other vertebrates, in the ovary. The first polar body is given off shortly after dehiscence of the egg, when it is lying in the body cavity. The second polar body is not given off until after fertilization, which takes place in the cloaca of the female. There is no copulation, but the female lifts into the cloaca the spermatophores previously deposited by the males on leaves and twigs on the bottom of the pools.

The eggs are laid shortly after fertilization in masses of varying size surrounded by a viscid jelly which becomes firm and swells considerably on exposure to water. Usually the jelly is glass-clear, but not infrequently is opaque white. Eggs from both kinds of jelly develop normally.

Each egg lies within a capsule which is embedded in the jelly. The capsule consists of an outer homogeneous, slightly opaque whitish layer, 0.15–0.20 mm thick, and a series of concentric layers of fibrous appearance, of which there may be from five to nine. The capsule is filled with a clear fluid in which the egg lies covered by a thin hyaline vitelline membrane, which is separated from the egg by a narrow vitelline space.

Eggs laid in the night or early morning may still be unsegmented when collected early and brought into the laboratory. Their cleavage begins at different times according to the period elapsed since spawning, so that their age would have to be reckoned from the first cleavage, or Stage 2. However, observations made by Dr. J. C. Lipsett (unpublished) make it possible to fill in this gap. In artificially fertilized eggs the interval between fertilization and first cleavage is 6 hours and 40 minutes at 20°C and 10 hours and 20 minutes at 15°C. The second polar body is given off about 3 hours and 20 minutes after fertilization at 19–20°C. The polar bodies are loosely attached to the egg near the animal pole but often become entirely free and may be found on any part of the egg after cleavage is underway.

There is considerable variation in the sequence of cleavages as well as in the actual pattern at different stages. For instance, the first cleavage may cut the egg in the plane of symmetry, at right angles to it or in intermediate planes. The first four cleavages may be quite regular or so irregular as to be almost random.

Outline of Normal Stages of *Amblystoma Punctatum*

Stage	
1	Unsegmented egg (polar bodies, pigment streaming, gray crescent).
2–7	Cleavage stages.
8–9	Early and late blastula (formation of blastocoele).
10–12	Gastrula (invagination, changes in contour of dorsal lip, formation of archenteron and blastopore).
13–20	Neurula (neural groove and plate, gradual rise of neural folds, closure of neural folds with fusion mainly in the trunk region at stage 20).
21–29	Rapid growth of the head process with gradual extension beyond the ventral surface of the embryo. Enlargement and rounding of the optic vesicle. Other sense organs evident: nasal pit and ear vesicle. Placodes of lateral line sense organs established anterior and posterior to ear vesicle. Branchial system (branchial grooves and clefts). At stage 26, definite prominence of the midbrain and slight hump of the hindbrain. At stage 27, protrusion of forebrain at ventral tip of head. At stage 28, midbrain region enlargement in dorsal bend of the head.
30–35	Gradual straightening of head curvature with elongation of trunk and reduction of trunk curvature. Progressive lengthening of tail bud. Beginning migration of trunk lateral line sense organs from postauditory lateral line placode. Development and increase in depth of the pericardial cavity with the growth of the heart tube, followed by definite twist to the right of the heart tube. Initiation of slow pulsation of heart between stages 34 and 35. Circulation of blood in gills and yolk at stage 37– .
36–40	Flattening of the ventral surface of the head with rounding of anterior region (forebrain). Elongation of balancer. Further development of gills. Progression of forelimb to paddle form. Beginning of differentiation in anterior part of trunk with liver and pancreas becoming definite posterior to the heart.

41–46 Further development of mouth and forelimb (mouth open at stage 43). Elongation of gills. Shift in position of nasal openings. Development of digestive system with absorption of yolk.

Description of Normal Stages of *Amblystoma punctatum*

Stage 1. A single cell, the upper hemisphere of which is dark brown and the lower a light gray, almost white. Viewed from the side, the egg is quite oblate above, owing to the fluidity of its contents. Changes take place in the distribution of pigment in this stage as have been described in many species of amphibia. These changes consist in a streaming of the cytoplasm involving to some extent the cortical pigment. As seen from the outside they are variable in *A. punctatum* and do not lead to the formation of a definite gray crescent as in some anuran eggs. The pigment may form in meridional streaks, and a latitudinal dark bar may form on one side of the egg with a lighter area above as well as below. As a result of these movements the boundary between the dark and light portions of the egg does not run equatorially but is oblique, and at one side of the egg it may be as high as 45° from the equator. Viewed from above, the egg shows a light area crescentic in shape on one side. Marking experiments with vital dye[1] indicate that this corresponds to the dorsal region of the embryo and that the plane bisecting this area corresponds to its median plane.

Stage 2. Two cells. This stage is initiated at the beginning of the first cleavage furrow, which is vertical, and ends just before the appearance of the second furrow. While the furrow is cutting through the upper part of the egg, it is very deep and the upper surface of the two cells is rounded. As the furrow approaches the animal pole the whole of it flattens. The first cleavage plane may cut the symmetry plane of the egg at any angle. Twenty-five eggs from a single bunch showed the following relations: in 44% the first cleavage coincided with the symmetry plane (sagittal), in 16% it cut the symmetry plane at right angles (frontal), and in 40% it was oblique. Another bunch of eggs showed the first cleavage to be more frequently frontal than sagittal: 50% frontal, 25% sagittal, 25% oblique.[2]

Stage 3. Four cells. This stage begins with the second (also vertical) cleavage, which may occur shortly before the first furrow is complete. It ends with the begin-

1. Made by Sally Wilens.
2. Determinations made by Sally Wilens.

ning of the third, or latitudinal, furrow. At this stage one pair of cells is sometimes larger than the other. During the progress of this furrow the upper surface of all 4 cells again becomes rounded. Viewed from above the two furrows may cross in a point, or two cells diagonally opposite may touch along a short line. The latter is the condition of equilibrium and occurs more frequently.

Stage 4. Eight cells. This stage is initiated usually by the formation of a horizontal or latitudinal furrow above the equator of the egg. The cells of the upper quartette are smaller and usually are all darkly pigmented. The larger cells of the vegetative quartette show dark areas of different size owing to the pigment boundary running oblique to the equator. A small cavity enclosed by the 8 cells is the segmentation cavity. In some eggs, one or more of the furrows dividing the 4 cells of stage 3 may run obliquely or may all even be vertical.

Stage 5. Sixteen cells. In the regular form this stage arises from the vertical division of the 8 cells of the previous stage. Since, however, the cells of the animal portion of the egg are more precocious in dividing, there is often an intermediate 12-cell stage.

Stage 6. Twenty-four cells, arranged in a single layer around a central cavity. This stage proceeds from the previous one through the division of the 8 upper cells ahead of the 8 lower ones. The regular pattern is pretty well obliterated by this time.

Stage 7. Approximately 100 cells. Since some cleavage planes are parallel to the egg surface there are now cells that do not touch the exterior. This is the last of the cleavage stages, the designation being, however, purely arbitrary, since the cells of the egg continue to divide.

Stage 8. There are perhaps 200 cells at this stage, which is characterized as an early blastula.

Stage 9. Late blastula, with innumerable small cells. The upper hemisphere, consisting of several layers of cells, roofs over the segmentation cavity. The lower hemisphere is a solid mass of much larger yolk cells. The smaller pigmented cells extend farther below the equator than previously, and in one sector of the egg, probably corresponding to that of light pigmentation in stage 1, it extends farther toward the vegetal pole than elsewhere. In the polar area of the less pigmented yolk cells, there is an irregular network of pigment along the cell boundaries. This is irregular and variable, becoming more definite in stage 10.

Stage 10–. Dorsal blastoporic lip present with beginning of invagination along a curved line forming the dorsal quadrant, a circle surrounding the vegetal pole. The bottom of the groove is intensely pigmented, and there is also an irregular accumulation of pigment in the yolk cells. Sagittal sections show the invaginating bottle cells or flask cells with their narrow necks ending in the groove. At stage 10, the dorsal lip is narrower and the invagination deeper than in stage 10– and is open. Pigmentation is intense both in the archenteric groove and in the yolk cells. At stage 10+, the blastopore is now a horizontal slit with a sharp dorsal lip extending almost straight transversely. In the yolk field there are streaks of pigment centering toward the blastopore. The invagination is deep but the archenteric roof almost touches the floor.

Stage 11–. The blastoporic lip now forms a semicircle. The opening is very narrow, the lip touching the yolk below. The invagination (seen in section) extends in about 90°—i.e. to above the equator but does not communicate with the segmentation cavity. Pigment streaks on the yolk running toward the blastoporic lip are fewer. At stage 11, the lateral blastoporic lips have extended farther ventrally to about three quadrants of circumference. The radius of curvature is smaller. Two sharp pigment streaks generally stream ventrally from the ends of the blastopore lip.

Stage 12–. The blastopore is completed and forms an elliptical or circular groove surrounding a large "yolk plug." Ventral invagination is shallow—about the same as the dorsal in stage 10. The archenteric cavity is large with arched roof. Segmentation cavity much reduced and now nearly filled with loose round cells. At stage 12, the blastopore is a small round opening out of which a small yolk plug projects. The egg has rotated about 90°, so that the blastopore now seems to lie at the equator instead of the vegetative pole. The animal pole of the egg now lies at the opposite end of the diameter which cuts the blastopore. In profile the egg, no longer round, is slightly ovoid. The blastocoele is almost obliterated.

Stage 13. The blastopore is a narrow vertical opening, somewhat dumbbell-shaped. Extending from the blastopore in the midline, the neural keel runs anteriorly over the dorsal surface of the egg in the midline. Embryo now slightly elongated in its blastoporic (anteroposterior) diameter; length:height = 1.17.

Stage 14. The embryo is now more elongated anteroposteriorly and is more flattened on the dorsal side; length:height = 1.31. Neural plate shows, with sharp median groove, but there are no distinct neural folds. Embryo more pointed posteriorly with prominences on each side of blastopore.

Stage 15. Shield-shaped neural plate surrounded by prominent neural ridge which is, however, not much lifted from the general surface of the embryo; there is a swelling on each side of the blastopore. Length:Height = 1.43.

Stage 16. Neural folds raised. Spinal portion of neural plate narrowed so that whole structure is racquet-shaped. Length:height = 1.49.

Stage 17. Neural folds higher, especially in head region, which is now narrowed. One pair of somites present. A short transitory stage not much used in experiments.

Stage 18. Neural folds again higher. Head groove narrow, fore and hindbrain vesicles outlined. Two somites. Length:height = 1.6.

Stage 19. Neural folds approximating. Brain curvature now distinct in profile. Fore-, mid-, and hindbrain vesicles outlined. Optic vesicle shows from surface. Mandibular arch marked off by shallow groove posteriorly. Three somites.

Stage 20. Neural folds completely fused in the trunk region. In the head there is a narrow cleft between them. At the ventral level of the hindbrain the ectoderm is drawn in, forming a sharp groove. Four somites.

Stage 21. Neural folds completely fused throughout. Optic vesicles are more distinct. The mandibular arches are bounded posteriorly by a shallow hyomandibular groove, which gradually has become more distinct. The head does not project much beyond the general surface of the egg in profile—only in the region of the neural keel around the dorsal semi-circumference. The pronephros scarcely shows from the surface, nor does the mandibular arch. When the ectoderm is removed, a number of myotomes show. The mandibular mesoderm appears as a band running behind the optic vesicle and curving ventrally to the midline where it ends in a blunt curved border. Four somites.

Stage 22. In this stage the head with the optic vesicles has become more distinct, and the mandibular arch with its hyomandibular groove becomes faintly apparent. Five somites.

Stage 23. The head becomes still more prominent as do the mandibular arch and hyomandibular groove. The mandibular arch forms a low ridge running from the side of the neural cord ventrally behind the eye. The hyomandibular groove runs far ventrally, gradually becoming more shallow and less distinct. Dorsally, just below the dorsal keel, it is a deep pit at this stage. The pronephric swelling is not prominent but is nevertheless clearly evident in the region of the third and fourth somites, which also are visible through the skin. Six somites.

Stage 24. (This embryo, though figured as 24, perhaps more nearly approaches 25, certainly as regards the ganglion crest.) The head is more prominent than in 23 but the head in front of the eye is still very blunt. The hyobranchial groove appears at this stage. Like the hyomandibular it is more distinct dorsally, where it is almost a pit running ventrally as a groove converging slightly toward the hyomandibular, but it does not extend so far ventrally. There is also a slight groove, very broad and gradual, behind the gill region in front of the pronephric swelling, which is now more distinct. This marks off the branchial swelling as an oval prominence, but it is still only slightly marked. The pronephros is more distinct—has assumed its pear-shaped form—and the ridge extends over four or five segments.

If the ectoderm is removed at this stage, the downgrowth of the neural crest is easily noted. The mandibular arch is covered to midway past the eye (see Stone 1922, p. 431, stage 25). The middle and posterior processes of the neural crest are not as far along as in this figure.

A break is seen in the mesoderm at this stage just behind the mandibular arch due to the out-pushing of the hyomandibular pouch.

The ear is now a distinct depression in the deep ectodermal layer running between two processes of the neural crest. It is usually visible from the outside. Nine somites.

Stage 25. This is distinguishable from 24 by the greater prominence of the head, the mandibular arch being now in the free portion. The tail bud is not yet distinct. The hyobranchial groove is elongated but does not extend to the ventral end of the gill swelling. When the ectoderm is removed the mandibular arch stands out as a sort of collar. It is covered largely with mesectoderm. Now the mesoderm of the hyoid arch is segregating and there is a strip dorsally where it is separated from the branchial mesoderm. Here the endoderm of the hyobranchial reaches out to the ectoderm.

When the pharynx is opened, the hyomandibular pouch shows as a groove which is fairly sharp and the hyobranchial as a very shallow not sharply defined depression.

The ear spot shows, at least in many cases, dorsal to the hyobranchial groove. Nine somites.

Stage 26. The body of the embryo begins to lengthen, the head becomes still more prominent, and the tail bud begins to show. On the surface of the pharyngeal region the most marked changes are the further extension ventrally of the hyobranchial groove and the appearance of the first branchial groove, which shows as a shallow pit on the dorsal part of the branchial swelling. The gill swelling itself is more definite than in the previous stage and appears as an oval mound owing to the deepening of the depression between it and the pronephros. The ear spot is more definite and shows in practically all ears.

The mesectoderm from the neural crest extends down over the mandibular arch nearly to the ventral border of the eye—that of the hyoid about one third of the way and that of the branchial almost the same distance.

From the inside of the pharynx two visceral clefts are distinct, the hyomandibular being a long sharp groove, the hyobranchial also sharp, extending about half as far. The first branchial is not yet present. Ten somites.

Stages 26–27. If the skin is removed the neural crest is seen to have made further progress in its downward growth (see Stone's figures, 1922). The mandibular arch is now almost entirely covered; the hyoid arch extends halfway ventrally and the two processes of the branchial region nearly as far. The first branchial depression, like that of the hyobranchial in earlier stages, is probably caused by the flanking of the two branches of the neural crest mesectoderm. The mesoderm of the first branchial region has not yet broken through; i.e. the first branchial pouch does not reach through to the ectoderm. The pharynx when opened shows three visceral pouches: the hyomandibular is longest, then the hyobranchial, and finally the short first branchial, limited to the dorsal part of the wall of the pharynx.

In stage 27 the stomodeum begins to show, the endodermal plug showing through faintly. Twelve somites.

Stage 28. This is marked by still further elongation of body, greater prominence of the head, and distinctness of the tail bud. The first branchial groove is also more distinct. The nasal area begins to show faintly. The endodermal plug shows more distinctly through the stomodeal membrane.

Removal of ectoderm reveals further downgrowth of the neural crest mesectoderm. The hyomandibular and hyobranchial pouches are slightly fused with the ectoderm, which when torn away may leave a slit into the pharynx. This feature however becomes more distinct in the next stage.

The inner wall of the pharynx shows the three visceral pouches still more distinctly. They are sharp grooves and the first branchial is about as long as the hyobranchial, but neither is more than half as long as the hyomandibular, which extends to the ventral midline. Fourteen somites.

Stage 29. The external features marking off the gill region are all more distinct—the tail bud is definitely more prominent. This stage may be distinguished from the preceding chiefly by the tail bud and the length of the head, together with the head curvature. After this stage the latter begins to straighten.

In removal of ectoderm it is found that the mesectoderm is farther ventral, this being especially noticeable in the posterior branchial region where it has lagged in the previous stages. The contact between the endoderm of the first branchial pouch and the ectoderm is now definitely established, though it is not so intimate as that of the hyomandibular, which, when the ectoderm is torn away, stands out as a slit leading into the pharynx. The hyobranchial slit at this stage may also show the same relation.

The internal wall of the pharynx has changed somewhat. The first branchial pouch has deepened, and from the posterior part of the recess the second branchial pouch may be seen. It is not very distinct at this period.

The relation of the first branchial pouch to the ectoderm renders operations involving the mesoderm difficult. There is left an area in the graft which is devoid of mesoderm and the sticking of the endoderm often necessitates rather rough treatment, with tearing of the mesoderm and mesectoderm, especially of the first branchial arch. Sixteen somites.

Stages 30–34. These are characterized by the gradual straightening of the head curvature by the elongation of the trunk, resulting in the dorsal trunk curvature being gradually reduced, and by the increase in the tail bud. The changes here are very gradual and the intermediate stages are not clearly definable. Stage 33 is, however, definite for it is then that the first movement (simple contraction in the side stimulated) begins. Embryos which move and those which do not have the same external appearance. The former have been designated then as stage 33+. By stage 35 the body curvature is practically eliminated, the body axis being straight from the hindbrain back.

As regards the mesectoderm, it moves farther ventrally in these stages. The mandibular, hyoid, and the first branchial processes are from the outset independent, but behind the first branchial arch there is a plate of mesectoderm which becomes broken into three vertical strips as the other branchial pouches push through. In stages 30 and 31 this has not yet occurred, the last pouch to reach the ectoderm being still the first branchial. This is now quite definite. On opening the pharynx, pouches show very clearly that the first branchial has now elongated ventrally and that the second branchial forms a recess at the dorsal part of the line where the lateral wall meets the posterior wall of the pharynx. In stage 31 the fin folds, especially ventrally, become distinct. In stage 30 it is not yet clear except when viewed ventrally. The lens pit shows in stage 31 but not in 30.

Somites: stage 30, eighteen; stage 31, nineteen; stage 32, twenty; stage 33, twenty-one or twenty-two; stage 34, twenty-four or twenty-five. After stage 34 it is difficult to make an accurate count of the somites.

Stages 32–36. Heart development. In stages 32–34 the pericardial cavity becomes progressively deeper. In stage 32, pericardial cavity present. No definite heart tube. In stage 33, definite heart tube with slight twist to the right and ventral mesocardium. In stage 34, heart tube more distinct. In stage 34 or 35–, still greater size of heart tube with bulge to the right. Beginning of slow pulsation. No evidence of movement of blood from outside. In stage 35, distinct but slow pulsation. Very rapid development of the bulge and twist to the right side. In stage 36, heart is larger—shape not changed much—bend to the right more accentuated, auricular portion more expanded. No movement of blood in yolk visible.

Stages 32–33. In stage 32 on lifting the skin one can see that the second branchial pouch has been pushed through the mesoderm to the skin, but the contact is slight and not very definite. In stage 33 it is extended, and by 33+ it has become a definite slit up to the ectoderm, but of course none of the branchial slits is broken through. There is little change in the interior view of the pouches in these stages except that, by 33+, the dorsal extremity of the two branchial pouches appears as a distinct pit running ventrally into a groove. The two last pouches now run off from an alcove at the corner of the pharynx.

Stage 35. This stage is marked by the straightening of the embryo, which now extends in a straight line from the hindbrain to the base of the tail. The latter may have a slight ventral bend at this time or it may be straight or even bent dorsally, and at the hindbrain there is a slight dorsal hump. The straightening of the back is not exactly synchronized with changes in the head (gills and balancer). Some embryos that are bent ventrally or are straight may be further along than those with tail curved upward. In the head region the balancer bud has now become visible as a round nodule on the mandibular arch, dorsoposterior to the eye. Lifting the skin shows the third branchial projecting through mesoderm and mesectoderm. The three external gills show as nodules projecting from the gill mass running in a ventroanterior–dorsoposterior direction; the first is considerably smaller.

In the interior of the pharynx the three branchials show they are all offshoots from a recess in the dorsoposterior corner of the pharyngeal cavity.

The first chromatophores appear at this stage distributed sparingly along the anterior part of the flank. The lateral line is rather variable but may reach to the beginning of the sixth somite. Coil movement of the embryo is initiated.

Stage 36. In this stage there are definite gill sprouts that stick out to the side from the general gill swelling. The balancer bud is more distinct, and the three sprouts together with the nasal pit lie in an almost straight line. Blood vessels have begun to develop (in stage 37 the first trace of the fourth branchial), and a sluggish circulation may be seen in the yolk vessels. It is difficult to remove the ectoderm without tearing underlying tissues, and such preparations are unsatisfactory. Interiorly, the pharynx is beginning to flatten out dorsoventrally, probably owing to pressure caused by the development of the heart. Embryo shows *S*-curve movement response to stimulation. About 26 somites.

Stage 37. There are three distinct gills pushing ventroposteriorly. The balancer is a short, rounded, cylindrical process converging toward the first gill and about the same length. Blood can be seen readily circulating in the gills and yolk vessels, as indeed it is in stages intermediate between stages 36 and 37.

Interiorly the pharynx is much flattened dorsoventrally and the clefts now extend around to the floor. The beginning of the fourth pouch running off from the posterior wall of the recess can be seen. About 28 somites.

Stage 38. The balancer is longer and slightly narrowed in its proximal portion. Secretion of sticky mucus begins at stage 38– . The gills are long and project more horizontally posteriorly and less ventrally than before. They reach the base of the forelimb, which has now become quite distinct and is just beginning to "point." The gills are more slender and tapering distally, but on each a pair of filament sprouts from the ventral border are discernible. The operculum is faintly visible as a rounded ridge running on the hyoid arch from the level of the base of the first gill and converging anteriorly toward its fellow of the opposite side, which, however, it does not yet meet. It is not very definite until the next stage. One deep otolith mass shows in the ear vesicle. This is probably the posterior or saccular otolith. In a specimen which could be called stage 38+ there is a very small otolith mass anterolateral to the larger one. About 30 somites.

Stage 39. The balancer is longer and more distinctly club-shaped. The gills are also longer and reach back to the tip of the forelimb bud or a little beyond. More filament processes have appeared and those present are longer. The first gill is distinctly smaller than the others, and in a typical specimen it has two pairs of filament buds, and the others have three. Two otolith masses are now visible, the anterior one being farther from the midline than the other.

The operculum is now distinct and its border can be followed almost to the midline. The limb bud points definitely and dorsoposteriorly. There is a small portion quite free from the body.

The pancreas shows as a transversely directed organ just posterior to the liver, the vitelline vein bending to the right (ventral view) and running to the right of the pancreas to disappear under the liver.

Stage 40. The gills are much larger than in stage 39 and, owing to the length of the filaments, are more feathery. They curve around dorsally, the tips standing almost perpendicular. The number of filaments is also increased. There are about four pairs of filaments in the first gill and six or seven in the second and third. They differ greatly in length, some being only nodules. The operculum is now distinct across the midline, the median notch being curved.

The pancreas shows now more distinctly as a whitish organ just posterior to the liver, the vitelline vein bending to the right at its posterior border. The liver is reddish, owing to the development of the portal circulation.

The anterior limbs are longer and the distal part is flattened slightly. The interdigital notch does not show, but the end is squared off. The cornea of the eye is cleared and the golden pigment cells of iris and choroid shine brilliantly.

Stage 41. The anterior end of the embryo is more rounded than before, since the forebrain does not project as in the earlier stages.

Viewed from below, the median notch of the operculum is a little deeper and sharper. The angles of the mouth are more distinct. Viewed from the side the overhang of the upper jaw is beginning to show, though it was hardly noticeable in stage 40. The gill filaments are longer and somewhat more numerous. The gills themselves are longer. The vitelline vein now takes a more distinct bend to the right, only faintly shown in the preceding stages. Sometimes the gall bladder is visible in this stage but usually not.

The forelimb now shows a distinct interdigital notch, and the first digit is a little more distinct and longer than the second.

Stage 42. The head is still further rounded and is broader and more blunt. The nasal openings are farther forward. The otolith in the saccus endolymphaticus first appears at this stage. The mouth is now a distinct groove, with the upper lip a little more prominent than the lower. It has not yet broken through.

The forelimb shows a much deeper bifurcation, and a slight lateral bulging marks the beginning of the elbow joint. The limb is considerably longer than in the previous stage. Viewed from the side, it shows a very slight enlargement in the upper or ulnar border, which will later form the two ulnar digits.

Viewed ventrally, the bend to the right of the vitelline vein is still more pronounced. The gall bladder may now be distinctly seen, a little to the left of the midline, filled with greenish bile. It is still small. It is to the left of the midline at the left end of the pancreas, which now is sharply defined from the much more reddish liver.

Stage 43. This stage is characterized by the breaking through of the mouth, which also has shifted forward relative to the anterior border of the head. The nasal pits are no longer on the ventral surface but have shifted to the lower slope of the anterior surface. The endolymphatic otolith is now larger and more distinct than before.

Viewed ventrally, the gall bladder is seen to be larger and the bend to the right of the vitelline vein is now quite sharp. There is now a deep fold running from right anterior to left posterior starting just behind the pancreas. This marks off an intestinal loop from the main portion lower down which now contains most of the yolk. This becomes still more marked in stage 44.

The forelimb is still longer, the interdigital groove deeper, and the elbow bend and the ulnar prominence (3rd and 4th digit rudiment) a little more distinct.

The hindlimb rudiments are just beginning to show as very slight nodules anterodorsal to the anus.

Stage 44. The changes which mark this stage concern principally the forelimb. The first digits are longer, the ulnar prominence more distinct. The whole limb is longer and is usually bowed slightly. The first movements of the limb begin in this stage at the shoulder joint but they are very slight.

In the head, the mouth is more widely open and the nasal openings have shifted farther around to the anterior surface.

The bend, to the right, of the vitelline vein is sharper, and more of the blood can be followed into the more superficial vessels of the liver. The balancer is not so plump—secretion of mucus abates.

Stage 45. Characterized by changes in the forelimb and in the topography of the abdominal viscera. The forelimb is longer than in stage 44 and the elbow bend is more distinct. The hand is further differentiated in that the third digit is distinct, the notch between it and the second making a right angle. The fourth digit is barely visible as a slight prominence on the ulnar border. The animal may now rest on its forelegs, which may be abducted and the elbow bent slightly forward. The hindlimb is now a distinct bud pointing posteriorly.

The ventral surface of the liver is much more extended anteroposteriorly, particularly on the left side where there is a long flap. The pancreas, which lies along the posterior border of the liver, thus has its long axis shifted from the transverse plane to an oblique direction running from anterior right to posterior left. There is still yolk in the intestinal wall. The part that will become the large intestine is now thrown into deeper folds than previously. Peristalsis may sometimes be seen.

The vitelline vein is no longer prominent and often becomes lost in the folds of the intestine. The transverse part runs across the (small) intestine, and below the border of the pancreas it crosses dorsal to it and breaks up into the smaller vessels in the liver. This is very variable. Arteries in the abdominal wall are now very distinct.

The mouth is now almost terminal and the nose, which is on the anterolateral curvature, is barely visible from the ventral side. The balancers are still attached.

The above describes what seems to be the usual condition. Sometimes, however, the limbs seem relatively further advanced than the viscera, and vice versa. Sometimes one or both balancers are shed while there is still considerable yolk left in the intestine—sometimes not until later.

Stage 46. This stage is marked by the completion of the absorption or elimination of the yolk. The intestines have no longer the white appearance that still persists in stage 45, but are evidently filled with a meconium often of a greenish hue owing to the bile.

The balancers have both been shed. There is much variation in this. The hand is much more fully differentiated, the third digit being from one third to one half as long as the second. The nodule on the ulnar border which will become the fourth digit is more distinct than previously. The movements of the forelimb are much freer and the animal now uses them in crawling and balancing. Both shoulder joint and elbow are movable. The hindlimb bud is longer and now projects to a slight extent free toward the tail.

The larva should begin to feed between stages 45 and 46; if not, the ventral contour of the abdomen is somewhat hollow.

Plate I

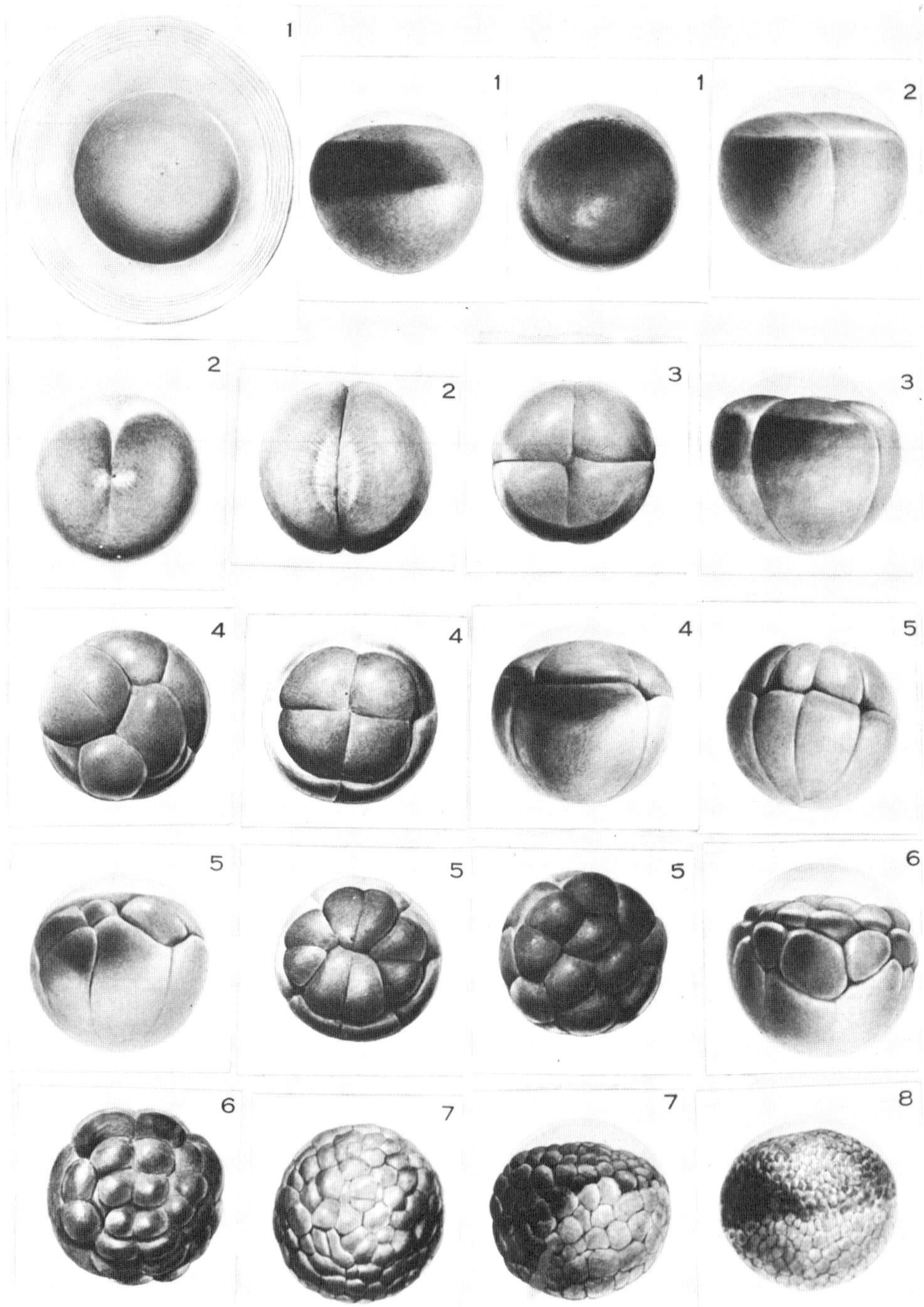

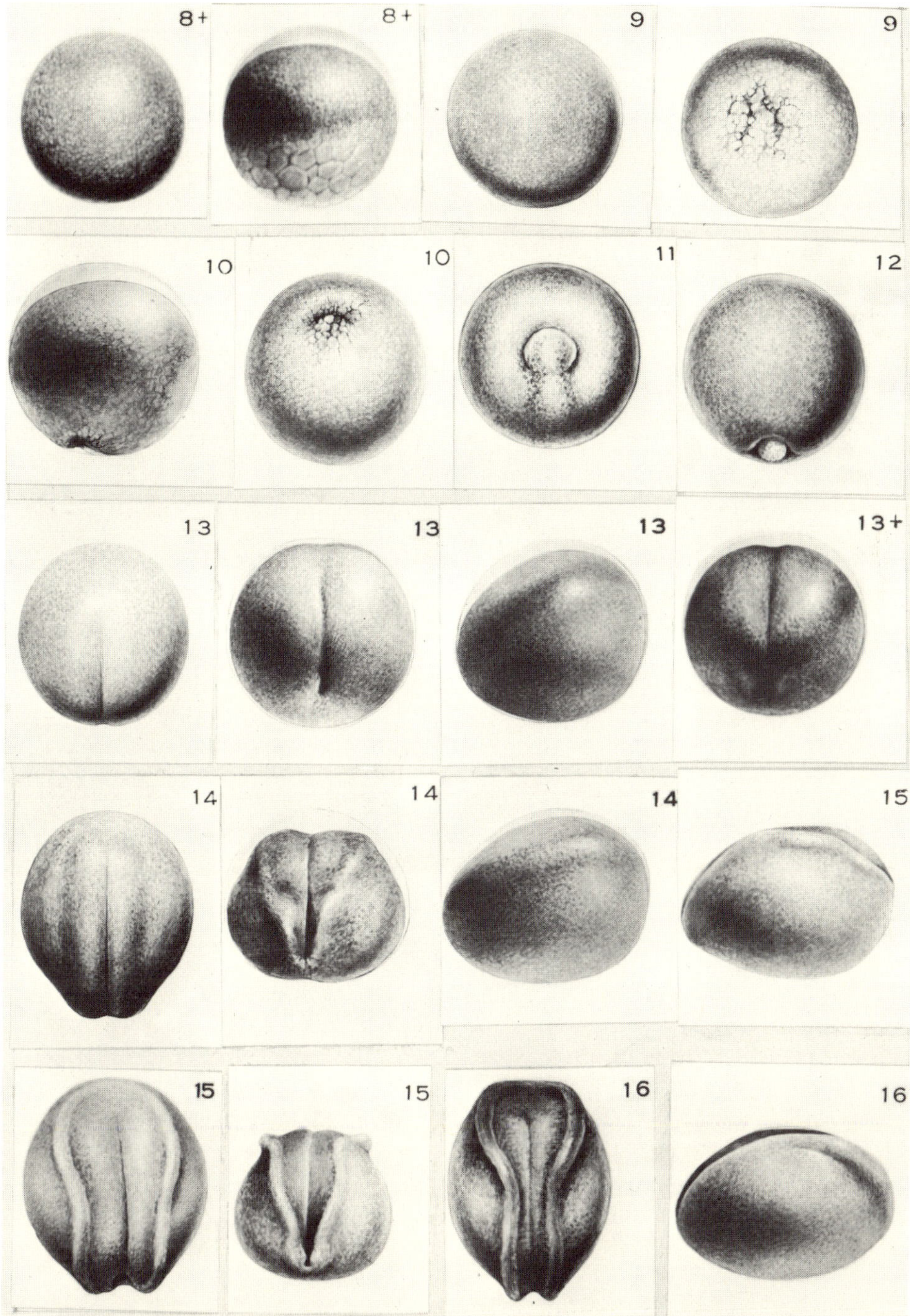
8+
8+
9
9
10
10
11
12
13
13
13
13+
14
14
14
15
15
15
16
16

Plate III

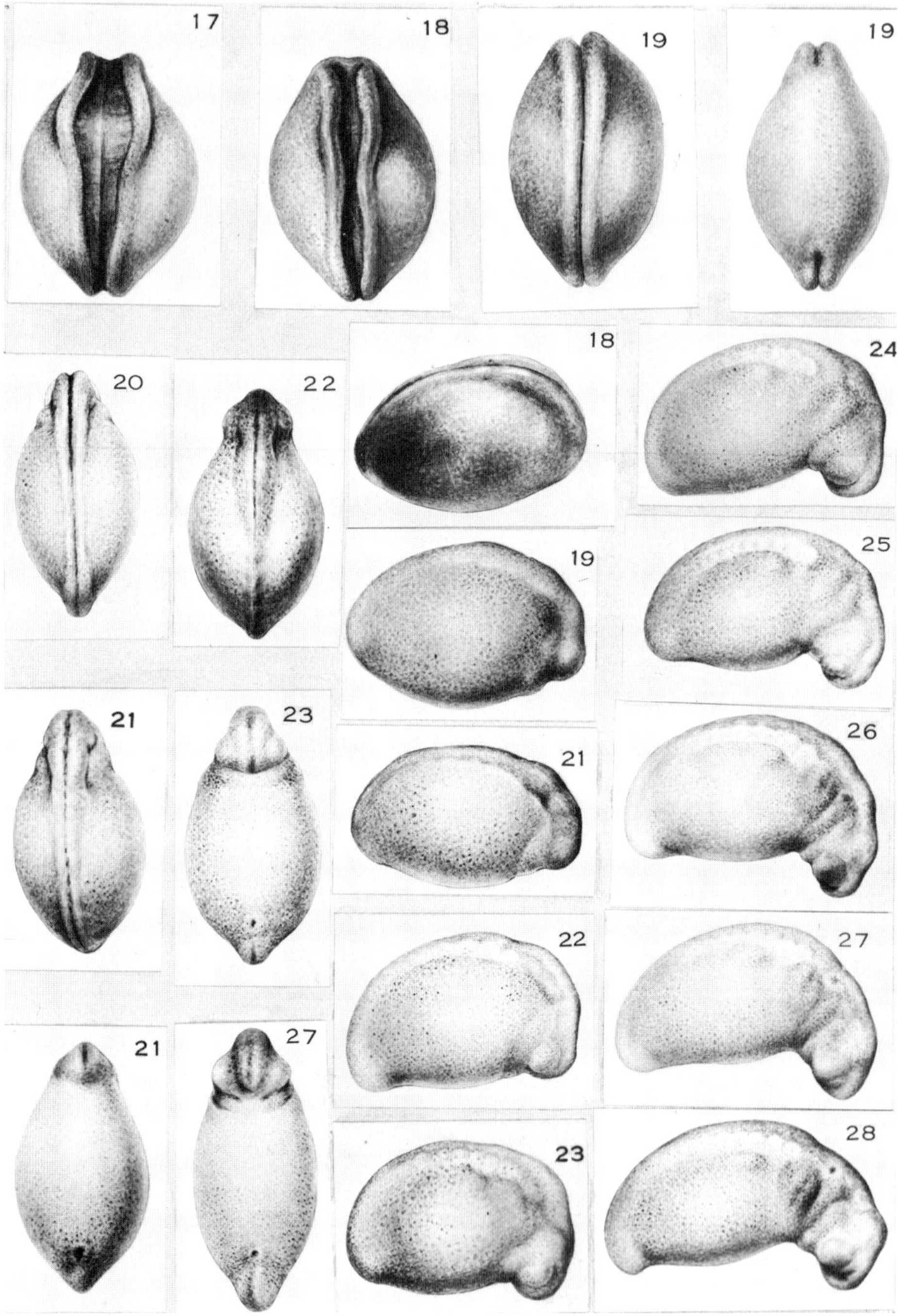

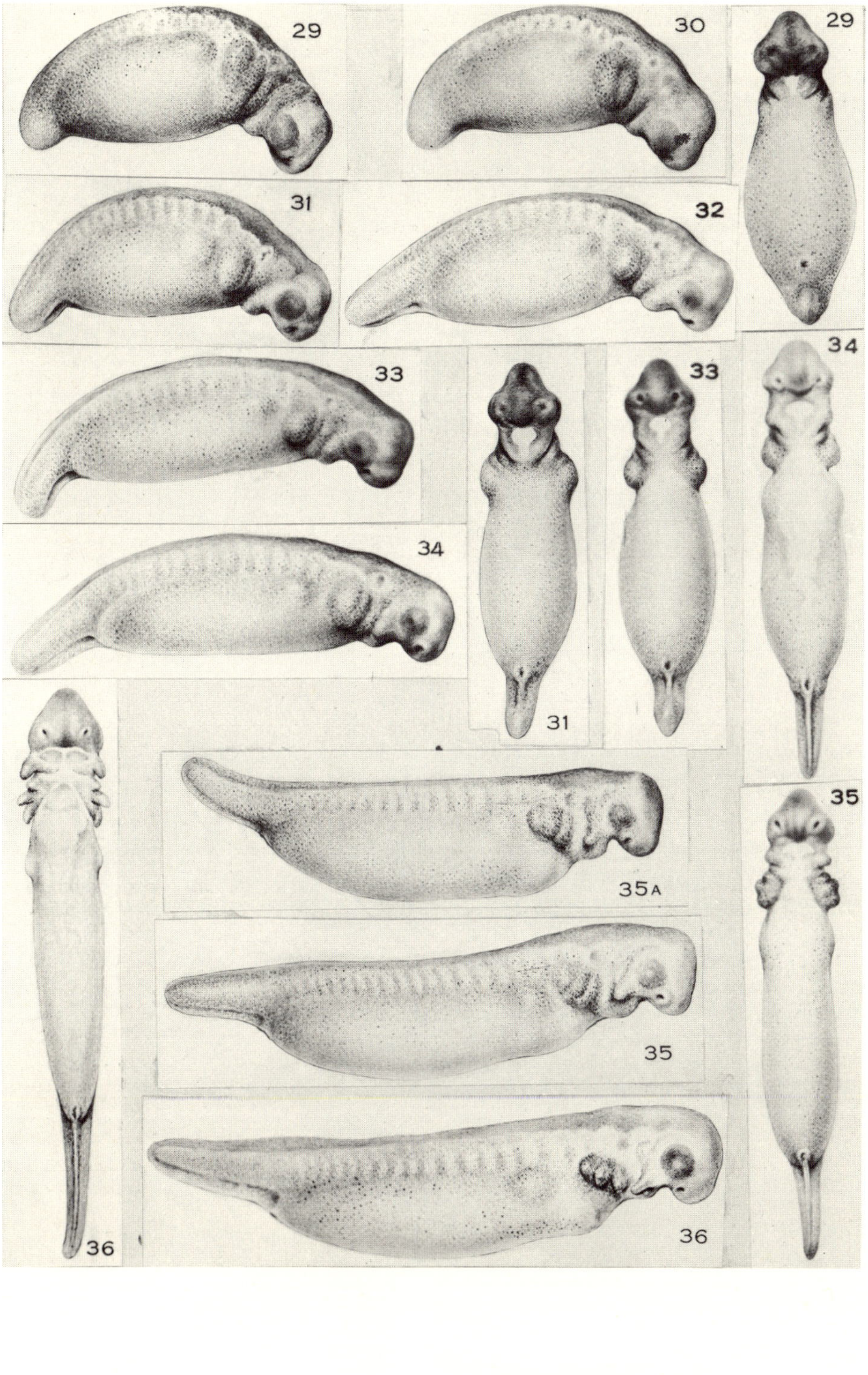
29
30
29
31
32
33
31
33
34
34
35A
35
35
36
36

Plate V

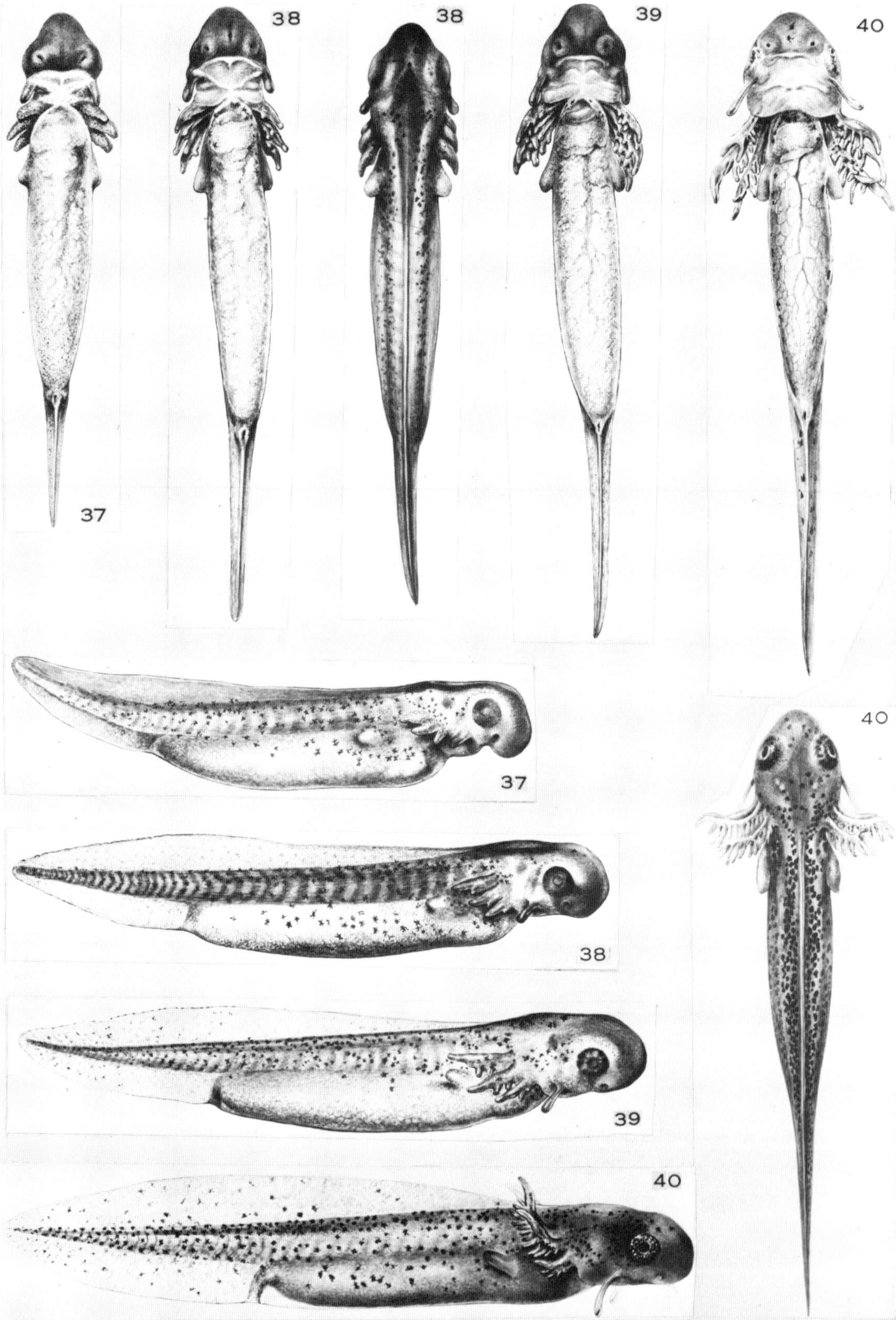

Plate VI

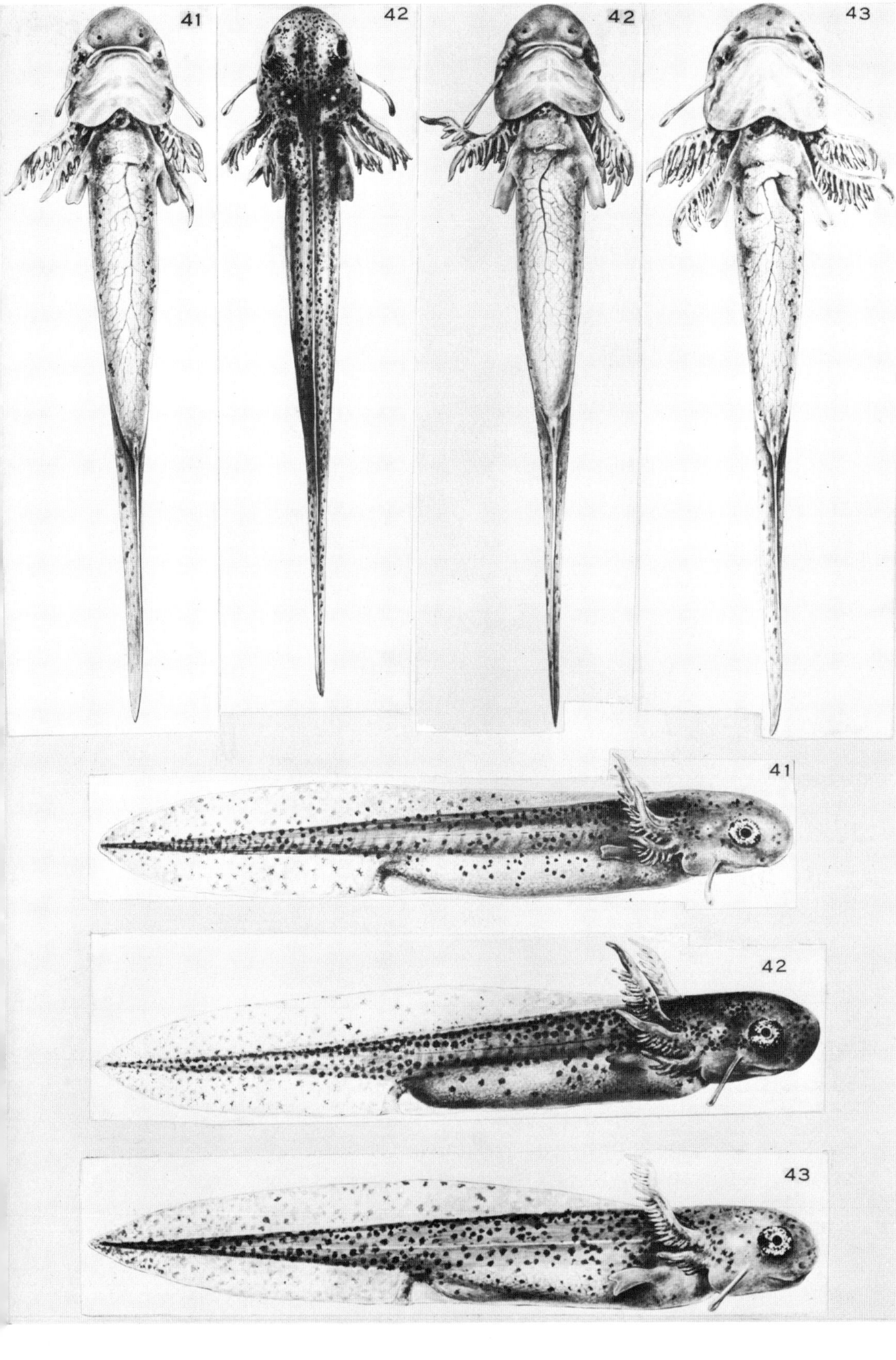

Plate VII

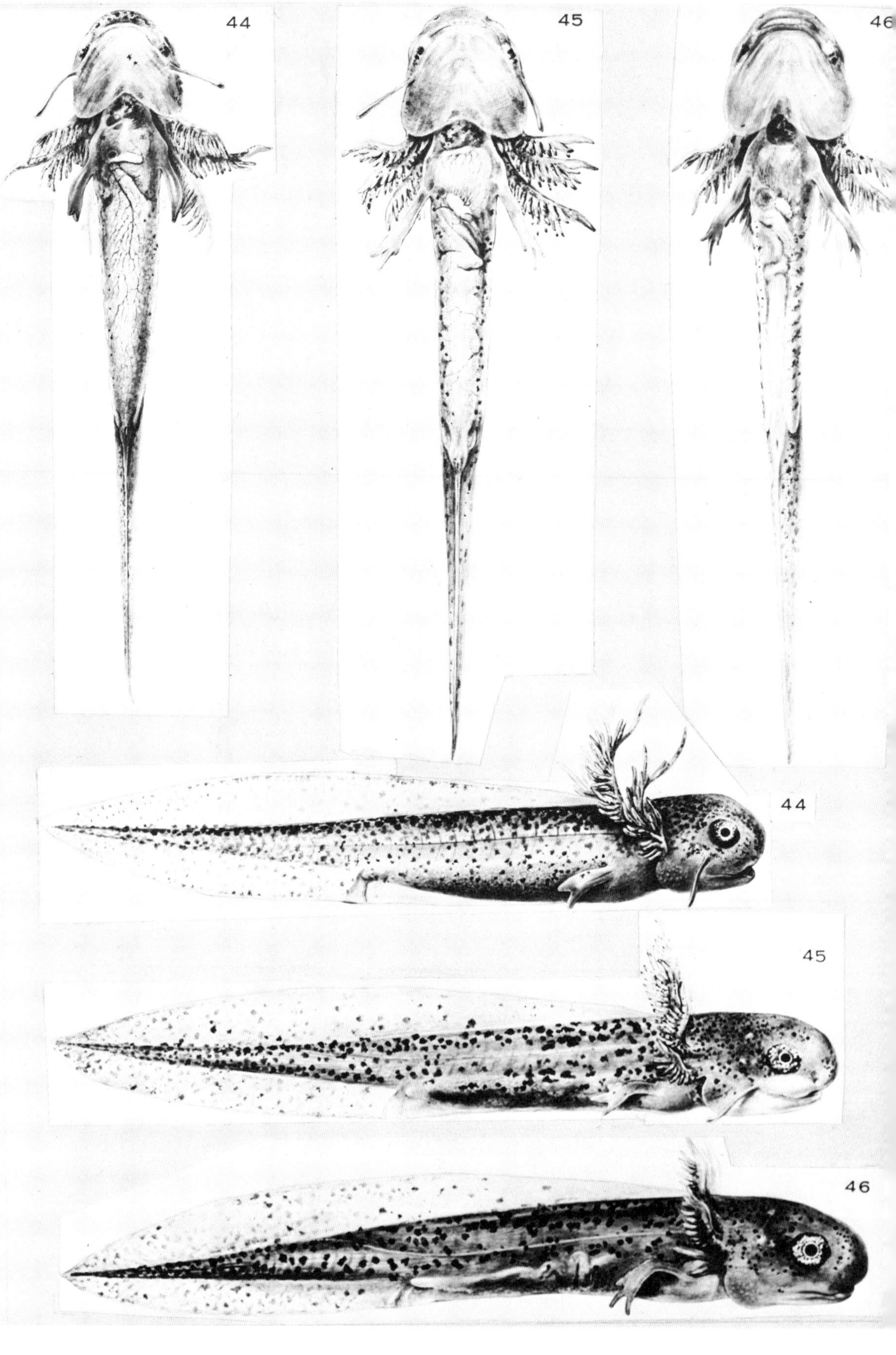

3

On the Status and Significance of Tissue Culture[1]

The fact that animals are made up of parts, each with a certain role to play in the economy of the whole, has been recognized from the very beginning of biological science and was long ago made familiar in the fable of the belly and the members. Modern anatomy and physiology are still based largely upon this fundamental concept, for although in recent years we have heard much of the organism as a whole, living creatures are not studied so much in their entirety as by processes of analysis in which each organ or part is taken, as far as possible, by itself.

Very early in the history of biology it was pointed out that the parts of animals are of two kinds. Thus Aristotle in his *Historia Animalium*[2] says: "Of the parts of animals some are simple: to wit, all such as divide into parts uniform with themselves, as flesh into flesh; others are composite, such as divide into parts not uniform with themselves, as, for instance, the hand does not divide into hands nor the face into faces. . . . All those parts that do not subdivide into parts uniform with themselves are composed of parts that do so subdivide, for instance, hand is composed of flesh, sinews, and bones."

This distinction, which characterizes the difference between what we now call tissues and organs and which is expressed in language scarcely less precise than that of

1. Address given in abridged form at the opening session of the Tenth International Zoological Congress in the Hall of the Hungarian National Museum, Budapest, September 4, 1927. Reprinted from *Archiv für experimentelle Zellforschung,* 6 (1928), pp. 4–27, Gustav Fischer Verlag, Stuttgart.
2. Translated by D'Arcy W. Thompson. Oxford. Clarendon Press.

the biology of to-day, is said to have antedated even Aristotle and to have been borrowed by him from Anaxagoras.

The same idea is formulated again by Aristotle in "The parts of Animals,"[3] in which he recognizes three degrees of composition:

1. Composition out of the elements (protoplasm and cells).

2. Composition of the parts uniform with themselves, such as bones, flesh, ligaments (tissues out of cells).

3. Composition of the parts not uniform with themselves (organs out of tissues).

Handed down through the ages, this idea became of greater significance through the revelations of the microscope in the seventeenth and eighteenth centuries, but it was not until Bichat published his *Anatomie Générale* at the outset of the nineteenth century that accurate systematic descriptions of the tissues were given. Bichat, however, worked without the aid of the microscope and enumerated as tissues twenty one different structures, including many now recognized as more complex. It remained, therefore, for Schwann[4] in the cell theory to lay the foundation of modern histology. Since then this branch of biologic science has divided its attention fairly between the investigation of the simple tissues and that of the more complex organs, and its very name places emphasis on the former rather than on the latter.

Physiology, on the other hand, busied itself more with the larger entities of the body. Its methods were better adapted to the study of organs than to that of the more recondite functions of cells and tissues. Nevertheless Claude Bernard[5] based physiology on the concepts of cellular differentiation, division of labor, and autonomy of the cellular elements, all subordinated to the whole by means of the *milieu intérieur*, and modern general physiology, following the principles so lucidly stated by that great master, has sought to study directly the activities of protoplasm and cells. Still the work in this field has depended upon the use of cells that are naturally isolated, such as unicellular organisms, blood cells, and the eggs of higher forms. The technique for a general study of cells and tissues was lacking.

The discovery that it is possible to isolate and keep growing in culture media very small fragments of almost any tissue either of embryonic or of adult organisms has in some measure supplied this need and has opened a new field now generally known as "tissue culture" or "explantation," through which, it is hoped, a deeper insight may be gained into what is meant by the autonomy of the structural elements and their relations to their *milieu*.

3. Book 2, Chap. 1.
4. *Mikroskopische Untersuchungen*, 1839.
5. *Leçons sur les phénomènes de la vie*, Paris, 1878.

We have already seen that the blastomeres or early cleavage cells may be separated from one another without injury and that their subsequent behavior varies. In some cases (ctenophores, molluscs, segmented worms, and ascidians) the isolated cells do little more than they would have if they had been left together with the rest of the egg, and sometimes even less: Strict preformation. In other cases the isolated cells, usually after some delay, may reconstitute themselves and give rise to a whole organism: Epigenesis.

This antithesis originally led to much controversy which is not yet by any means settled: Whether the organism consists of a large number of individual units or cells or whether it is consolidated into a whole which has more than the combined capabilities of its constituents.

So there began in the second half of the last century, the study of the behavior of isolated cells. For many years this was confined to blood cells (begun by Max Schultze, 1865). Amoeboid activity was discovered in the white blood corpuscles and these cells were often compared with the free-living protozoa such as the amoeba. Their properties as phagocytes or "devouring cells" were noticed and gave further ground for comparison with the amoeba. This property of devouring micro-organisms that had invaded the body and produced disease was recognized as one of the means the organisms had of fighting off infection. Another property is antibody formation which produced immunity. Later Ranvier and especially Jolly kept amphibian blood corpuscles outside the body for months and observed their repeated division. This work was not extended though numerous observations were recorded on the survival of cells of other tissues after removal from the body or after the death of the organism as a whole.

After the experimental method had become established in embryology, W. Roux (1895, 1896) made a study of the behavior of the cells of the morula and blastula stages of the frog's egg (*Rana fusca*). He claimed to have found that when separated in dilute egg albumen or salt solution, cells that were close together (ca. 50–60 μ or the diameter of a cell apart) would gradually approach each other and ultimately come into contact. He called this "cytotropism" and discussed it fully in connection with movements and behavior of micro-organisms and with tropisms in general. He also observed in many cases that when small groups of cells were in contact they would change their relative positions, which act he called self-arrangement (*Selbstordnung*) of the cells or "cytotaxis."

These were stimulating results, and Roux discussed them very ably in their general bearing. However, there is serious doubt as to their special significance in the development of the egg, and especially whether the observations on cell movements are statistically significant, i.e. whether they are directed or merely random movements. According to Vogt (1913), Voigtlander (1932), and Kuhl (1937) the movements are random. Time-lapse motion pictures show this, but the extreme view that cells with hyaline processes are moribund is untenable (v. Holtfreter, Harrison). [Lecture III notes.]

The expression "tissue culture," which has come into such general use and which has been employed in the title of the address assigned to me today, is in some degree a misnomer, for it fails to convey in full the idea it represents. Moreover the use of the phrase might be taken to imply that this method of research—in reality a technique—is an object in itself, whereas it is but a means of investigating the properties of cells under conditions not previously available. What was originally the designation of a method has in fact come to connote the field of knowledge it has brought to light.

The mode of experimentation in question has been employed in the study of both plant and animal tissues and has proved to be of particular value in the fields of embryology, histology, physiology, pathology, and immunology, but in spite of the wide divergence of the lines in which it has been applied, there is a realm of common interest, apart from mere technique, that holds together those who work with it and has brought here today this large group of investigators. This common ground centers about an interest in the activities of protoplasm, cells and tissues, as such, and rests upon the recognition that what goes on in isolated cultures gives some measure of information regarding living processes within the organism itself, and further, that the reactions of living elements in disease are not fundamentally different from the normal processes of development and function. Our method affords a means of studying these reactions where conditions are most simple.

The methods of tissue culture have been mothered by various necessities. In the case of my own work on the development of the nervous system[6] there was need for an experiment in which the behavior of certain cells could be observed when removed from the bewildering conditions obtaining within the embryonic body, for it was recognized that the ordinary methods of histology were inadequate to answer definitely the question of the origin of the nerve fiber. Here, then, was a problem that demanded for its solution some such radical form of isolation as that afforded by the method of explantation.

6. *Proc. Soc. Exp. Biol. and Med.*, *4*, 1907 (Also in *Anat. Rec.*, *1*); *Anat. Rec.*, *2*, 1908; *J. Exp. Zool.*, *9*, 1910.

Others approached the field in conscious search of a physiology of cells and tissues, and in this, as usual, the botanists anticipated the zoologists. Thus, Haberlandt, who was the first to cultivate isolated cells of higher organisms, sets forth very fully in his introductory paper the purposes and possibilities of modern tissue culture.[7] On the animal side Leo Loeb, in reviewing his earlier work said: "It [tissue culture] will I hope, ultimately lead to the building up of a physiology and pathology of tissues in a similar sense as we now possess a physiology and pathology of organs."[8] A similar thread has run through the work of Carrel,[9] who more consistently and perseveringly than any one else has aimed to go beyond the mere explantation of fragments and to carry on for long periods true cultures of cells in pure line.

The introduction of a method as radical as this was naturally met with skepticism on the one hand and with, perhaps, an oversanguine enthusiasm on the other. These opposite attitudes may still be occasionally observed, even though the skeptics have been met at most points and the enthusiasts have often encountered disappointment. The method has thus fairly established itself and the number of investigations involving its use has increased at a greatly accelerated rate within the past few years. Articles touching upon some phase of the subject are appearing at frequent intervals and are attracting wider and wider interest. In the latest bibliography published,[10] covering the years 1925 and 1926, about four hundred titles are listed; a number of extended treatises on the subject in at least four languages have appeared; and the journal devoted to this field and founded by Professor Rhoda Erdmann less than three years ago, has already achieved signal success. All of this, taken together with the

7. "Es sind meines Wissens bisher noch keine planmäßig angelegten Versuche gemacht worden, isolierte vegetative Zellen von höher entwickelten Pflanzen in geeigneten Nährlösungen zu kultivieren. Und doch müßten die Ergebnisse solcher Kulturversuche manches interessante Streiflicht auf die Eigenschaften und Fähigkeiten werfen, die die Zelle als Elementarorganismus in sich birgt, sie müßten Aufschlüsse bringen über die Wechselbeziehungen und gegenseitigen Beeinflussungen, denen die Zellen innerhalb des vielzellingen Gesamtorganismus ausgesetzt sind." *Sitz.-Ber. Math.-Nat. Kl. Kais. Akad. Wiss. Wien, 111,* Abt. 1, 1902, p. 69.

8. *Anat. Rec., 6,* 1912, p. 120. About the time that Haberlandt's first paper appeared Loeb published his experiments with pieces of epithelium from the guinea pig, imbedded in small blocks of clotted blood or agar which were placed for incubation in the body of another animal (Arch. Entw.-Mech., Bd. 13, 1902; Journ. Med. Res., Vol. 8, 1902). The technique was entirely different from Haberlandt's and had nothing in common with that of modern tissue culture, although the underlying purpose of the experiment was essentially the same. Loeb in an earlier paper, published separately (Ueber die Entstehung von Bindegewebe, Leukozyten und roten Blutkörperchen aus Epithel und über eine Methode isolierte Gewebsteile zu züchten, Chicago 1897), mentions having made experiments in which the agar blocks containing pieces of living tissue were incubated outside the organism. The results were not stated.

9. *J. Exp. Med., 15,* 1912; ibid., *17,* 1913; ibid., *18,* 1913; *Physiol. Rev. 4,* 1924.

10. Erdmann and Pinzke, *Arch. Exp. Zellforsch., 4,* 1927.

wide application of the method in biology and medicine, makes it clear that the subject has extended far beyond the bounds that any one individual may reasonably be expected to cover in a short address. I must, therefore, plead my natural limitations, if, in what follows, I should seem to ignore fields of work with which I have had no experience, and confine myself to the consideration of a few topics in the fields of embryology, histology and physiology where tissue culture and related methods have proved fruitful and are promising of further progress.

The culture method used was that known to the bacteriologists as the "hanging drop" culture [Fig. III–1]. The tissue to be studied was dissected from the embryo under aseptic precautions, placed in a drop of serum or lymph on a cover-slip which was then inverted over a depression ground in a glass slide, the preparation then sealed with paraffin and set aside to develop. The tissue could then readily be observed under high powers of magnification, just as bacteria or molds could be followed in their multiplication and growth. [Lecture III notes.]

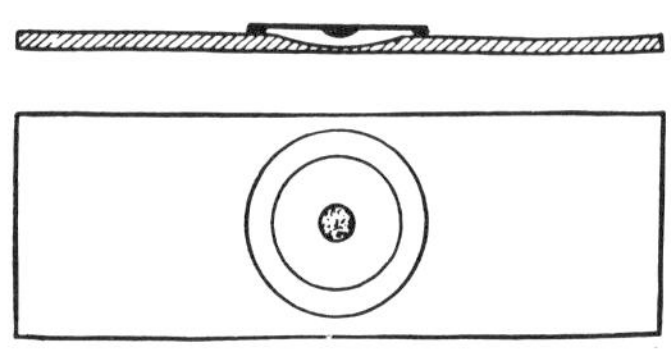

FIG. III–1. Hollow ground slide for hanging-drop cultures. After Muir and Ritchie, 1907.

The Nerve Elements

The first observation[11] on nerves growing in clotted lymph (the medium then used) [Figs. III–2 and III–3] gave ocular proof of the histogenetic theory of His[12] and Ramón y Cajal.[13] The fiber was seen to be formed by the spinning of a thread of protoplasm by means of the amoeboid activity of the free end of a cell process. In this protoplasmic filament neurofibrillae develop a little later.[14] These findings have been

11. Harrison, op. cit., 1907, 1908.
12. *Abh. d. math.-phys. Kl. d. Kgl. Sächs. Ges. d. Wiss.*, *13*, 1886; *Arch. Anat. u. Physiol., Anat. Abt.* 1887; Ibid., 1890.
13. *Anat. Anz.*, *5*, 1890; *La Cellule*, *9*, 1892.
14. Burrows, *J. Exp. Zool.*, *10*, 1911.

confirmed and extended by numerous investigators,[15] and the degeneration of injured nerves and their subsequent regeneration in the same manner have also been followed in vitro.[16]

FIG. III–2. Isolated cell from a piece of embryonic spinal cord growing in a drop of clotted lymph. The cell body, which is filled with yolk granules, is sending out a hyaline protoplasmic process which undergoes amoeboid movements. Drawn from a live specimen. After Harrison, 1908.

Subsequent studies by Levi[17] have greatly enriched our knowledge of the physical properties and the physiology of the protoplasm constituting the growing fiber and have demonstrated the value of microdissection methods in work of this kind. According to Levi the young fiber is tough and resistant, and may be considerably stretched without suffering permanent injury. At the same time it is exceedingly sensitive to mechanical stimuli. If the cell body of a neuroblast is compressed, instantly its visible character changes and the nerve fiber attached to it is seen to emit fine processes, thus demonstrating its conductivity. Again the shock of cutting a fiber is transmitted to the growing end, as shown by the cessation of movement for a short time after such treatment. Subsequently the amoeboid activity is resumed and may continue for hours, during which time the fiber increases in length, as if it were still attached to the cell body. Ultimately, of course, this movement ceases and the severed fiber degenerates. Furthermore, both stumps of a cut fiber may manifest amoeboid activity and the two may reunite *per primam intentionem,* whereby the whole

15. Cf. Burrows, op. cit.; M. R. a. W. H. Lewis, *Anat. Rec., 6,* 1912; Braus, *Münch. med. Woch.,* 1911; Id. *Verh. d. Ges. Deutsch. Naturf. u. Ärzte,* 1911; Levi, *Mem. R. Accad. Lincei, 12,* 1917.
16. Cf. Braus, op. cit.; Ingebrigsten, *J. Exp. Med., 17,* 1913 and *18,* 1913.
17. *Arch. exp. Zellforsch., 2,* 1925.

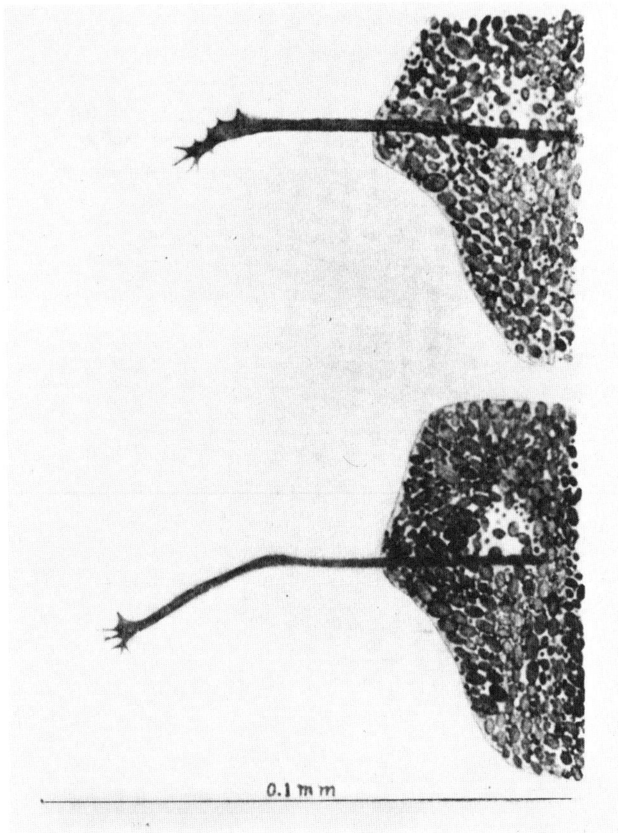

FIG. III–3. Two views, taken twenty-five minutes apart of the same nerve fiber growing from a group of embryonic spinal cord cells into the lymph. After Harrison, 1908.

is restored and the distal portion spared the fate of degeneration. None of these facts could have been discovered by means of the more conservative methods.

From the time when the first nerve fibers were observed in culture it was evident that through the study of the effect of various physical and chemical agents on the outgrowth process we might gain insight into the play of factors that produce the complex pattern of the nervous system within the organism. It was soon shown that the outgrowth of a nerve fiber, like other cell movements, requires a solid substratum,[18] and this, taken together with observations on the course of nerves in the embryo, renders it likely that the configuration of the solid structures in the way of a

18. Harrison, *Science, 34,* 1911 and *J. Exp. Zool., 17,* 1914.

developing nerve determines to a considerable extent its path.[19] In respect to responses to chemical stimuli, as postulated in Cajal's theory of chemotaxis, our knowledge is not so definite. Experiments made some years ago,[20] by a modification of Pfeffer's method, in implanting very small capillary tubes filled with various substances in proximity to the explanted nervous tissue gave conflicting results. Hoadley[21] on the other hand, working with tissues implanted on the chorio-allantoic membrane, has found evidence that nerve fibers from pieces of the brain do tend to grow toward other pieces of tissue. Likewise the preliminary experiments of Ingvar[22] point to the conclusion that nerve fibers respond as regards their direction of outgrowth to a very weak galvanic current. [Not confirmed by experiments of Weiss 1934; see p. 132]. This, if definitely established, would afford important experimental evidence in favor of the theories of Kappers[23] and Bok.[24]

Connective Tissue

Cells from the connective tissue play an important, frequently a dominating part in tissue cultures of almost all organs. They survive without difficulty, remaining usually without marked visible differentiation, and lend themselves readily to indefinite culture. These cells are usually referred to as fibroblasts, but it has not been definitely proved that they have anything more than a passive role in the formation of connective tissue, and none of the numerous investigators who have cultivated them has proved them actually to form collagenous fibers. Baitsell[25] has, in fact, brought forward strong evidence to show that, under certain conditions, connective tissue fibres may be formed by the transformation of the fibrin clot used in tissue cultures, just as scar tissue is formed in wound healing. The fibers would thus be a coagulation prod-

19. *J. Exp. Zool., 9,* 1910, p. 835.
20. *Anat. Rec., 6,* 1912, p. 192. Experiments made by the same method by Centanni and Ugurgeri (Accad. Fisiocrit. Siena, 1913) and later by Sanguineti (*Rev. Patol. Nerv. e Ment., 19,* 1914) are not relevant to this immediate question, since the outgrowths from pieces of peripheral nerves found by these authors are obviously, as they themselves admit, of connective tissue origin. Still this work, as well as the later study by Bisceglie (*Arch. exp. Zellforsch., 2,* 1926), is of great interest in showing that cellular outgrowth from explanted tissues is stimulated chemically, as indicated by the direction in which it takes place most actively.
21. *J. Exp. Zool., 42,* 1925.
22. *Proc. Soc. Exp. Biol. a. Med., 17,* 1920.
23. *Psychiatr. en Neurolog. Bladen,* 1916.
24. *Psychiatr. en Neurolog. Bladen,* 1915.
25. *J. Exp. Med., 21,* 1915; ibid., *23,* 1916.

uct rather than a direct secretion of the connective tissue cells. For this view much evidence has been advanced by those who have studied the histogenesis of connective tissue in the embryo,[26] for here fine fibrils formed out of a more homogeneous interstitial jelly seem to precede the immediate presence of cells. It is true that Levi[27] described delicate fibers which were formed in certain cultures of mesenchyme from the heart and skin of advanced chick embryos, but their identity with connective tissue fibers is not claimed to be proved. Likewise M. R. Lewis[28] has reported the formation of very delicate fibrillae in coverglass cultures of mesenchyme cells in purely fluid media, though in my opinion neither the exact nature of the fibrillae observed nor the direct participation of the cells in their formation has been shown. Thus there is not necessarily any inconsistency in the findings just referred to. It is likely that the further pursuit of the subject with the aid of improved technique will settle definitely this perplexing question. Certainly we may hope to find among the numerous investigators working with fibroblasts some one who will give us further information regarding the histogenetic powers of these cells.

Blood Cells

I hesitate to tread upon the difficult and hazardous ground of the formation and genetic relationship of the blood elements, for with all the diligence and ingenuity that has been spent in this field, uncertainty still exists with regard to even the most fundamental questions, and the enormous mass of publication that has been poured forth cannot possibly be mastered by an outsider. The work done by the method of tissue culture would alone fill several volumes, but, although it has made a definite impression and points in one direction, it cannot be said yet to have given a complete solution of the problems. However, the conditions obtaining within the organism are so complex that it is almost impossible to determine there the life history of cells capable of moving about by themselves and of assuming so many different guises. We must, therefore, have recourse to some such method of isolation as is afforded by

26. Cf. v. Ebner, *Zeitschr. wiss. Zool.*, *62*, 1896; Merkel, *Anat. Hefte*, *38*, 1909; Studnička, *Anat. Anz.*, *31*, 1907 and *40*, 1911; Nageotte, *C. R. Soc. Biol.* Paris, *79*, 1916; Id. L'Organisation de la matière etc., Paris, 1922; Isaacs, *Anat. Rec.*, *17*, 1919; Baitsell, *Am. J. Anat.*, *28*, 1921; Id. *Quart. J. Micr. Sci.*, *69*, 1925; Harrison, *J. Exp. Zool.*, *41*, 1925.
27. *Mon. Zool. Ital.*, *27*, 1916.
28. *Contr. to Embryol. Carnegie Inst. of Wash.*, *6*, 1917.

tissue culture, where transformations may be followed by direct observation. For this reason it seems clear that this method will have the last word in clarifying the questions at issue.

The problems that confront us here concern the interrelations of the cells of the blood, bone marrow, lymph and connective tissue, together with their behavior both in the normal body and in reaction to conditions produced by disease. Various types of cells are known, differing from one another in size, shape, nuclear structure, granulation, motility, and reaction to abnormal conditions, and occurring in the blood and tissues in relative numbers which fluctuate according to physiological and pathological conditions. Are these all merely different phases of one kind of cell or are at least two fundamentally different types involved which do not pass over into one another?

Work done with the aid of tissue culture methods generally favors the first of these alternatives. The old distinction between lymphogenous and myelogenous elements has been broken down by the discovery that cultures of the cortex of a lymph gland will yield a whole series of cells formerly thought to be characteristic of bone marrow.[29] The more recently drawn antithesis between monocytes and the various forms of leucocytes on the one hand, which are considered to be reticular in origin, and the clasmatocytes and erythroblasts on the other, thought to be derived from endothelium,[30] will also have difficulty in maintaining itself if the transformations which have been described by various investigators prove to be real.

All transitional stages between small lymphocytes, mononuclears and granular leucocytes and myeloctyes have been observed,[31] and although the direct transformation of lymphocytes has not been actually followed in the living cell, the early findings of Maximow have been abundantly confirmed by later workers.[32] The transformation of mononuclear blood cells into macrophages, epithelioid cells, and giant cells has been followed,[33] and it has been found that blood monocytes and tissue macrophages or clasmatocytes assume identical characters with regard to form, size, pseudopodial activity, mitochondria, and granulations, when cultivated in the same media, and that these vary with the medium employed.[34] Again the change of mononu-

29. Maximow, Arch. *Mikr. Anat.*, 97, 1923.
30. Sabin, Cunningham and Doan, *Contr. to Embryol., Carnegie Inst.*, *16*, 1925.
31. Maximow, op. cit.; Cf. also Arch. *Russ. Anat., Histol., Embryol.*, *1*, 1916; Awrorow and Timofejewsky, *Virchows Arch.*, *216*, 1914.
32. Cf. int. al. Fazzari, *Arch. Exp. Zellforsch.*, *2*, 1926; Caffier, *Arch. Exp. Zellforsch.*, *4*, 1927; Hirschfeld, ibid.
33. M. R. and W. H. Lewis, *J. Am. Med. Ass.*, *84*, 1925; *Contr. Embryol. Carnegie Inst.*, *18*, 1926.
34. Carrel and Ebeling, *J. Exp. Med.*, *44*, 1926.

clears from cultures of blood over into fibroblasts has been repeatedly observed by many workers since first described by Carrel and Ebeling.[35] The occurrence of this transformation, which is of prime importance in the organization of blood clots in wound healing and which could hardly have been proved crucially by the older methods, shows that fibroblasts are one of the end products in the metamorphosis of blood cells. Recently, however, it has been found that under certain conditions, not yet altogether understood, fibroblasts may be changed back into macrophages.[36]

From all of this the inference may be drawn that the various forms of blood and connective tissue cells may be produced by the transformation of a monocyte-like element. Some of these transformations are known to be favored by certain qualities of the medium, as, for instance, the change of lymphocytes into myelocytes and granular leucocytes by the presence of bone marrow extracts,[37] and the hastening of the change of blood monocytes into fibroblasts through the action of sarcoma extract added to the culture medium.[38] From such facts we are naturally led to view the mode of activity of these cells and their particular habitus at any given time as responses to factors in the environment rather than as characters indelibly stamped upon them. It remains to be determined, however, which of the transformations are reversible.

Muscle and Heart

That the cells of the myotomes are self-differentiating to the extent that normally striated fibers will develop from them in plasma was shown in my first work on the outgrowth of the nerve fiber *in vitro.* This was in fact one of the essential control experiments, for in order to be sure that the interpretation of the nerve preparations was correct, it had to be shown that other tissues would undergo characteristic differentiations under similar conditions.[39] For the study of the histogenesis of muscle the methods of tissue culture are, however, of no particular value. On the other hand they have been an important adjunct in the study of the development of muscle function, especially that of the heart.

35. *J. Exp. Med., 36,* 1922.
36. Fischer, *Arch. Exp. Zellforsch., 3,* 1927.
37. Maximow, op. cit.
38. Carrel, *J. Exp. Med., 43,* 1926.
39. *J. Exp. Zool., 9,* 1910, p. 812.

The pioneer work of Burrows[40] showed that groups of partially differentiated myoblasts, which had migrated out of explanted pieces of embryonic chicken heart, might after some days begin to contract rhythmically under cultural conditions that precluded contamination with any kind of nervous element.[41] Thus a further crucial test of the myogenic theory of the heart beat was given. An isolated cell was observed to beat with a rhythm of its own and two portions of the cellular net, formed by the migrated cells, which had become functionally separated from each other by the degeneration of the intervening portion, pulsated still in the same rhythm but with an alternation of phase, i.e. not synchronously.

Experiments made by M. R. and W. H. Lewis,[42] with different forms of contractile tissue of the chick embryo—smooth muscle from the amnion, heart muscle and voluntary muscle from the myotomes and limb buds—have shown that all may produce cells that extend into the medium and there contract rhythmically [Fig. III–4]. However, the cells which grow out from the explanted fragments undergo a considerable amount of morphological dedifferentiation, and while striated fibrillae are sometimes found, as, for instance, in some remarkable cases described by Levi,[43] this is the exception rather than the rule. Outgrowths formed from voluntary muscle often resemble the muscle buds formed in the regeneration of muscle *in vivo,* and similar buds growing from explanted mammalian heart muscle have also been described.[44] In culture, however, these buds usually remain protoplasmatic without characteristic fibrillae.

Considerable discussion has taken place with regard to the nature of the cells which grow out from explanted heart tissue, some referring to them as fibroblasts while others, particularly Levi,[45] emphasize their myoblastic origin. It seems clear, however, from the work of M. R. and W. H. Lewis[46] that both types of cells proliferate and migrate in cultures, myoblastic outgrowths occurring relatively more frequently from tissue taken from young embryos, than from older. Again, according

40. *Science, 36,* 1912; *Münch. med. Woch., 59,* 1912.
41. That fragments of heart from chicken embryos could pulsate in culture media was shown by Burrows in his first work (*J. Exp. Zool., 10,* 1911). Braus (Münch. med. Woch., 1911; Wiener Med. Woch., 1911) demonstrated the same for the heart of the amphibian embryo (*Bombinator, Rana*) and further that such fragments responded to electrical stimuli. Carrel (*J. Exp. Med., 15,* 1912) found pieces of embryonic chicken heart beating after sixty-five days of isolation in culture media.
42. *Am. J. Physiol., 38,* 1915; ibid., *44,* 1917; *Am. J. Anat., 22,* 1917.
43. *C. R. Assoc. Anat., 19.* Réun., Strasbourg, 1924.
44. Cf. Erdmann, *Verb. Anat. Ges.,* Halle, 1924.
45. *Arch. Sci. Med., 50,* 1916; *Rend. Reale Accad. Lincei, 32,* 1923.
46. *Behavior of Cells in Tissue Culture.* In *General Cytology* edited by E. V. Cowdry, Chicago, 1924.

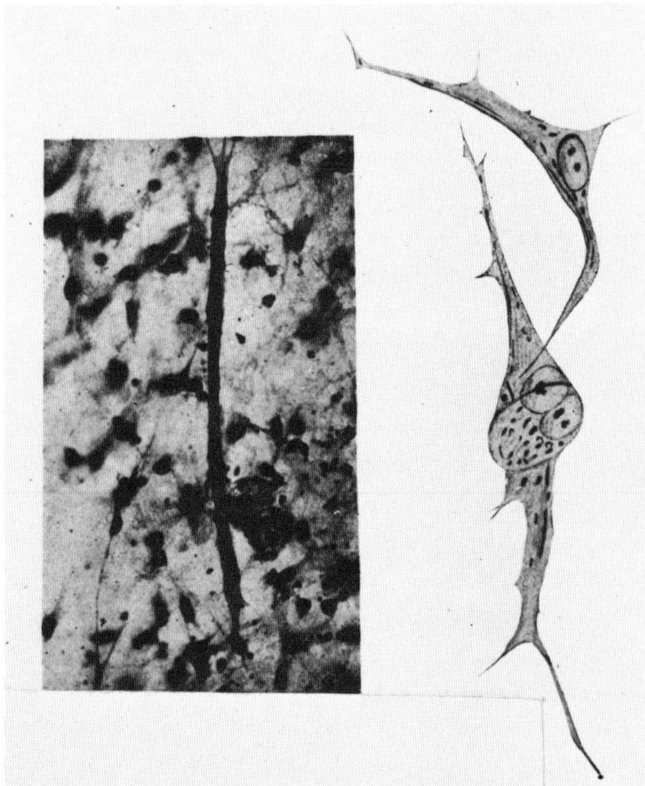

FIG. III–4. Variation in contraction rhythm of voluntary muscle from chick embryo leg. Left, isolated fiber which contracted 120 times a minute for 8 hours; right, two living myoblasts which contracted slowly (once in 15 and 23 seconds) for one-half hour. After M. R. Lewis, 1915.

to Erdmann, in preparations of the guinea pig's heart myoblasts tend to predominate at first, but if the piece is replanted a few times with removal of the cells that have migrated, pure cultures of fibroblasts are obtained. Myoblasts seem therefore to be more active and fibroblasts to have greater powers of survival.

Pieces of embryonic heart planted out in culture media beat frequently with different rhythm even when taken from the same individual and subjected as far as possible to identical conditions. Experimenting with such fragments, Fischer[47] found that after two had established close organic union by outgrowth of cells from one to

47. *J. Exp. Med.*, *40*, 1924.

the other, which was completed after from 48 to 72 hours, the two pieces beat in complete unison. No such connection occurred between pieces from chick embryos and those from the duck. According to Olivo,[48] however, pieces of heart tissue from chicken and pigeon embryos do coalesce and in such cases complete synchronicity is established in the pulsation of the two components.

The mode of connection between the elements in cultures of cardiac muscle has been considered fully by Levi[49] as part of the general question of the occurrence of syncytia. Many cells that appear fused with one another often move apart later, and contact between cells seems to be all that is necessary for the transmission of stimuli. Even in cells that are actually fused each nucleus seems to have a definite sphere of influence. Lewis[50] takes a still more radical stand against the existence of syncytia, especially in the mesenchyne, with regard to which there had been perhaps most general agreement as to the reality of cellular fusions. The study of tissue cultures has therefore tended to restrict very much our notion of the real occurrence of syncytia.

In contrast to the foregoing, which concerns small fragments of heart tissue and isolated cells, the heart rudiment of the amphibian embryo, when removed from the body with a small amount of overlying ectoderm, will, as shown by Ekman,[51] develop into a heart having the main features of the normal organ. Sinus, atrium, ventricle and bulbus may be formed, the whole structure with a certain degree of curvature[52] and enclosed in a cavity inside an epithelial sphere. Such hearts, which lack altogether a nervous system, contract rhythmically and afford a valuable means for the study of the action of drugs.[53] Explanted without the covering ectoderm, the heart rudiment forms a pulsating organ, but it is less perfect in form than when it develops in a pericardial cavity inside an epithelial sack. Such aggregates, which are intermediate between tissue cultures proper and whole explanted organ complexes, are of great interest in relation to problems of organization and differentiation.

48. *Arch. Exp. Zellforsch.*, *2*, 1925.
49. *Arch. Exp. Zellforsch.*, *1*, 1925.
50. *Anat. Rec.*, *23*, 1922.
51. *Oversikt. Finn. Vetensk.-Soc. Förhend.*, *63*, 1920–21. *Ann. Soc. Zool.-Bot. Fenn.*, *2*, Helsingfors, 1924.
52. Stöhr, who has repeated and extended Ekman's experiments, emphasizes the defective nature of most of the explanted hearts and the dependence of this organ on its surroundings for the development of its normal form (*Arch. Mikr. Anat. u. Entw.-Mech.*, *102*, 1924; *Verh. d. Phys.-Med. Ges. Würzburg*, *N.F. 50*, 1925).
53. Cf. experiments by Schübel and Stöhr on heterotopically transplanted hearts. *Arch. Exp. Pathol. u. Pharmakol.*, *104*, 1924.

Explantation of Organ Rudiments into Media and to the Chorio-allantoic Membrane

The study of the development of organs has not been so greatly advanced by the method of explantation as that of histogenesis. Their growth is not perfectly correlated in the unusual environment of the plasma culture and, while normal differentiations do occur, the ensemble is usually atypical. The most readily interpretable cases are those in which the epithelial component of the organ rounds itself off into a covering for the rest of the cells, as in Ekman's hearts above referred to. Fischer's[54] intestinal organisms, formed from small pieces of the gut of chick embryos are of this character, and show that at least the epithelium and connective tissue components undergo marked progressive differentiation. Explanted limb buds of chick embryos give rise to typical cartilage, connective tissue, and epithelium according to Strangeways and Fell[55] and under certain conditions they show in gross form the early stages of a normal limb skeleton. Typical differentiation of the retina occurs in explants of the optic vesicle of the chick[56] and an optic cup with lens is formed when the amphibian optic vesicle is planted out with its covering ectoderm.[57] One of the most notable examples of the constructive development of organ units is that found by Rienhoff[58] in the formation of tubules in explanted pieces of the mesonephros and permanent kidney of the chick embryo.

There is a procedure, akin to that of tissue culture that will probably prove to be of greater value in this field, even though it is of but limited application. This is the method of explantation upon the allantois of the embryo, whereby the explanted tissue is given a substratum from which it may obtain a continuous supply of nutriment through the circulation, and which, though living, has but very slight morphogenetic powers of its own and is very tolerant of intrusion.

First developed by Murphy and Rous[59] for the study of transplantable tumors, this method has yielded interesting results with pieces of normal organs of either embryonic or adult animals. It has formed the basis of investigations on haematopoiësis,[60]

54. *J. Exp. Med.*, *36*, 1922.
55. *Proc. Roy. Soc. London*, *B*. *99*, 1926.
56. Strangeways and Fell, *Proc. Roy. Soc. London*, *B*, *100*, 1926.
57. Filatow, *Arch. f. Entw. Mech.*, *107*, 1926.
58. *Bull. Johns Hopkins Hosp.*, *33*. 1922.
59. *J. Exp. Med.*, *15*, 1912; Murphy, *J. Am. Med. Ass.*, *59*, 1912; id., *J. Exp. Med.*, *17*, 1913.
60. Danchakoff, *Anat. Rec.*, *10*, 1916; id., *Am. J. Anat.*, *24*, 1918.

on sex differentiation[61] and on the phenomena of determination and differentiation of various embryonic organs.[62]

From all of these experiments it is seen that in the chick embryo at the stages studied most parts are capable of a high degree of self differentiation. According to Hoadley, for example, the primordium of the eye is already partly determined in the blastoderm of four hours incubation. More complete determination follows rapidly, as shown by the degree of development attained by grafts taken at closely succeeding stages. Again, the specific sex characters of the gonads are maintained, when growing on the allantois, in spite of the antagonistic sex hormones of the host that are present in the circulation when the latter is of opposite sex to the graft (Corinaldesi).

Removal from normal surroundings can, however, have a very different effect, and may bring to light unexpected powers of growth which are ordinarily held in abeyance by the inhibiting action of the organic environment, as in the case of the pronephros. Normally undergoing regression and disappearing early in development, this organ when planted out on the allantois persists and grows materially.[63]

Microdissection

The technique of micromanipulation,[64] thanks to the skill of Schouten, Barber, Kite, Chambers, and Péterfi, has reached a state of almost incredible precision, which is commensurate with the delicacy of the objects it has to deal with. The adaptation of these methods to the study of cells in tissue cultures[65] opens to view many interesting possibilities. It is feasible thus to investigate not only the physical properties of living cells and their constituent parts but also their reactions to mechanical, electrical, and chemical stimuli locally applied, the effect of injecting foreign substances, and the consequences of definite mutilations.

61. Cf. Minoura, *J. Exp. Zool., 33,* 1924; Willier, *J. Exp. Zool., 46,* 1927; Corinaldesi, *Arch. Exp. Zellforsch., 4,* 1927.
62. Danchakoff, *Zeitschr. f. Anat. u. Entw.* 74, 1924; *Contr. to Embryol. Carnegie Inst., 18,* 1927. Hoadley, *Biol. Bull., 46,* 1924; *J. Exp. Zool., 42,* 1925; ibid., *43,* 1926. Murray and Huxley, *J. Anat., 59,* 1925; *Brit. J. Exp. Biol., 3,* 1925.
63. Hoadley, op. cit., 1925.
64. For the history of the development of "micrurgical" methods see article by Péterfi in *Abderhalden's Handb. Biol. Arbeitsmeth.*
65. Péterfi, *Arch. Exp. Zellforsch., 4,* 1927.

On the other hand, it is doubtful whether the cell nucleus[66] will tolerate much interference, as for instance the mutilation necessary to remove individual chromosomes. Its organization is so delicate that any attempt to manipulate by direct mechanical action the carriers of the genes is likely to fail. Such effects are for the present best obtained indirectly by the action of chemical and physical agents, such as ultraviolet light and X-rays, on the mechanism of cell division, whereby an abnormal distribution of the chromosomes to the two daughter cells may be brought about, as in the case of non-disjunction, or selective action upon individual chromosomes may be produced.

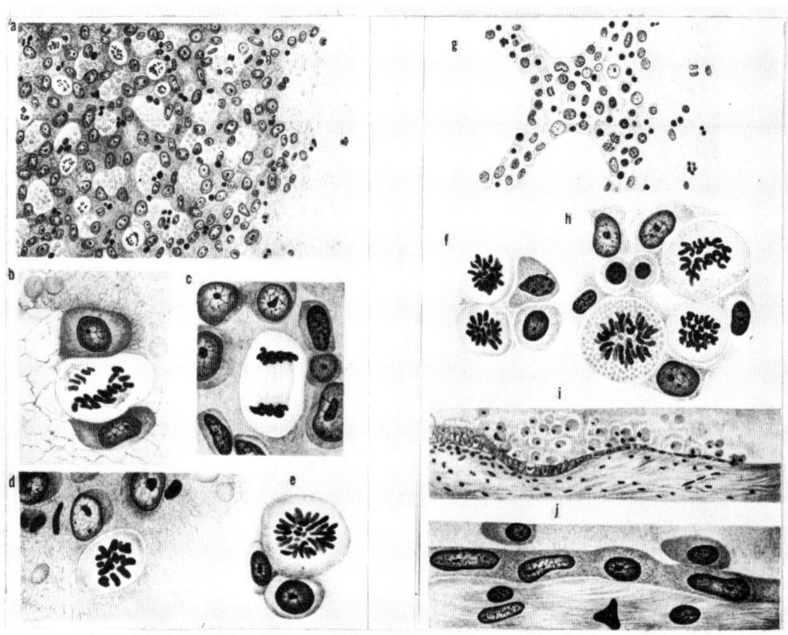

FIG. III–5. Mitotic figures in plasma cultures of cat spleen (a–f), bone marrow (g, h), and mucous membrane of the larynx (i, j). After Oppel, 1912.

The Prolonged Culture of Pure Strains of Cells

One of the earliest and most persistent criticisms of the method of explantation was that tissues did not actually grow under such conditions but merely survived—

66. Péterfi und Olivo, *Arch. Exp. Zellforsch.*, 4, 1927.

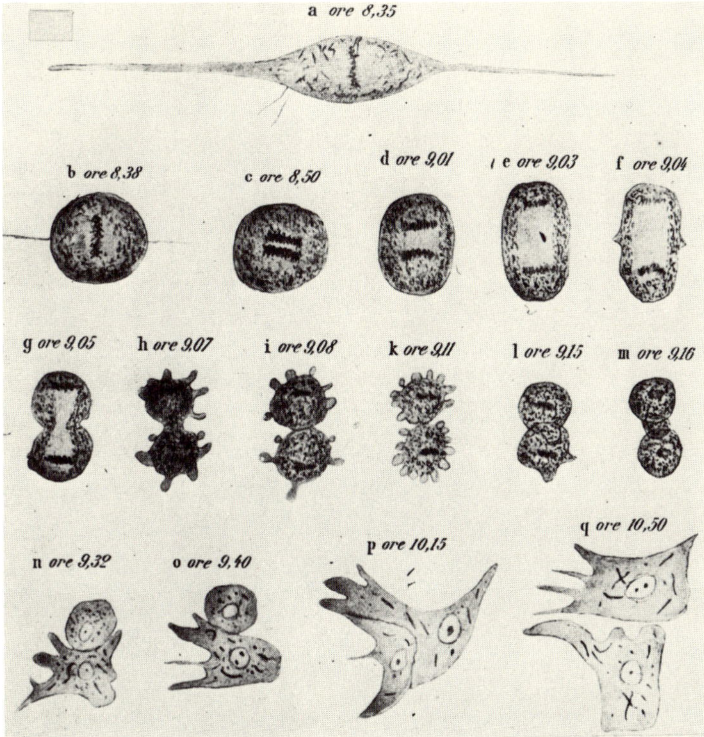

FIG. III–6. Cell division in fibroblasts of chick embryo (97 minutes). After Levi, 1919.

that, in other words, there was no true tissue culture in the sense that we speak of the culture of bacteria.[67] Never well founded, this objection has been shown to be meaningless in so far as a large number of investigations are concerned, and furthermore, it has been completely refuted in other cases where it would have been of crucial importance. In the first place, many of the problems studied by the method do not require for their solution that the explanted tissue remain alive and reproduce indefinitely. This is quite obvious from the work on the outgrowth of the nerve fiber, on the histogenesis of other tissues, on the nature of the connections between cells, on the course of mitosis, and in many other problems of cell structure and behavior. In the second place, it is now known that permanent cultures are possible to a degree once thought to be altogether fantastic [Figs. III–5 and III–6].

67. Jolly, *C. R. Soc. Biol. Paris, 69,* 1910. Gargano, *Publ. Staz. Zool. Napoli, 4,* 1923.

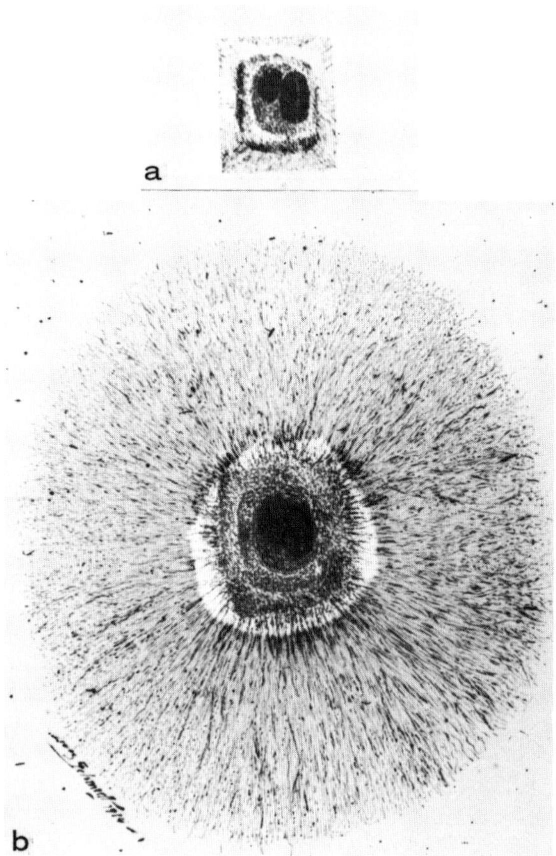

FIG. III–7. Culture of a strain of connective tissue twenty-eight months old, derived from a piece of 7-day chick embryo heart: (a) fragment extirpated and cultured, after one hour; (b) same fragment after forty-eight hours. After Carrel, 1914.

This was first demonstrated by Carrel and Ebeling[68] with the now celebrated culture of fibroblasts from the heart of the embryonic chick, which was set out originally in February 1912 and is still growing with undiminished vigor [Figs. III–7 and III–8]. Since the beginning of the experiment the mass of growing cells has been divided and reimplanted, usually at intervals of forty eight hours, in which period of time it about doubles its volume. The life of this line of cells has therefore exceeded by a number of years the span of life of an ordinary chicken and the amount

68. Carrel, *J. Exp. Med.*, *15*, 1912; ibid., *20*, 1914; Carrel and Ebeling, *J. Exp. Med.*, *34*, 1921; *C. R. Soc. Biol.* 90, 1924; Ebeling, *J. Exp. Med.*, *17*, 1913; ibid., *30*, 1919; ibid., *35*, 1922.

FIG. III–8. Fibroblasts, from heart of embryo chick, cultivated for 10 years. After Ebeling, 1922.

of living material produced, potentially at least, has been far greater in mass than the body of a single individual. In fact, had it been possible to provide nutriment for all the cells, and to allow them all to multiply, they would now greatly outweigh the terrestrial globe. The general significance of this indefinitely growing line of somatic

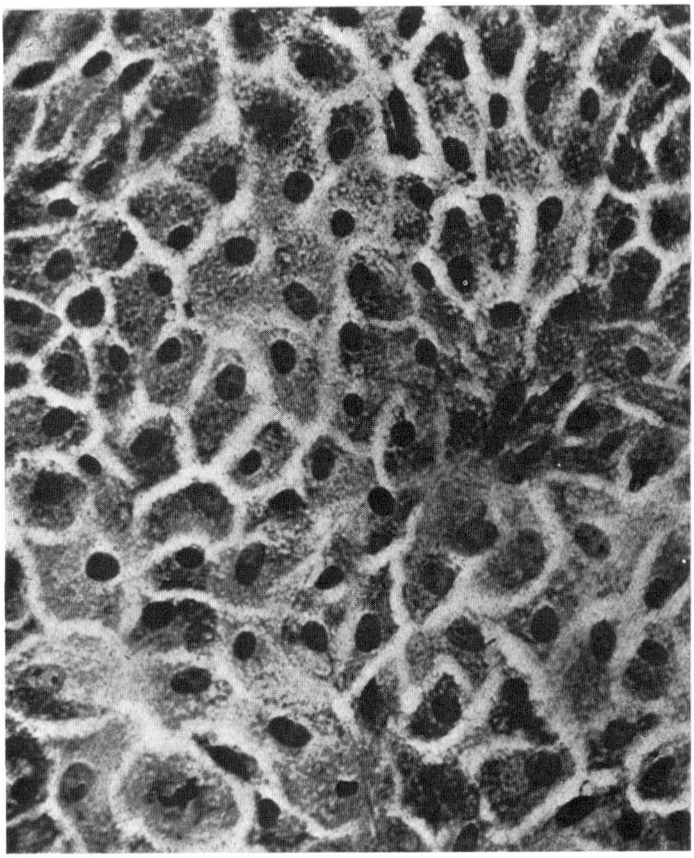

FIG. III–9. Pure culture of epithelial cells derived from the iris. After Fischer, 1922.

cells, not to speak of the perseverence and skill required to obtain the result, renders this discovery the greatest single achievement in the field of tissue culture.

A number of other lines of cells, mostly avian, have also been grown for considerable periods of time in pure culture. This list includes mononuclear leucocytes,[69] cartilage,[70] and a number of different strains of epithelium, such as that from the iris[71] [Fig. III–9] and the thyroid gland.[72] Each type of cell, though losing some of its visible differentiation, nevertheless retains certain characteristics of form, mode

69. Carrel and Ebeling, *J. Exp. Med.*, *36*, 1922.
70. Fischer, *J. Exp. Med.*, *36*, 1922.
71. Fischer, *J. Exp. Med.*, *35*, 1922.
72. Ebeling, *J. Exp. Med.*, *41*, 1925.

of growth, and physiological activity. Ebeling's culture of thyroid epithelium, after months of growth, formed follicles filled with colloid. Mammalian cells are also capable of indefinite cultivation, as shown by Erdmann's lines of connective tissue and endothelium from the guinea pig embryo.[73]

As is now well known, the longevity of cell cultures can be extended beyond a short time only when the medium is properly constituted. A small piece of tissue left to itself in a drop of plasma, serum, or artificial medium will cease growing actively after a few days, although some cells may remain alive for some weeks. If, however, embryonic tissue extract is added to plasma[74] very active growth is set up which may continue indefinitely, provided the growing tissue is planted out again at short intervals or the medium is otherwise renewed.

In addition to their general interest, these strains of cells maintained over a long period of time are of great value in affording a uniform material, already acclimated to conditions of culture outside the body, for the analysis of the factors which influence the rate of growth. By measuring the halo of cells formed through the migration of elements from the implanted piece out into the surrounding medium the amount of growth may be determined to within an error of ten per cent or even less,[75] which is sufficiently accurate for use in a great variety of comparisons in which differences are large. Much has already been accomplished in this line, for the details of which those interested are referred to the original papers.

Differentiation, Dedifferentiation and the Influence of Cells upon one Another

It will already have been gathered from what has been said that explanted tissues undergo a considerable amount of visible dedifferentiation. In fact some authors, as for instance Champy,[76] have gone so far as to claim that cells in culture revert to a primitive condition unless there are two types of cells side by side, for example connective tissue and epithelium, to react upon one another.[77] While a considerable

73. *Verh. Anat. Ges.*, Halle. *Ergänz.-Heft, Anat. Anz. 58,* 1924.
74. Carrel, *J. Exp. Med., 17,* 1913; ibid., *18,* 1913. An interesting suggestion in this direction is made in Haberlandt's first paper (op. cit., p. 89–90).
75. Cf. Ebeling, *J. Exp. Med., 34,* 1921, and Carrel, *Physiol. Rev., 4,* 1924.
76. *C. R. Soc. Biol. Paris, 75,* 1913; ibid., *76,* 1914. *Arch. Zool. Exp., 53,* 1913; ibid., *54,* 1914; ibid., *55,* 1915; ibid., *60,* 1920.
77. Cf. Drew, *Brit. J. Exp. Pathol., 4,* 1923 (not seen in the original).

body of evidence has been brought forward in support of this thesis, it can scarcely be maintained in last instance, for although cells under cultural conditions do lose some of their physical and functional characteristics, they nevertheless retain others, as shown by their form, habit of growth, behavior in various culture media and by their mode of reaction when recombined. Embryonic cells not only do not dedifferentiate immediately on explantation but exhibit marked powers of differentiation, as shown by the formation of nerve fibers, the differentiation of cilia, and the cuticular border in explants of ectoderm and the formation of striated fibrillae in cells from the myotomes.[78] Notable examples of differentiation in tissues of more advanced stages are the formation of follicles filled with colloid in Ebeling's[79] culture of thyroid and the complete course of spermatogenesis in cultures of the testis of the cecropia moth, as observed by Goldschmidt.[80]

While it may be accepted as demonstrated that pure lines of cells under culture do not by any means lose all of their differential characters, it is nevertheless true that mixed cultures behave differently from pure. It is this fact that has lent support to the more extreme views of Champy and Drew just referred to. Now this very circumstance is of essential importance in our study of organization. So long as we merely separate the different kinds of cellular elements found in the metazoan body and study their properties as isolated strains, we gain no deeper insight into the nature of those interactions that constitute organization. It is as if the chemist should content himself with obtaining pure substances and observing their properties, all the while neglecting the study of their reactions with one another. It is necessary here as there to investigate the phenomena of recombination—in other words to synthesize as well as to analyze.[81]

Certain things have already come to light that seem to me to be of significance in this direction. It is, for example, a very interesting and at present inexplicable fact that single somatic cells isolated in culture media do not proliferate. Experiments to this end made in my own laboratory some years ago but not published did not succeed and other workers[82] have reported similar experience. As Fischer puts it, a colony of fibroblasts cannot arise from a single cell even when the nutrient conditions are most favorable. Likewise small groups of cells if isolated do not undergo division

78. Harrison, op. cit., 1910.
79. Op. cit., 1925.
80. *Proc. Nat. Acad. Sci.*, *1*, 1915; *Arch. Zellforsch.*, *14*, 1917.
81. Cf. Harrison, *Anat. Rec.*, *6*, 1912, p. 185.
82. Cf. Fischer, *J. Exp. Med.*, *38*, 1923; *Arch. Mikr. Anat. u. Entw.-Mech.*, *104*, 1925. Policard, *C. R. Soc. Biol. Paris*, *92*, 1925; Laser, *Arch. Exp. Zellforsch.*, *1*, 1925.

and their growth remains at a standstill. On the other hand certain tumor cells (Rous chicken sarcoma) are capable of multiplying and producing colonies when isolated singly.

The earliest and at the same time most extensive experiments on this question were made by Haberlandt[83] with plant cells. It was found that while leaf cells of *Lamium purpureum* and hair cells of *Tradescantia* would undergo some growth, that is increase in size, when isolated singly, cell division never occurred. Experiments with small pieces of tissue from the potato showed a direct relation between the size of the piece or number of cells present within it and the number of cell divisions, and also that the presence of vascular bundles or sieve tubes stimulates cell division in the parenchyme of the explant. From his more recent studies Haberlandt[84] has been led to attribute the power of inciting cell division to substances emanating from injured cells. These he terms wound hormones.

Again in the matter of differentiation such fragmentary experiments as we have indicate that it is stimulated or at least accelerated by the presence of two kinds of cells in close contact. The best example of this is found in the mixed cultures of pure strains of fibroblasts and epithelial cells in the experiments of Ebeling and Fischer.[85] Here a ten year old strain of fibroblasts was combined with a two months strain of epithelium. After cultivation together for some time it was found that the epithelium had rounded itself off into a sort of glandular tissue lying in a stroma of fibroblasts.

It will be possible undoubtedly to make further progress in the study of the interactions between different kinds of cells and thus bring to light various manifestations of dependent determination. That functional stimuli are transmitted from cell to cell in culture has been amply demonstrated, and it is hardly too much to expect of the method of explantation that it will give similar proof of the passage of formative stimuli.

Holtfreter's study of larger masses of germ layers taken from the early amphibian embryo (blastula and gastrula) has served toward an understanding of the movements and foldings that take place in the embryo.* Isolated endoderm

83. *Sitz.-Ber Math. Naturw. Kl. Kais. Acad. Wiss., Wien. 111,* Abt. 1, 1902; *Sitz.-Ber. Kgl. preuß. Akad. Wiss.,* Jahrg. 1913, 1. Halbbd.

84. *Sitz.-Ber. Preuß. Akad. Wiss.,* Jahrg. 1921, 1. Halbbd.; *Biol. Zentralbl., 42,* 1922.

85. *J. Exp. Med., 36.*

* Editor's note: The account of Holtfreter's experiments was edited by Prof. V. C. Twitty, who also amplified the notes concerning his own work on pigmentation that concludes Silliman Lecture III. In both instances Twitty conformed to the content of the original lecture notes.

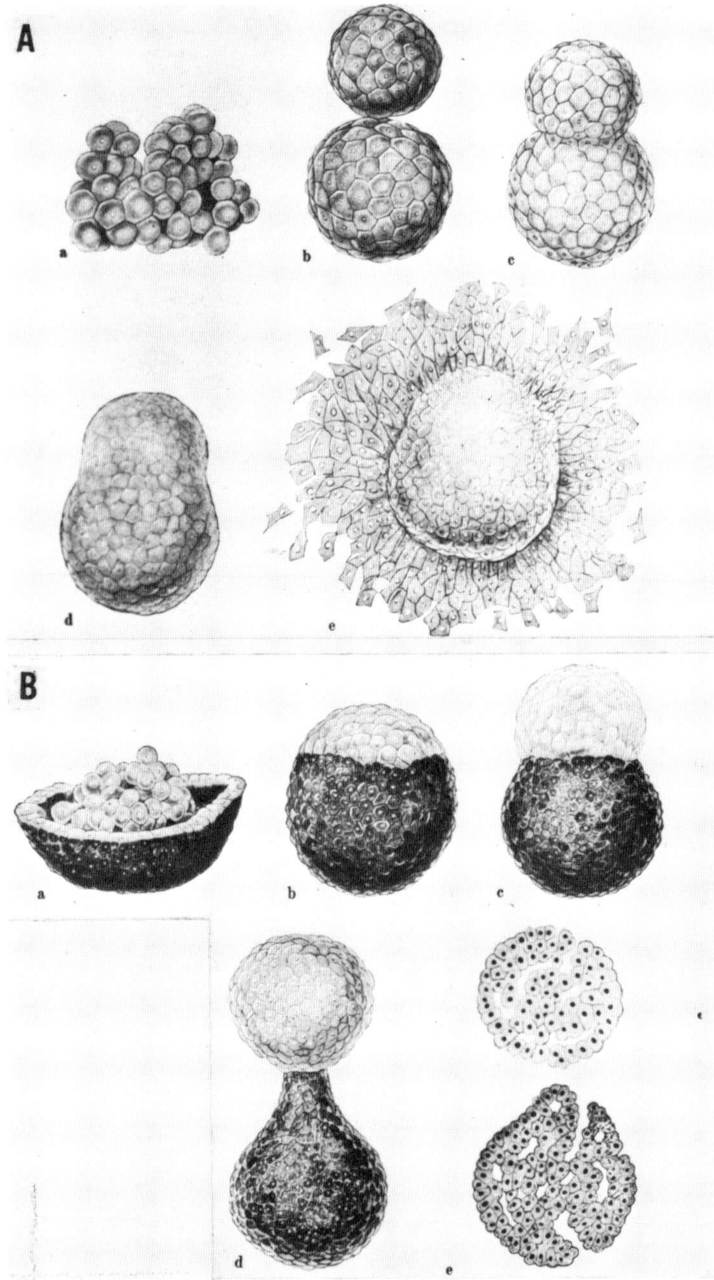

FIG. III–10. A. Behavior of isolated endoderm cells from the amphibian gastrula: (a) the cells shortly after isolation; two rounded aggregates placed in juxtaposition (b) form a single mass (c and d) which then (e) flattens with cells spreading from the mass.

B. Combination of endoderm and ectoderm: (a) initial arrangement; (b) formation of a large, rounded mass of cells; (c, d) gradual separation of component tissues; (e) the isolated pieces consisting of unorganized gut or epidermis cells. After Holtfreter, 1939.

(precursor of gut) forms little balls of tissue which tend to fuse with other similar balls if present, and flatten out later [Fig. III–10A]. There is a tendency of cells to spread. This coincides with the time of spreading in the embryo to close off the archenteric roof. On glass, cells tend to spread in all directions, but in the embryo local configurations guide the cells.

A cup of blastula ectoderm containing loose endoderm cells at first heals into a cohesive spherical mass, but a constriction soon develops and the ectoderm and endoderm separate cleanly from one another [Fig. III–10B]. This "negative affinity" between the two germ layers is prevented from manifesting itself, however, if some mesoderm is present in the explanted complex; the ectoderm and endoderm now remain attached as parts of a continuous sheet enveloping the mesoderm, which breaks up into loosely scattered mesenchyme cells [Fig. III–11A]. If the amount of ectoderm is increased, so that it completely envelops the mesoderm and endoderm at the time of healing, the three germ layers assume the concentric relationship characteristic of the normal body plan, with an outer ectodermal envelope separated from an internal endodermal vesicle by an intervening layer of mesoderm [Fig. III–11B]. The positive affinity of mesoderm with both ectoderm and endoderm thus overcomes the antagonism manifested by the ectoderm plus endoderm combination and, in addition, appears to foster the arrangement of the constituent cells of the latter into orderly epithelial layers.

Positive and negative affinities also operate within the germ layer, and help account for its subdivision and regrouping into derivative structures. If the anterior portion of the medullary plate is isolated together with an extensive piece of the adjoining epidermal ectoderm, the plate closes normally and differentiates into a well-organized neural complex inside an epidermal vesicle [Fig. III–11C]. But if a smaller amount of epidermis is included, so that it cannot readily envelop the neural tissue, the epidermal and neural components of the explant express their inherent negative affinity by separating from one another and forming independent vesicles. The neural tissue and its derivative optic vesicle, now deprived of their normal association with mesenchyme and epidermis, are unable to undergo normal morphogenesis.

The effect of inducing agents on the form of the neural tube should also be considered briefly here. Lehmann showed that when the notochord is absent the floor plate remains thick, and when muscle plates are removed on one side the neural tube is lopsided. Holtfreter has confirmed and extended this by means of his explantation method of combination, as shown in [Fig. III–11C]. He carries this further to development of the central nervous system and eye, showing both nega-

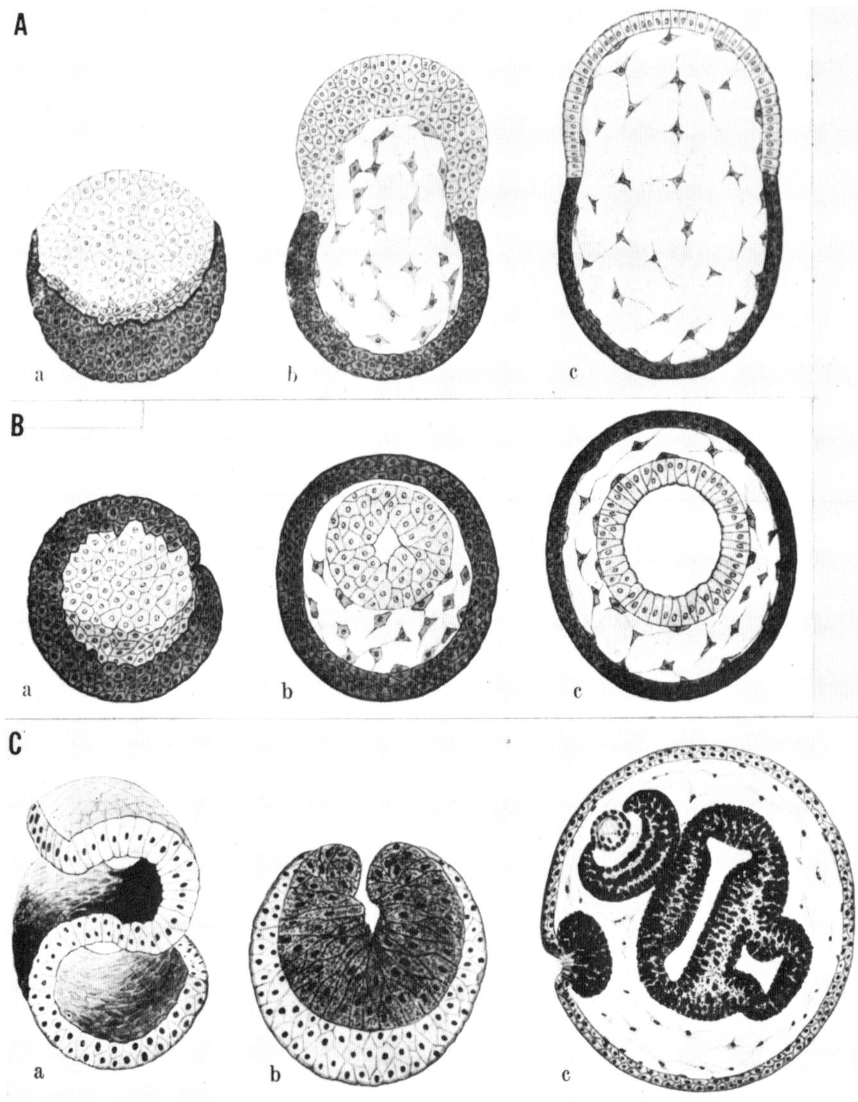

FIG. III–11. A. (a) Combination of endoderm and ectoderm separated by a layer of mesodermal cells, shown in section. (b, c) Upper expanded endoderm and lower ectoderm form a common epithelial wall with mesenchyme enclosed (see text).

B. (a) Endoderm and mesoderm covered by ectoderm. (b, c) Inner endoderm with lumen surrounded by mesenchyme and ectoderm.

C. Isolated neural plate (head region) with adjoining ectoderm. (a) Concave curling of medullary material above, epidermis below. (b) Neural tissue within the isolate partially enclosed by epidermis. (c) Differentiated brain, eye, and nose within the epidermal vesicle. After Holtfreter, 1939.

tive and positive affinities. Why these affinities exist is what has to be explained and the general application of a name for the process is not an explanation of the factors involved. Surface phenomena of some kind are involved; possibly a molecular fit or misfit between the patterns of cell surfaces (Schmitt 1941). [Lecture III notes.]

Reconstitution of Organisms from Dissociated Cells

Leaving now the results of the more usual methods of tissue culture, we may next consider a somewhat different mode of isolation in relation to ontogeny and regeneration or reconstitution; for here, it seems to me, is a field that has been relatively neglected and one in which, through recombination of known elements, syntheses of viable organisms may be obtained more readily than by the combination of pure lines of cells from the most specialized animals.

It is well known that most organisms show some power of restitution of lost parts and that in some the whole can be restored by very small fractions, as in hydroids, some worms, ascidians, etc. This takes place either by the formation of a blastema at the wound surface, from which the missing parts develop, or by a process of rearrangement of the constituents of the fragment (morphallaxis), as shown by Morgan[86] years ago.

Now it has been found possible to go much further than the students of regeneration had done in the attainment of isolation by a method introduced in 1907 by H. V. Wilson.[87] If small fragments of certain marine siliceous sponges are squeezed in fine bolting cloth, and the material which passes through is stirred into sea water, a fine suspension of isolated cells may be obtained, which after a time settle to the bottom. From this deposit small groups of living cells segregate and in due course reconstitute themselves into orderly arranged masses out of which new sponges develop [Fig. III–12]. Some of these have been raised to sexual maturity.

The most important elements in this process are the archaeocytes (amoebocytes) which move about by amoeboid activity, unite by means of their ectoplasmic layer with other similar cells, incorporating also into the small masses thus formed other cellular elements of the sponge. There has been some difference of opinion between

86. *Arch. Entw.-Mech.*, *9*, 1900, p. 563–586; *Regeneration*, New York, 1901.
87. *J. Exp. Zool.*, *5*, 1907.

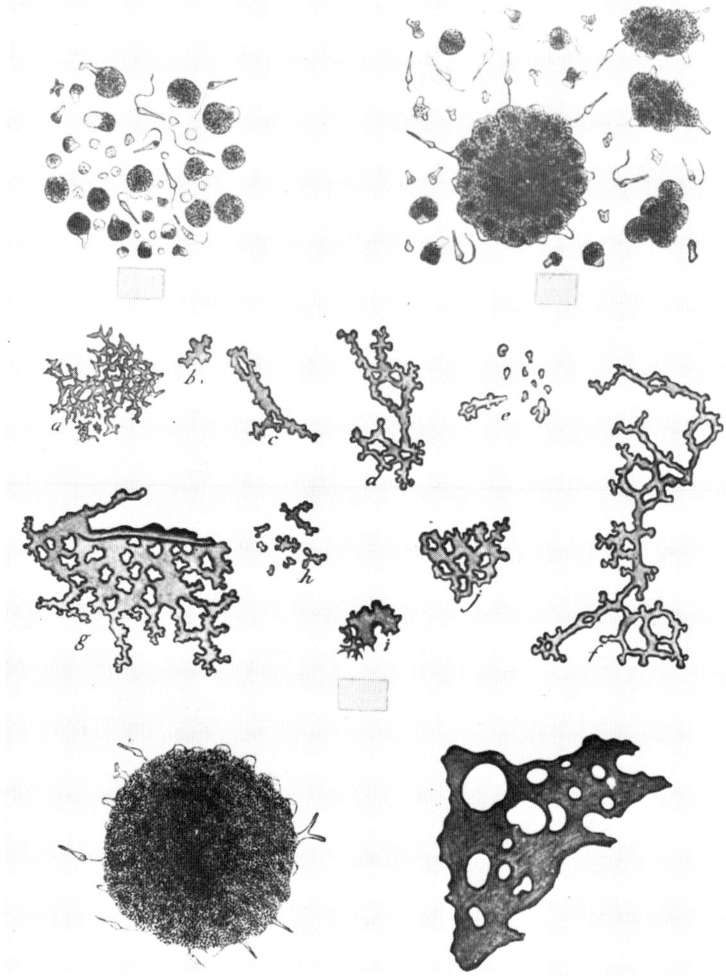

FIG. III–12. Restitution from dissociated cells of sponges (*Microciona prolifera*). After H. V. Wilson, 1911b.

Wilson and those who have followed him with regard to the potencies of the different kinds of cells and the extent to which dedifferentiation occurs. According to Galtsoff[88] the dissociated cells do not revert to an indifferent condition, as Wilson was inclined to think, but each retains some of its morphological characters, and no cell is totipotent. At least one kind of cell other than archaeocytes, namely pinacocytes,

88. *J. Exp. Zool.*, *42*, 1925.

which constitute the dermal layer and the lining of the canals, is necessary to form a viable aggregate that will develop into a sponge. The only cells that probably do not take part in the reconstitution of the sponge masses are the collar cells, but in the calcareous sponge (*Sycon*) studied by Huxley[89] the collar cells do enter into the composition of the masses and form the new flagellated chambers. Development of the restitution masses, according to Huxley, is primarily a process of sorting out, by which the dermal cells cover the surface and the collar cells pass to the interior. There are obviously important specific differences in the mode of reorganization but it is of equal interest, whether brought about by redifferentiation of indifferent material or by local sorting of the elements.

It is interesting to note, as Wilson points out, that the archaeocytes have great regenerative power under other circumstances. They play the chief role in reproduction by means of gemmules and out of them develop also the sexual cells. We have, then, in these cells elements which are most active in all three of the modes of reproduction occurring in sponges.

Another important fact brought out by these studies is that no coalescence takes place between cells of different species [Fig. III–13]. Wilson has made mixtures of cells from two sponges that could be recognized by means of differences in pigmentation, but always with the result that each cell aggregate formed was of homogeneous origin. Galtsoff obtained similar results with mixtures of other species. This is odd in view of the fact that in vertebrates there is no difficulty in getting tissues of different species to unite and to grow and function in harmony, as in numerous well known cases of heteroplastic grafting in amphibian embryos.

Wilson[90] later extended his experiments to the hydroids, showing that here, too, a similar power of reconstitution is present [Fig. III–14]. In both *Eudendrium* and *Pennaria* new colonies of polyps grew out of masses formed by the reintegration of cellular sediment. The masses of cells readily come together, secrete a perisarc, and later send out coenosarcal outgrowths on which typical polyps develop. Microscopic examination shows that at first a syncytial mass is formed in which cells of both ectodermal and entodermal origin are represented but without definite arrangement. Later an outer layer segregates, forming a definitive ectoderm and a more or less solid mass representing the yolk entoderm, which finally divides into a definite layer of entoderm closely applied to the ectoderm, and an irregular inner yolk mass, which gradually is used up. This stage closely resembles the planula stage in the normal ontogeny of *Pennaria*.

89. *Phil. Trans. Roy. Soc. London*, 202 (*B*), 1912; *Quart. Journ. Micr. Sci.*, 65, 1921.
90. *J. Exp. Zool.*, 11, 1911.

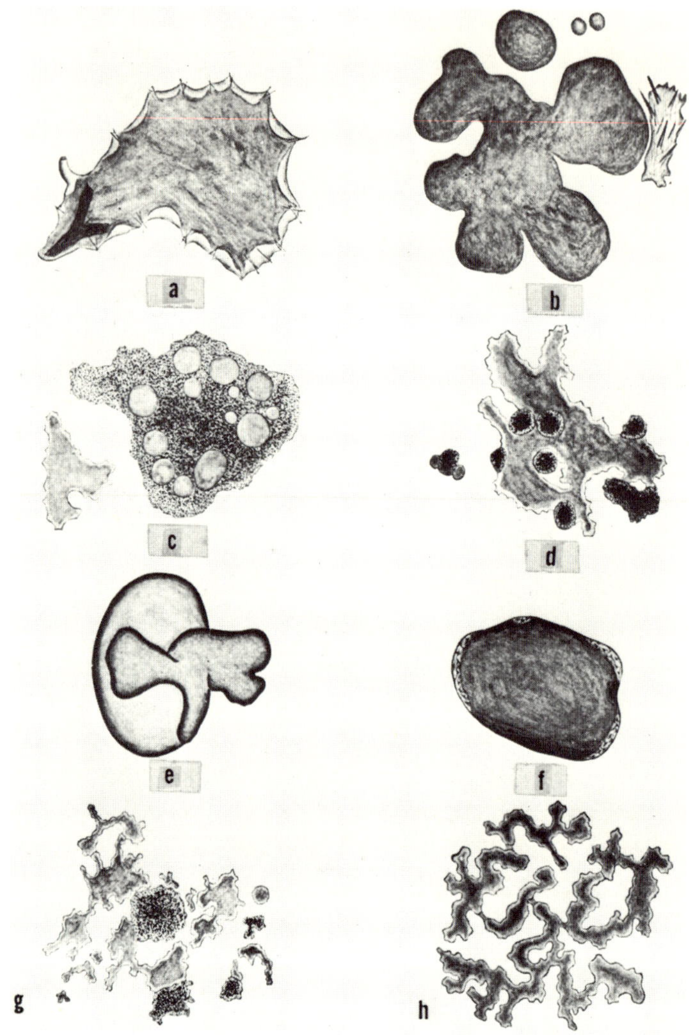

FIG. III–13. Restitution from dissociated cells of sponges (a, b, f, *Microciona;* h, *Stylotella*) and segregation of mixed cells (c, d, e, g, *Microciona* with *Lissendoryx* or Stylotella). After H. V. Wilson, 1911b.

Experiments by Hargitt[91] yielded similar results in *Podocoryne* and to some extent also in two species of *Eudendrium,* but in the latter the polyps, which are formed only in exceptional cases, are small and weak, and reconstitution usually does not go

91. *Biol., Bull., 28,* 1915.

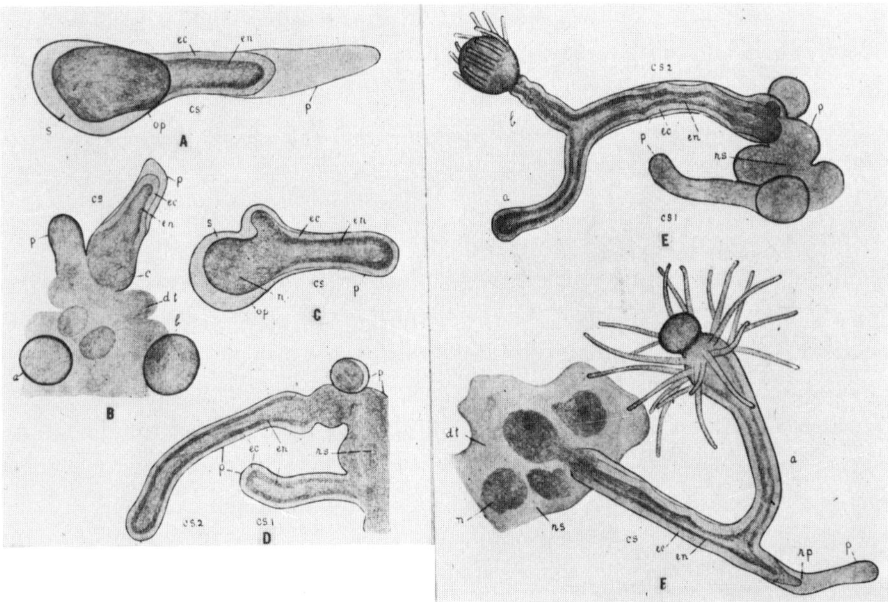

FIG. III–14. Polyps of *Eudendrium* which have grown from dissociated cells (see text). After H. V. Wilson, 1911a.

A. Restitution mass four days old; cs, coenosarcal outgrowth; ec, ectoderm; en, endoderm; op, perisarc of original mass; p, perisarc of outgrowth; s, space between original mass and its perisarc.

B–D. Parts of restitution masses showing coenosarcal outgrowth and nodule formation.

E. Restitution mass of (D) twenty-four hours later.

F. Restitution mass with coenosarcal outgrowth bearing a vertical branch which ends in a hydranth.

beyond the stages of encystment and stolonization,[92] as appears always to be the case in certain other hydroids. In *Clava* according to Föyn,[93] polyps are readily produced, the original sex of the colony being preserved, and hermaphroditic aggregates can sometimes be made by mixing cells from colonies of each sex.

Wilson again raises the question of the dedifferentiation of specialized elements, pointing out that the cell masses consist of both ectoderm and endoderm cells. The

92. De Morgan and Drew (*J. Marine Biol. Assoc.*, *10*, 1914) repeated Wilson's experiments, using two species of *Antennularia.* Restitution masses were formed, some of which lived as long as sixty days, but none of these went so far as to form polyps. Hargitt (op. cit.) reports similar failure in several species of hydroid polyps and medusae.

93. *Zool. Anz.*, *61*, 1924.

formation of the layered mass must take place, then, either by differentiation *in situ* irrespective of origin or by the wandering of the original ectodermal elements to the surface. There is no histological evidence for the latter and Wilson therefore adopts tentatively the former view which means that already specialized cells may dedifferentiate and later become transformed into something else.

In his earlier papers Wilson inclined to the view that the cells completely dedifferentiate before reconstituting a new organism, but it now seems that certain types of mesenchyme cells and collar cells make up the bulk of the reconstitution masses and that the gray mesenchyme cells in *Microciona* actually move around collecting other cells into masses. The collar cells then line small cavities which become the flagellated chambers and the other cells become the main mass of the sponge and its covering. That is, to a considerable extent each kind of cell seeks out its place in the mass to take part in new organism formation without either de- and re-differentiation (Wilson 1932, Wilson and Penney 1930). [Lecture III notes.]

As far as is known this mode of reconstitution takes place only in sponges and hydroids. However, no exhaustive efforts have been made to find out to what extent in the animal kingdom it may occur. Wilson[94] has made some experiments with the pressed out ovarian cells of the starfish (*Asterias*) and with cells from the stolon of *Perophera,* an ascidian, but with negative results. His experiments with *Leptogorgia,* a gorgonid, nevertheless lend encouragement to the view that under proper conditions successful syntheses may be obtained with the cells of this or related forms. It would seem worth while to experiment also with the regenerating tissues of forms in which the whole may be restored by very small fragments, such as the starfish *Linckia,* planarians, ascidians, *Ctenodrilus,* and some of the fresh water oligochaetes. By the use of this method, if further perfected, it will be possible to study not only the types of cellular combination that will produce new individuals but also to investigate the external physical and chemical conditions that influence the process. In the latter direction Galtsoff[95] has already broken the way.

By employing dissociation in conjunction with the more usual procedures of tissue culture we may be able to study the transition between the process of reorganization and the simple propagation of cells in pure or mixed cultures without the production of individual organisms. Of much interest in this connection are the recent experi-

94. *J. Exp. Zool., 11,* 1911; *J. Elisha Mitchell Soc., 39,* 1923.
95. *J. Exp. Zool., 42,* 1925; *J. Gen. Physiol., 10,* 1926.

ments of Murray[96] in which it was found that small fragments of planarian tissue (*P. dorotocephala*) isolated in culture media would generally behave like tissue cultures and send out into the medium cells of various kinds. Some of the fragments, however, showed a certain amount of individuation either of an axiate or a spherical type. Here, then, we have the actual border line between the apparent anarchy of an ordinary tissue culture and the subordination of elements requisite to produce a higher organization. The investigation of the conditions under which these different types of reaction occur should, as Miss Murray points out in her program, throw light on the whole question of reconstitution.

In fact much more should be done in cultivating invertebrate tissues. The work with vertebrates has, no doubt, been pushed not only because of the convenience of obtaining suitable culture media but also because of its greater practical and theoretical interest in medicine. Nevertheless, with the less specialized invertebrate elements we are more likely to obtain those syntheses which will help us to gain insight into the meaning of what we call organization.

Significant in relation to the above experiments is the fact that in nature many organisms undergo periodically a kind of reduction or disintegration during which they are able to survive detrimental environmental changes that they could not withstand in their fully developed state. This occurs in some marine sponges and in freshwater sponges as well. The sponge tissues shrink in volume, leaving the skeleton bare and in the form of strands and small oval bodies. When such "reductions" are placed again in better conditions, they form new sponges much like those from aggregates of dissociated cells. Probably only the archaeocytes (amoebocytes) are involved in the restitution.

This kind of reproduction leads into gemmule formation, which also occurs widely in several marine sponges. Aggregates of mesenchyme cells form spherical groups which become surrounded by other cells, and a horny capsule may be secreated around the mass. These gemmules may withstand drying and other adverse conditions, and when more optimal conditions are restored they develop into swimming larvae which settle and form sponges (Wilson 1894). The ordinary mode of reproduction by means of single egg cells fertilized by sperm may then occur. In the sexual mode of reproduction the resulting larvae are much like the gemmule larvae.

The case of some of the ascidians is particularly interesting in view of the char-

96. *J. Exp. Zool.*, 47, 1927.

acter of their development. According to Conklin and others (Chabry et al. 1887) the egg (*Styela*) follows a very definite cleavage pattern, each cell giving rise to a certain organ or tissue even when displaced or when its visible inclusions are disturbed by centrifuging. There are, it is true, certain evidences of regulations [see pp. 12–13]. These are, however, relatively minor departures from the general mosaic character of an ascidian's embryonic development.

Let us now contrast the latter with vegetative reproduction or budding, whereby in the case of the compound ascidians large colonies are formed. Some species protect themselves, like the sponges, from adverse conditions by involution and the formation of "winter buds" consisting of epidermal sacs filled with loose cells. One of the most thoroughly studied species is *Clavelina lepadiformis*. This form is of special interest also because it was used by Driesch (1929) as one of the bulwarks of his vitalistic theory. The colonies are built up by stolon budding. Furthermore, regeneration will take place from a cut surface of the zooid in either direction at any level (Spek 1927). Also, isolated pieces of stolon will regenerate zooids.

This was one of Driesch's best examples of an "harmonic equipotential system." Under adverse conditions, either in the laboratory or in nature (winter in the north or summer in the south), *Clavelina* zooids undergo involution. Cells laden with droplets of nutrient material appear in the branchial basket and in most other parts of the organism. These wander to the stolon, which puts out diverticula filled with these cells and perhaps other mesenchyme cells (Brien 1927–1937). The diverticula become sealed off and remain as winter buds until conditions get better, when blastozooids develop in them. The buds consist of loose cells without any organization (droplet cells and mesenchyme—perhaps just different phases of the same type of cell covered by a layer of ectoderm). Here we have, then, amorphous sacs filled with loose cells, totipotent and capable of giving rise to organisms like the oözooid developed from the egg, where development is strictly determined [Fig. III–15].

In the blastogenetic forms the organization of the animal cannot be represented by a spatial configuration of cells but must be in their finer structure.

The fact that the mesenchyme cells in the buds are totipotent and the facts of involution and reconstitution seem to lead to a paradox when taken in connection with the determining of egg development. One explanation would be that the mesenchyme cells which are derived from the anterior pair of vegetative cells in the 8-cell stage are totipotent but that they are unable to realize their potencies at that period of development or in later embryonic stages because of their small number and because conditions for their protection are inadequate. Only later

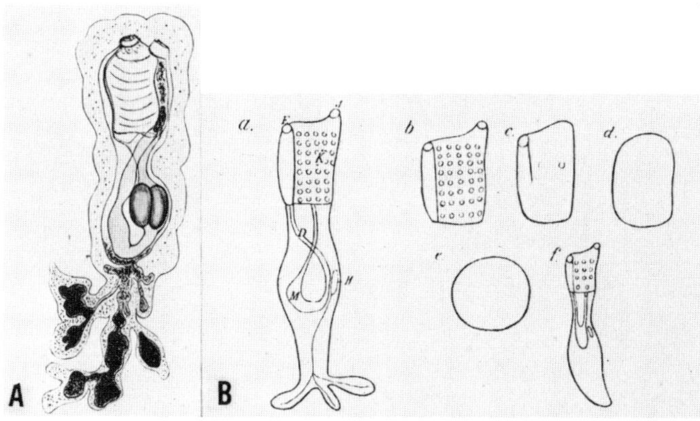

FIG. III–15. A. Degenerating *Clavellina* with greatly enlarged winter bud ampullae into which the trophocytes are wandering from the lining of the anterior regions of the alimentary canal, prior to detachment of zooid masses. After Spek, 1927.

B. Regeneration in *Clavellina.*

(a) Diagram of the normal animal; openings, E and J; branchial apparatus, K; intestine, D; stomach, M; heart, H. (b) The isolated branchial apparatus. (c–e) Different stages of reduction of the branchial apparatus. (f) The new whole little ascidian. After Driesch, 1929.

when there are much larger numbers of them are they capable of multiplying further under the protection of the stolon test (cellulose), ultimately developing into a blastozooid.

Berrill has made a comprehensive study of asexual and sexual development in ascidians. He points out that the cleavage pattern and the limitation of cell numbers lead to a larval form and that a number of cell types—muscle, notochord, nervous system—are significant only for the larva. "The process of determination or differentiation discernible in the early development of an egg refers in many cases not to the final differentiation of miniature adult structure but to a precocious differentiation of peculiar larval structures" (Berrill 1935, p. 421). "The egg of an ascidian is never so large or so yolky that it is able to give rise to a sufficient number of cells for the full expression of specific character" (ibid., p. 422). In *Ecteinascidia* ca. 132,000 cells are formed by the egg; ca. 330,000 are necessary as shown by bud development. In some ascidians (*Molgulidae,* e.g.) the larval stage is dropped (9 species out of 21), and the egg develops directly into the ascidiozooid.

The *Myxomycetes* or slime molds are claimed by both botanists and zoologists,

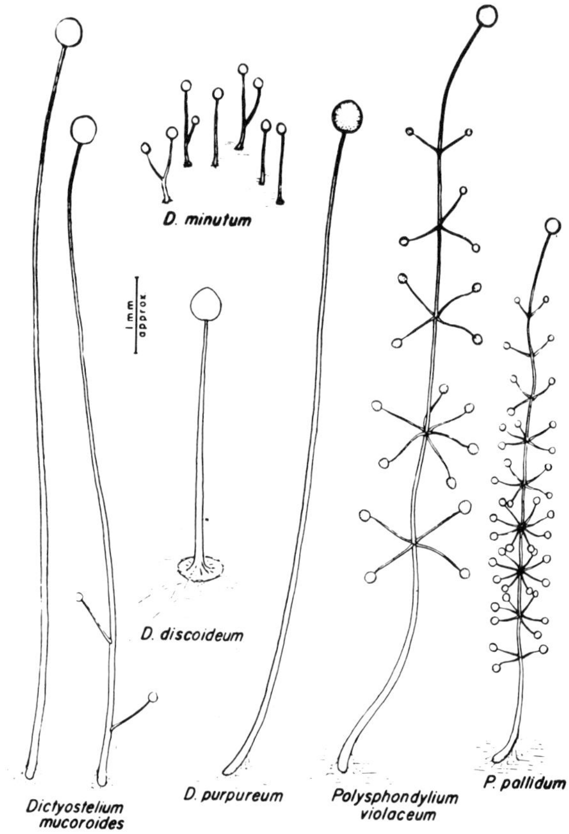

FIG. III–16. Semi-diagrammatic figures of the mature fructifications, or sorocarps, of various slime molds. After Raper, 1941.

the latter calling them Mycetozoa, or fungus animals. Whatever they may be, they are very interesting primitive organisms. They live mostly on decaying vegetable matter and form large masses of slime plasmodia without form or structure except that the masses are polynucleate (Raper 1935–1941).

In one suborder (Acrasida or Acrasieae) the organism consists of a colony of amoeba-like cells, which remain apart. Each contains a nucleus, contractile vacuoles, and food vacuoles, a more fluid granular endoplasm, and a clear ectoplasm which sends out pseudopodia. They feed on bacteria (phagocytosis) and as long as there is a plentiful food supply they keep on multiplying and move around actively. Various species of *Dictyostelium* and *Polysphondylium* are impossible to

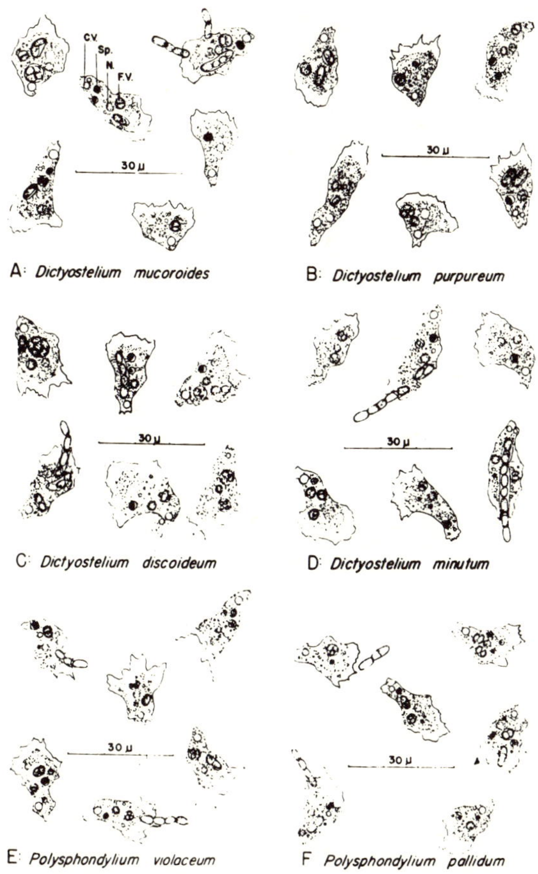

FIG. III–17. Camera lucida drawings of feeding myxamoebae of each of the six species shown in Figure III–16. Cultures grown in association with *Bacillus megatherium* upon hay-infusion agar at 22° C ±. All figures drawn to a constant scale as shown. F.V., food vacuole; N., nucleus; Sp., undigested bacterial spore; C.V., contractile vacuole. After Raper, 1941.

distinguish from one another in this stage. During this vegetative stage there is no trace of organization; the individual amoebae are very evenly distributed.

When the food supply is depleted these colonies go into the fruiting stage [Figs. III–16 and III–17]. First, small clusters of amoebae appear. What starts these is unknown. Once started, however, the amoebae within range orient toward the center, and flow toward it in radial streams forming a wheel-like pattern. Breaks may occur in some of these radial streams and more than one pseudoplasmodium may arise from an originally single aggregate, but often all go into one. They form ultimately an elongated mass like a sausage with a tapering point at

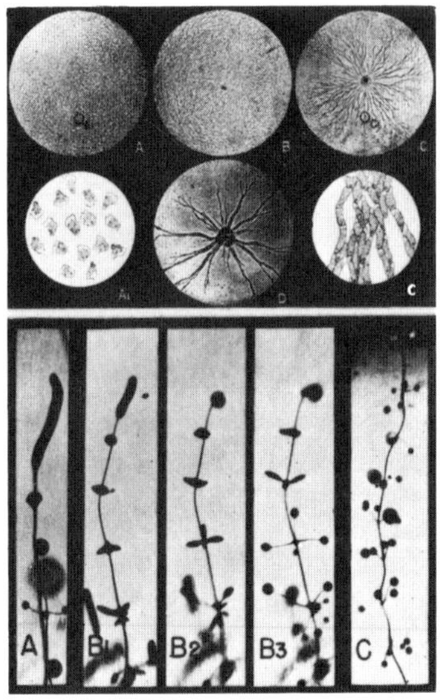

FIG. III–18. Upper: Progressive stages in pseudoplasmodium formation in *Dictyostelium discoideum:* (A) close of vegetative stage; (A_1) enlarged view of small portion of colony, myxamoebae unoriented; (B) early stage in aggregation, pseudoplasmodial center prominent; (C and D) progressive stages in orientation and aggregation of myxamoebae; (C_1) detail of aggregating myxamoebae showing their elongate form and consistent orientation.

Lower: Progressive stages in sorocarp formation in *Polysphondylium violaceum:* (A) young sorocarp showing elongate terminal sorogen and the first deposited mass of myxamoebae. In the lower portion of the photograph are shown the terminal portions of two completed sorocarps. (B_{1-3}) successive stages in the development of a single sorocarp. (C) greater portion of a mature sorocarp showing general habit; upon touching the agar surface, sori immediately spread as shown at the fourth and fifth whorls. After Raper, 1941.

the forward end and somewhat more blunt at the other. This raises itself in the air above the medium and the first individuals to elevate themselves form the stem, which the others climb to elongate it and to form the fruit at the upper end, or in some species, on side branches as well (*Polysphondylium violaceum*) [Fig. III–18].

The individual amoebae in the aggregates, instead of fusing into a multinucleate mass, maintain themselves separately with definite cell membranes. The conditions responsible for this behavior must be contained mainly in the protoplasm of the amoebae. The number involved in each sorocarp (fruiting body) varies enormously. In a series of cultures Raper found migrating pseudoplasmodia varying in number of myxamoebae from 770 to 189,000. Thus a structure of the same form may construct itself out of cells in which the minimum number is

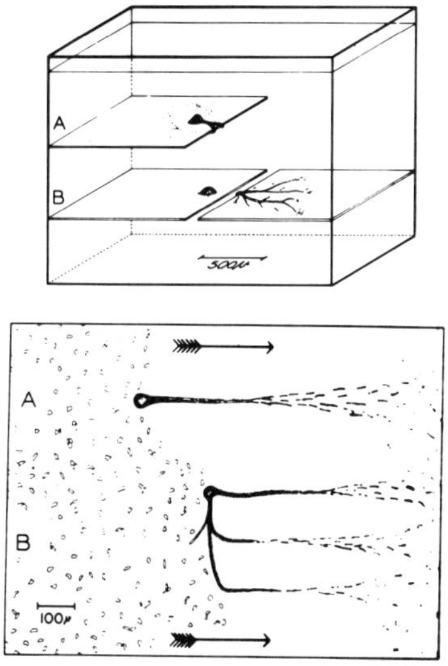

FIG. III–19. Upper: A semi-diagrammatic representation of 2 experiments done on aggregation in *Dictyostelium* using coverslip shelves held under water. A, the myxamoebae previously at random under the coverslip are attracted around the edge to the center on the upper surface; B, the myxamoebae previously at random on the right-hand coverslip are attracted to the center on the left-hand coverslip, across the substratum gap.

Lower: A semi-diagrammatic drawing showing the effect of a moving stream of water (the arrows indicate the direction of movement) over an aggregation pattern in *Dictyostelium*. A, An aggregation pattern that formed while the water was in motion; B, an aggregation pattern that formed in still water and was subsequently subjected to a stream of moving water. After Bonner, 1947.

only 1/245 of the maximum. Raper states, "The final result is that the proportions, and therefore the pattern, of the sorocarp are maintained, while individual cells contributing to stalk formation assume whatever shape is necessary to accomplish this result" (1941, p. 67).

What starts the first concentrations? This is not known, but Bonner (1947) has made some interesting experiments indicating that it may be due to attraction to a chemical substance secreted at a certain point. He tentatively calls the substance secreted at the center "acrasin." He has made very ingenious experiments showing that the myxamoebae do react to diffusible substances, and has ruled out such things as agglutination, electric field, magnetic field, contact, and radiation [Fig. III–19].

These Mycetozoa, while in a certain sense showing the same kind of behavior as the sponges and ascidians previously considered, bring out in a more striking way the contrast between the cells as a unit or individual and the cell as part of a unit of higher order. These two roles alternate. The cells when free constitute an "equipotential system" according to Driesch's definition and the part that each plays in constituting the whole is a function of its position. We accept these phenomena as facts, which will ultimately be explained. There is no ground for taking them as proof of either vitalism or holism. [Lecture III notes.]

Relation to the General Problems of Embryology

The relation of tissue culture to general questions of embryology has not been fully considered. With regard to the problem of determination, the method has served par excellence in proving that certain elements, such as muscle cells, neuroblasts and pigment cells are self-determining so far as their histogenetic processes are concerned. On the other hand, as already pointed out, no adequate experiments have been made to show the interaction of cells in development, though enough has been found to show that here is a rich field for future study.

The determination problem has dominated experimental embryology ever since the first embryological papers of Roux and Driesch were published. To ascertain whether a given differentiation is due to activities inherent within the differentiating element itself, or whether it is dependent on influence from without, and if the latter, according to what pattern the actions take place, is a task that all will admit must, in some measure at least, take precedence over the study of the more hidden qualities of the activities themselves. However, the nature of the protoplasmic changes involved in determination and differentiation, and the differences between various cells in respect to the kinds of change they undergo are questions that demand equal attention if we are to give a rational description of development in general terms. The growth energy of different types of cells, changes in chemical composition as shown by vital staining and other microchemical methods, the nature of the changes going on in any transformation as suggested by the study of temperature characteristics, the character and intensity of the respiratory activities of cells and their change during differentiation, regional differences in metabolic activity, changes in the physical state of protoplasm as shown by the methods of microdissection and by the measurement of viscosity—all such processes must be taken into account in any full consideration of ontogeny. While at present the ways are not altogether clear, it is probable that the value of tissue culture in embryology will lie in the investigation of phenomena

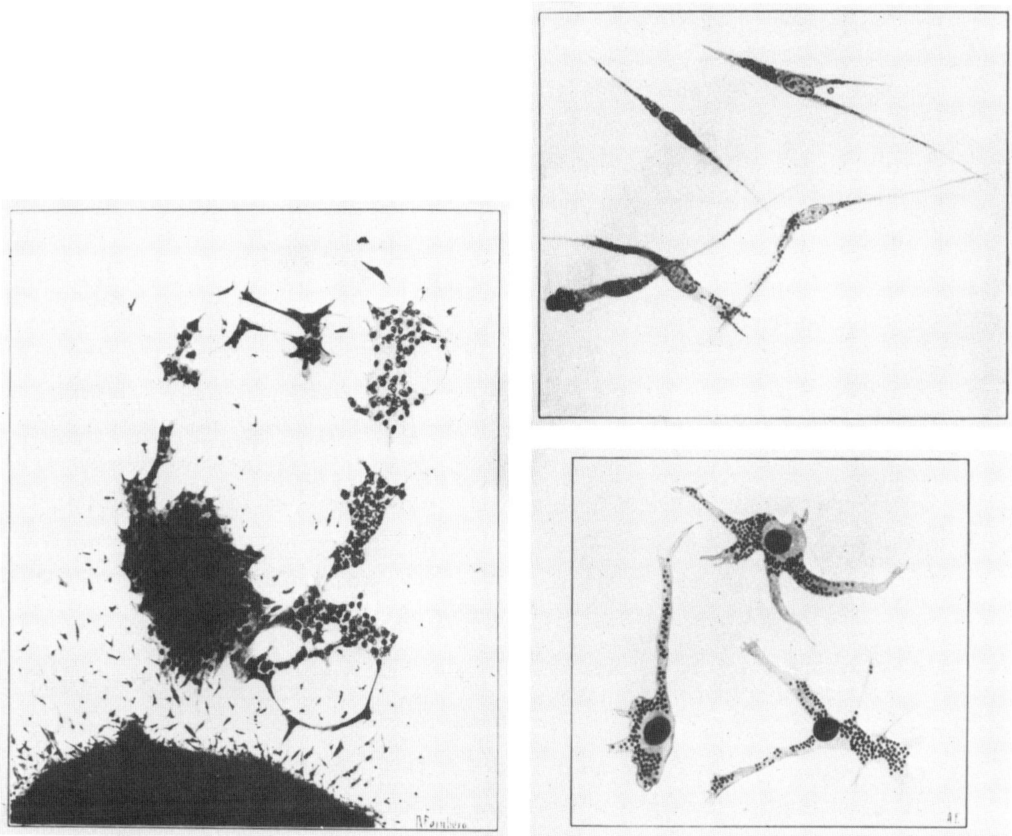

FIG. III–20. Left: Amoeboid wandering carcinoma cells growing *in vitro*.
Right: Carcinoma—stained with iron hematoxyliyn and Sudan III. Amoeboid wandering cells similar to type of growth of mesenchymal tissue. After Lambert and Hanes, 1911.

of this kind, rather than in the working out of a topographic map of these changes, which is involved in the concept of determination.

Of other great fields in which tissue culture has found application, particularly in relation to medicine, time does not permit me to speak at any length. In pathology its most important contribution has been to the study of tumors [Fig. III–20]. Early in the history of the subject it was shown that tumor cells grow well in culture media, exhibiting distinct differences from normal tissue cells. At the same time each neoplastic type was seen to grow and behave in characteristic manner. Certain tumor cells, at least, after long cultivation outside the body lose none of their virulence, for, injected back into a living animal, they readily produce new tumors of the original character. Now evidence is accumulating to show that normal body cells subjected to certain treatment while growing *in vitro,* may be capable of producing malignant metastasizing tumors after reimplantations into the body of a healthy individual. These facts, which could scarcely have been revealed by other methods, give witness to the unique value of the method of explantation and lend encouragement to the expectation of further important discoveries.

Résumé and Conclusion

Summarizing briefly what has been said and including certain topics not considered here at length, we may point to the following substantial contributions to biological and medical science made through the application of tissue culture and closely related methods of research:

1. A direct ocular demonstration of the embryological basis of the neurone concept.
2. A crucial proof of the myogenic theory of the heart beat.
3. A demonstration of the potential immortality and indefinite power of reproduction of somatic cells.
4. The discovery that dissociated somatic cells of animals having at least the degree of organization found in sponges and coelenterates can reconstitute themselves into normal viable individuals by a process resembling ontogeny.
5. A means for the exact study of growth in pure lines of cells, conditions influencing the process and the chemical nature of the medium required.
6. The discovery of numerous facts with regard to cell transformations that go a long way toward the solution of the vexed question of the genetic relations of blood and connective tissue elements in favor of the monogenetic or unitarian theory.
7. Important evidence regarding cell movement, the behavior of mitochondria, the mode of connection between tissue elements and their differentiation and dedifferentiation.
8. The opening of the process of cell division to direct observation, through which the peculiar movements of the cytoplasm, the duration of nuclear phases, the periodicity of the process and the influence of various external conditions upon it have been studied.
9. The discovery that tumor cells have the power of indefinite multiplication out-

side the body without losing their properties and that normal somatic cells may through proper treatment be rendered capable of producing malignant tumors when reintroduced into the body of an animal.

10. The direct observation of the behavior of tissue cells toward microorganisms and the conditions under which certain immune reactions occur.

11. The opening of tissue cells to the direct action of drugs whereby the influence of the latter on the living protoplasm may be studied.

Let me, however, not place too much emphasis on what has been accomplished. We are here not merely to review the past and take stock of the present but to look toward the future. From what has been said it must be clear that the actual accomplishment is but little in comparison with what remains to be done. In all of the subjects just enumerated, and in many others, further intensive investigation is needed. The time has gone by, however, when the amateur may describe a few casual experiments and still expect a hearing, for the novelty of the results and the manual dexterity involved can no longer be a cause for wonderment. The method, if it is to continue fruitful, must be applied with perseverence, thoroughness and discretion, with all necessary controls, and with data sufficiently numerous to bear statistical scrutiny. The ideas underlying the method are sound, but in its application vision of the great problems awaiting solution and their precise formulation for experimental test must hold sway over mere technical skill.

The last topic of today's discussion is pigment cells and animal coloration. The connection with the other topics may seem remote. It is brought in here, however, to show how cells may comport themselves with a considerable degree of independence within the body of a complex organism where they are nevertheless subjected to a considerable extent to the influence of internal environment.

In the vertebrate embryo pigment cells originate from the neural crest. This source was suspected more than fifty years ago by several observers (Harrison 1910, 1914; Weidenreich 1912), though not put to an experimental test until the 1930s by Graham DuShane, then a student in the Osborn Zoological Laboratories. The neural crest, which is formed along the crest or ridge of the paired neural folds, cuts itself loose from the neural tube as soon as it closes (sometimes even before), and forms a mid-dorsal wedge of tissue between the central nervous system and the covering epidermis. The neural crest cells soon begin to disperse widely, and form or at least contribute to the formation of the corium or dermis, the connective tissue of the fins, spinal ganglia, nerve sheath cells, pigment cells, and also the visceral skeleton of the head.

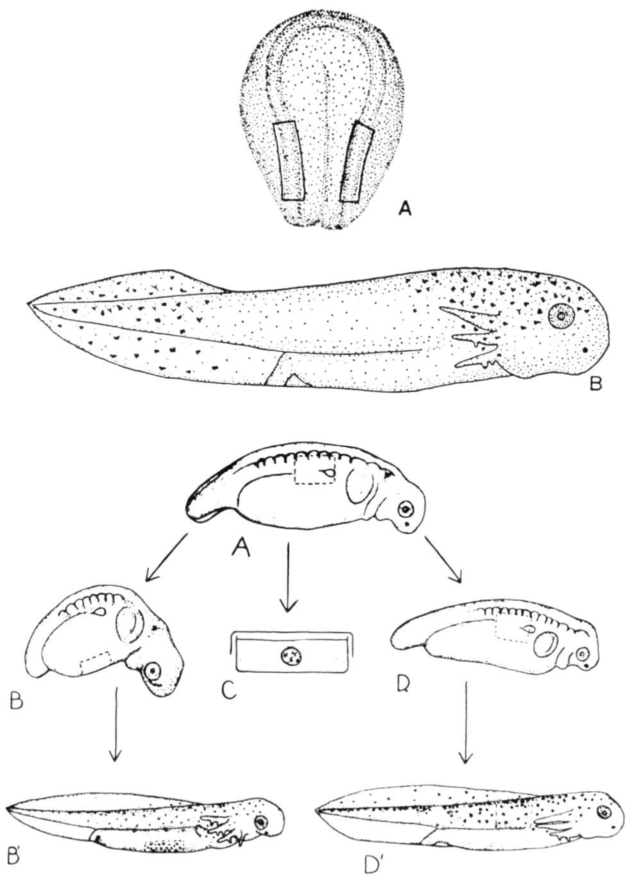

FIG. III–21. Upper: The effect of the removal of trunk neural crest on pigment cell formation. (A) Neurula stage (*Amblystoma tigrinum*) showing the area that was removed; (B) larva resulting from the operation in A. The dorsal fin and melanophores are absent in the trunk region.

Lower: Transplantation of flank ectoderm and mesoderm from older embryos (*A. punctatum* stages 32–37), showing effect of migration of neural crest. A. Homoplastic transplantation to the abdomen (B) produces melanophores (B′). Explanted pieces (C) form melanophores. However, when this tissue is transplanted to *A. tigrinum,* melanophores of graft type do not appear in the graft (D, D′). After DuShane, 1935.

Three types of pigment cells are recognized: melanophores, containing black or brown granular pigment; xanthophores, containing a yellow substance—lipochrome, which looks homogeneous in the cytoplasm; and guanophores having granular pigment with metallic luster, silvery or golden. We shall consider principally the melanophores, but there is evidence that all three types come from the same source.

Pigment cells are subject to many different intrinsic influences—the nervous system, the endocrine system through the circulation—and to extrinsic factors in the environment such as light and temperature. The responses of the melanophores consist in clumping or dispersal of the dark granules whereby color changes may be brought about. Other changes are of longer duration, as when an animal changes from a summer to a winter coat. When the pigment cells move out from the neural crest they are at first without color and cannot be distinguished from other cells derived from the mesenchyme or connective tissue. For that reason their origin can not be directly observed.

In DuShane's original experiment he found that removal of the neural folds on the two sides of the trunk region resulted in a larva devoid of pigment cells in most of the trunk and the base of the tail [Fig. III–21]. It was also shown that two factors are involved in the formation of melanin pigment, presumably an enzyme and a substrate, since heteroplastic grafting of ectoderm from a pigmented form to the white axolotl called forth pigment development in the host (Harrison 1935). The melanophores which developed under these circumstances were of axolotl type, not of the type of the graft species. Further variations of the experiments confirmed these conclusions.

Twitty has made an exhaustive analysis of the behavior of pigment cells and the factors involved in the development of pigment pattern in several species of California newt (*Taricha*). The whole study of Twitty and his collaborators is elaborate—involving inter-specific grafting, hybridization, and tissue culture of pigment cells. The following account is merely an outline of this important work (Twitty 1936, 1944, 1945; Twitty and Bodenstein 1939; Twitty and Niu 1948).

The species of *Taricha* present striking differences in pigmentation that lend them to analysis of the developmental basis of pigment pattern. In *T. torosa,* for example, the melanophores are almost all concentrated into a pair of compact dorsal bands extending along the dorsal margins of the myotomes, whereas in *T. rivularis* the melanophores are widely and uniformly dispersed over the entire sides, with no trace of banding. Hybridization results in a "compromise distribution" of the pigment cells. When a *rivularis* egg is fertilized by a *torosa* sperm the paternal influence brings about an increase in concentration of dorsally placed melanophores, while in the reciprocal hybridization there is some diffusion of the paired bands [Fig. III–22].

These hereditary differences in pigment pattern are referable primarily to factors

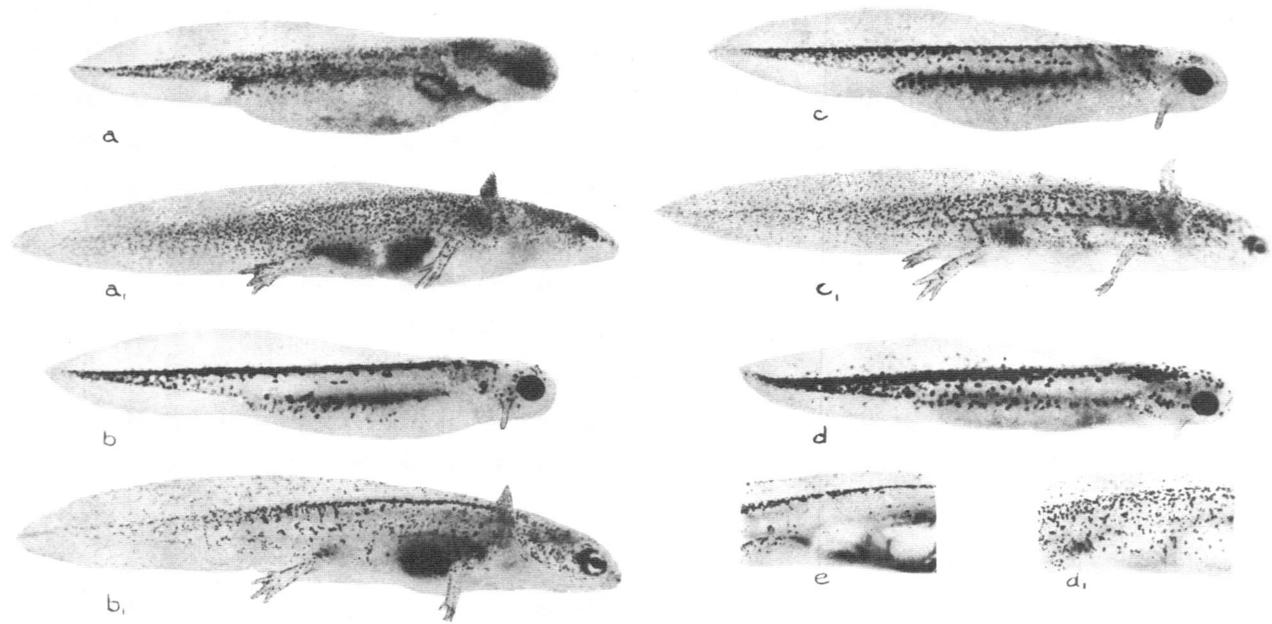

FIG. III–22. Normal and F_1 hybrids of T. torosus and T. rivularis. a, a_1 normal rivularis, younger and older larval stages; b, b_1 normal torosus (Santa Lucia Mountains variety); c, c_1 rivularis ♀ × torosus (S. L. variety) ♂ hybrids; d, d_1 torosus (ordinary variety ♀ × ♂; e, normal torosus (ordinary variety) larva for comparison. After Twitty, 1936.

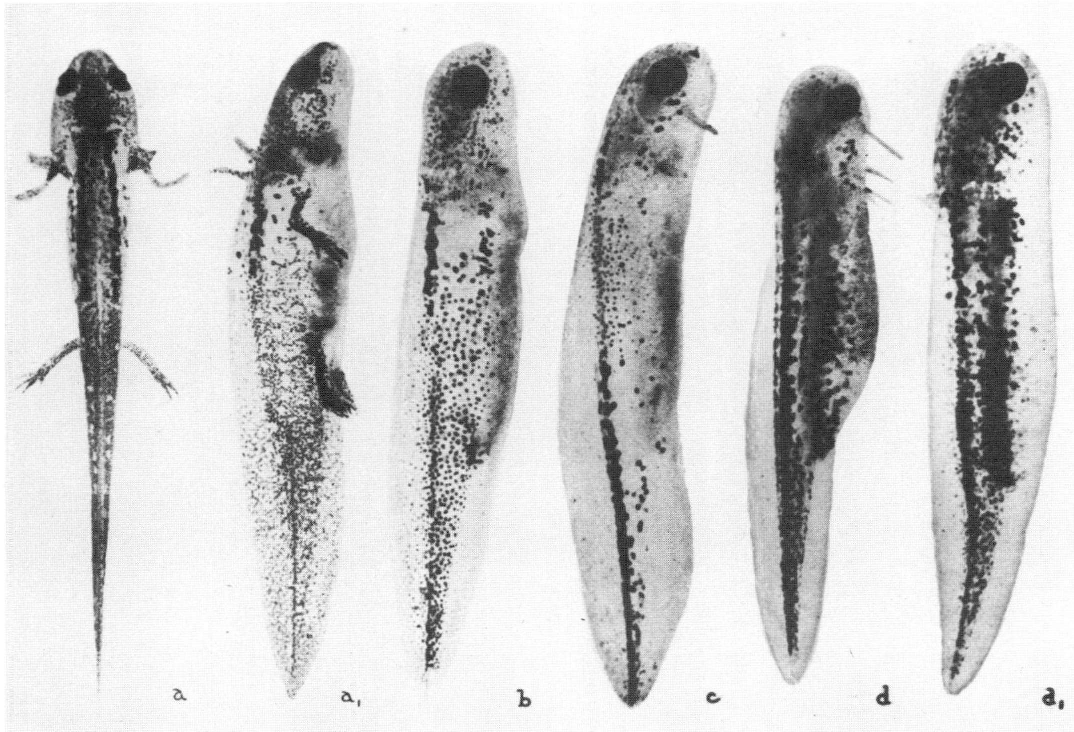

FIG. III–23. Orthotopic, heteroplastic transplantation of neural crest: a, a_1 transplantation of neural crest from torosus to similans; a, dorsal view, a_1 lateral view of two different cases; b, transplantation of neural crest from torosus to rivularis; c, transplantation of neural crest from rivularis to torosus; d, normal rivularis ♀ × torosus ♂ hybrid; d_1, transplantation of neural crest from rivularis ♀ × torosus ♂ hybrid to rivularis. Note in all cases that the pigmentation formed by the graft is of the donor pattern. After Twitty, 1936.

inherent in the chromatophores themselves; exchange of neural crest between the species leads to reproduction of donor-type patterns in the host environment [Fig. III–23].

An analysis of the banding behavior of the *torosa* melanophores revealed that the bands are formed through a secondary reaggregation of the pigment cells following an initial period in which they are more widely dispersed on the flanks. This was especially clear from studies *in vitro,* in which the cells could be observed to disperse freely from explanted pieces of neural crest and then subsequently withdraw into increasingly compact groupings. This reaggregation occurs only in a medium that fosters normal pigmentary synthesis in the melanoblasts (coelomic fluid), and is believed to result from shortening of the adhesive filamentous processes that interconnect the cells of a differentiating melanophore population. Choice of the dorsal somite margins as the site for band formation is attributed to the fact that melanophore differentiation, and hence shortening of their interconnecting processes, is initiated earlier here than at more ventral levels on the flank. If the myotomes are surgically inverted prior to crest cell migration, so that their true dorsal margins now lie ventrally, the site of band formation is correspondingly shifted (Twitty 1945).

The behavior of embryonic pigment cells under especially devised conditions *in vitro* led to the conclusion that their dispersal from the neural crest is motivated by mutual chemical stimulation mediated by diffusible substances emanating from the chromatoblasts. Cells migrating under conditions conducive to accumulation of these substances in effective concentrations, as when a crest outgrowth spreads into a low-ceilinged chamber beneath a fragment of coverslip, move more rapidly and become more widely spaced than under ordinary conditions of culture (Twitty and Niu 1948). [Lecture III notes.]

4

On the Origin and Development of the Nervous System Studied by the Methods of Experimental Embryology

Permit me first to thank the Royal Society for the honour of the invitation to deliver the Croonian Lecture, which I appreciate as one of the highest distinctions that may be conferred upon a biologist. I am reminded, too, of a personal relation that may have something to do with my presence here today. Fifty years ago, in 1883, the Croonian Lecture was given by H. Newell Martin, who on that occasion described his epoch-making work on the isolation of the mammalian heart. It was but a few years later that, as an undergraduate at Johns Hopkins University, I attended Martin's lectures on general biology and there received the first inspiration to follow my chosen calling. The awakening of my serious interest in biology was thus due to one who has filled this place before me, one whom I shall always remember with gratitude and respect.

Although many of my predecessors have concerned themselves with the nervous system, no one has considered it in this place from the standpoint of embryology. It therefore seems appropriate to direct your attention to this aspect of the subject today, more especially to the part that experiment has played in its advancement. The field covered is almost entirely a development of the twentieth century and is still in the rough. No well-rounded account of it is at present possible, so that I shall not have the satisfaction of showing "wie wir's dann zuletzt so herrlich weit gebracht." On the contrary, I approach the task with full realization of its perplexities, but

The Croonian Lecture, delivered June 29, 1933. Reprinted from *Proceedings of the Royal Society of London,* Series B, no. 808, *118* (August 1935), pp. 155–96.

nevertheless with the conviction that the new lines of investigation are leading in the right direction.

Recent advances in experimental embryology, particularly those gained from the study of the amphibian egg, have taken us further and further away from the preformationist point of view. The cells destined normally to give rise to the central nervous system may be located in the blastula by means of Vogt's vital staining method and may even be projected back upon the unsegmented egg. They require, however, the stimulating action of cells of the substratum after gastrulation, or some other agent, before their potency to form brain and spinal cord can be realized. Furthermore, transplantation experiments show that other cells than those which normally develop into the nervous system may be made to do so by bringing them into contact with this substratum, or "organizer" (Spemann 1918, 1927). Thus from the earliest stages of development the nervous system is highly regulable, a state which is retained for a long period, possibly as long as the organism is capable of growing and learning.

The nervous system is the principal means by which the organism is kept in touch with its surroundings and also one of the principal coordinating mechanisms within the organism, the other being the internal circulatory system. These two agents work entirely differently. The blood, a fluid with a complex of substances in solution and living cells in suspension, bathes all organs and tissues, which respond selectively in respect to what they take from the blood and what they give to it.

The nervous system, on the other hand, consists of fixed cells definitely arranged and connected by communicating paths or nerve tracts. Each nerve fiber is a part of a particular cell and is dependent upon the integrity of the connection with that cell for its continued functioning and existence. Also the neuroglia or supporting cells are essential components of the nervous system.

The cells are located in nerve centers or ganglia which in lower vertebrates are diffuse but in the higher phyla are more and more sharply localized and in the aggregate constitute the central nervous system, i.e. the brain and spinal cord. The nerve tracts serve to connect centers in the central nervous system and extend as peripheral nerves to all parts of the organism. They conduct impulses in a centripetal direction from the sense organs, integument, muscles, etc. and also efferent or centrifugal impulses to muscle and secreting cells. There are also certain nerve centers or ganglia outside the central nervous system which constitute the elements of the autonomic or sympathetic system.

Nerve units have the properties of irritability and conductivity which though common to protoplasm in general are, however, exhibited in high degree by the nerve cell and fiber, the fibers conducting impulses with facility. The speed of the nerve impulse varies with temperature, size of fiber, medullation, and species of animal. In man nerve fiber conductivity is calculated to be about 100 meters per second, in the frog, 30 meters per second, and in the alga, *Nitella,* protoplasmic conductivity is 0.02–0.2 centimeter per second—i.e. 5,000 to 500 times slower than nerve impulses in man.

The task of the embryologist is to find out how this elaborate system of conductors becomes established out of a compact mass of ordinary cells. The localization and differentiation of the nervous system involves a series of the principles encountered in the study of development. The first is the localization of the rudiment of the system which in the vertebrates is found to be dependent on influences emanating from the subjacent material turned in during gastrulation. The last is those changes in structure involved in habit formation and intellectual processes in man. Between these there is a variety of other processes which have been more or less thoroughly studied. On the whole, however, in spite of much and varied effort our knowledge remains fragmentary.

The induction of the neural plate and associated organs was described briefly in the second lecture. The problem is simplified in that it has been shown that the ectodermal structures are dependent upon the pattern of the underlying material and no self-differentiating mechanism need be postulated in the former. Previous to gastrulation—i.e. the folding in of material to form the lining of the archenteron, notochord, and mesoderm—the cells that will make up the nervous system are interchangeable with those that will form skin (epidermis) and sense organs. The cells of the outer layer differentiate according to location. After gastrulation, which involves the underlying of the superficial cells by the invaginated layer, this is no longer the case. The outer layer material then differentiates according to site of origin.

We shall not consider in detail the process of induction since it has been fully treated in Spemann's Silliman Lectures in 1933. The conclusions presented by Spemann are based upon a great variety of experiments conducted by him and his school and by many independent investigators.

The first visible change in the rudiment of the central nervous system, the neural plate, is a thickening of the cell layer due to change in shape of the cells which arrange themselves in a palisade. Next the thickened area becomes more sharply demarcated by a raised rim and a line of pigment. In contour these appear racquet-shaped with the enlargement anteriorly situated. The rim rises into folds (neural) which bend over and become apposed, then close to form a tube. The walls of the tube give rise to the substance of the brain and spinal cord. By the time the tube is closed, the main divisions of the brain and cord are demarcated (see Normal stages 14–21).

For some time after the completion of the neural tube, although the definitive parts can readily be made out, there is still no histological tissue differentiation. The walls which give rise to the substance, i.e. the cells and fibers of the spinal cord and brain, are not very different from other embryonic cells, except for their general columnar arrangement and the extension of each cell from the inner to the outer surface of the nerve tube. This primitive nervous system touches the skin on the dorsal midline and on each side the segmental axial musculature, which is also still without histological differentiation. Ventrally in the midline it is underlain by the notochord. The cells are in direct contact solely with adjacent cells, either internally with those of the spinal cord and brain or externally with the skin and muscle segments.

The next problem is to find out how the truly nervous connections, which put the central nervous system into direct association with all parts of the body, are established. This involves changes in the individual nerve cells which lead to the formation of fibers, to movements of single cells and cell masses especially in the formation of the nerve nodes or ganglia. [Lecture IV notes.]

There is no system of organs in which proper function is so dependent upon the minute arrangement of its cellular elements as is the nervous system, and none that approaches it in the complexity of this arrangement. In no other constituent of the body have studies of structure and of function gone so closely hand in hand. Yet many of the phenomena of development have no obvious relation to function as understood by the physiologist. The relative size of the nervous system in vertebrate embryos reaches its maximum before nervous function begins. Complicated neuro-muscular mechanisms in higher vertebrates are essentially complete structurally before they become active, and in those forms, such as the Amphibia, where activity seems to develop concomitantly with the development of structure, nervous function may be suppressed by chloretone without interfering at all with the development of correlated structures or with their normal functioning as soon as the effect of the anaesthetic passes off (cf. p. 158).

The embryo carries on the ordinary functions of organisms, such as respiration and metabolism, but the peculiarly developmental processes are apart from or superimposed upon these and are for the most part continuously changing and irreversible. Nothing is as yet known of the ultimate nature of these changes, but their loci may be ascertained and they may be modified in definite manner by known agencies. In other words, they are amenable to experiment.

Scientific theory has advanced through the interpretation of observed facts in terms of physical units. By this means physics, chemistry, and crystallography have reached a high degree of precision and, in a certain sense, perfection. With far less exactness, something of the same kind has been attained in biology, first in the cell theory and later in the theory of the gene, by means of which the phenomena of heredity may be stated in general terms. Whether we have faith in the reality of these particular units or not, there seems to be something in the human mind that makes it cling to devices of this kind.

In respect to the development of the nervous system, the theory of the gene has at present no direct practical bearing, but the cell theory has, since the nervous system is itself composed of cells. These nervous elements, according to the neurone concept, which is but a special phase of the cell theory, are essentially autonomous in their development, function, and structural arrangement. The embryonic nerve cells, or neuroblasts, appear at first like ordinary cells, but later form long processes or axones (nerve fibres) and dendrites, which connect all parts of the organism according to a definite arrangement.

The cell theory, however, is not without its critics, and the neurone theory, in particular, has been subject to most violent onslaughts. In general, its opponents maintain that cells are not independent but are frequently fused to form syncytia; they stress the view that an organism must be considered in its totality, being more than a mere summation of its parts, and that one of its most characteristic properties is its regulable structure, by virtue of which a part may function as a reduced whole. The latter has been emphasized by one school of experimental embryology and, in the field of neurophysiology, by Bethe in his studies on locomotion in arthropods and vertebrates,[1] and by Lashley (1926, 1929) in connection with his work on cerebral localization.

This raises a much-discussed philosophical question that need not be entered into here, since the unit theory remains in practice as the most effective instrument of analysis, provided the experimenter recognizes certain complications in the mode of action of the units and regards them as constituents of an integrated whole rather than as altogether independent entities. While, therefore, the neurone theory in its

1. Reviewed in full by Bethe and Fischer (1931).

cruder form may be taken as an oversimplification, the fog in which the opposite assumption of a diffuse nerve net is enveloped involves much greater difficulties.

The earlier phases of the development of the nervous system, such as the formation of the neural plate and the closing of the neural folds to form spinal cord and brain, will not be considered here, since they involve general processes of development that are not specifically nervous. The main topics which will be taken up are the following:

(1) The constitution and physical properties of the single nervous unit and its activities in forming the nerve fibre.
(2) Responses of nerve fibres to conditions in the medium in which they are growing.
(3) Influences that determine the nerve paths in the embryo, particularly those of peripheral nerves, and the terminal connection between nerve fibre and end organ.
(4) Responses of nerve centres to changed conditions at the periphery and in other parts of the central nervous system.

Neuroblast and Nerve Fibre

The three views regarding the development of the nerve fibre that have been held till recently, and even now have adherents, were all foreshadowed in the middle third of the last century. The oldest, or *cell-chain theory,* according to which the nerve fibre or axone is formed by the same chain of cells that gives rise to the neurilemma sheath, goes back almost a hundred years to Schwann (1839) and has been advocated by a long line of embryologists. The *outgrowth theory,* first enunciated by Bidder and Kupffer in 1857 but much more firmly established by His (1886–1890) and Ramón y Cajal (1890),[2] constitutes the embryological basis of the neurone concept, according to which the nerve fibre is an outgrowth of a single cell or neuroblast. In the *plasmodesm theory* of Hensen (1864) the nerve fibre is assumed to be a product of differentiation *in situ* of preformed protoplasmic bridges which takes place under the influence of functional activity. As alternatives, the first and second theories concern

2. Cajal's papers on the development of the nervous system have been collected in a volume (text in French) published in Madrid, 1929.

themselves with the constitution of the nerve fibre, while the contradistinction between the second and third rests rather upon the manner in which nervous connections are supposed to be established. Both pairs of alternatives have been subjected to experimental tests.

The decision between neuroblasts and sheath cell was reached by removing the source of each element in turn and finding that neurones with naked axis cylinders developed when the ganglion crest, the source of the sheath cells, was removed and the neuroblasts were left intact (Figs. 1 and 2), while no nerve fibres developed at all when only sheath cells remained (Harrison 1904, 1906, 1924; Müller and Ingvar 1921, 1923; Van Campenhout 1930).

The question of the role of protoplasmic bridges was a more difficult one to dispose of by experiment, for these structures are found everywhere in the interstices between the primitive organs, where the first nerve fibres run, and their relations to the latter are difficult if not impossible to resolve by purely histological methods. The question, whether the plasmodesmata are actually incorporated into the substance of the axones, or whether they constitute merely a lattice upon which the young nerve fibres are supported in growing out, could be decided only by substituting something else for them.

To test this it was necessary to devise an experiment in which the growing neuroblast was brought to a foreign medium lacking the protoplasmic bridges. A preliminary experiment, in which a small cylindrical blood clot was introduced into the body of the embryo in place of the excised spinal cord, showed that fibres from the rhombencephalon grew out into the clot. The next and final step was to remove the embryonic nervous tissue from the organism entirely by means of the explantation experiment, the method now known as tissue culture (Harrison 1907).

Pieces of medullary cord from frog embryos, taken before nerve fibres had differentiated, were placed each in a hanging drop of frog's lymph under aseptic precautions and the preparations sealed. After an interval varying from one to three days, fibres of hyaline protoplasm with finely branched amoeboid endings were found extending from the implanted tissue into the clotted lymph (Fig. 3). The fibres, at first coarse, were drawn out to fine threads by the amoeboid movement at the ends. The rate of progression and the independence of each fibre from its fellows and from other cells could be readily observed (Harrison 1910). Detailed comparisons with the earliest nerves in sections of embryos (Fig. 4) showed that the hyaline threads were in fact nerve fibres, and appropriate controls made with cultures of other tissues proved that their formation was due to the specific activity of presumptive nervous tissue. These observations have since been greatly extended and are now generally accepted.

Only one conclusion is possible. The substance of the axone originates in the neuroblast and is spun out by means of the amoeboid activity of the end without material contribution from the surrounding protoplasmic bridges, which may serve, however, as a framework to support and guide the fibres. The outgrowth of the nerve fibre is, therefore, a mode of protoplasmic movement and not a mere progressive differentiation *in situ*.

With most other cells of the embryo the neuroblasts share the property of being actively amoeboid. In plasma cultures they send out hyaline processes terminating in fine pseudopodia, which, according to Mossa (1927), are seen to be very long and delicate when observed under dark field illumination. Here the close similarity ends, for the processes of the nerve cells have a different consistency, with extreme ductility. The usual result of the movement of the amoeboid end is, therefore, not the locomotion of the cell as a whole but the extension of the process into a fine filament. Tension is thus exerted and this may be supplemented by the flowing of the cytoplasm of the perikaryon into the fibre itself or, as Olivo (1927) has occasionally observed, by the actual movement of the whole neuroblast through the culture medium. Both of these types of movement have their counterpart within the embryonic body.

The velocity of outgrowth varies within wide limits. In the amphibian nerve it may attain the rate of 1 μ per minute, though usually not over 15–20 μ per hour. In cultures from the warm-blooded chick embryo Burrows (1910) found that the rate of extension may reach 90 μ per hour. The available data on nerves developing within the embryo or regenerating in adult organisms, which have been tabulated by Williams (1930), indicate velocities of about the same order of magnitude, with varia-

---

Plate I

FIG. 1. Ventral ramus of a lumbar nerve of a larva of *Rana palustris* from which the ganglion crest was removed 7 days prior to the preservation of the specimen; no sheath cells are present, and the mesenchyme cells enter into no definite relation with the nerve fibres. Exp. Z65. × 350.

FIG. 2. Similar nerve from a normal larva (*R. sylvatica*), showing numerous sheath cells closely attached to the nerve fibres. Exp. Z77, control. × 350.

FIG. 3. Group of nerve fibres growing from a piece of medullary cord of a frog embryo (*R. palustris*) in clotted lymph, about 36 hours after explantation. Exp. Is135. × 350.

FIG. 4. End of a sensory nerve lying between the ectoderm and the muscle plates, from a parasagittal section of an embryo of *Rana pipiens*, 4 mm long. × 750.

FIG. 5. Piece of duodenum of a 9-day chick embryo cultivated 2 days in chicken serum supported by spider web. Note chains of cells closely applied to the web fibres. Exp. C48. × 300.

FIG. 6. Two cells from a preparation of the medullary cord of a frog embryo (*R. palustris*), growing in serum (from *R. clamitans*) and attached to spider web fibres, 8 days after explantation. Exp. SW39. × 300.

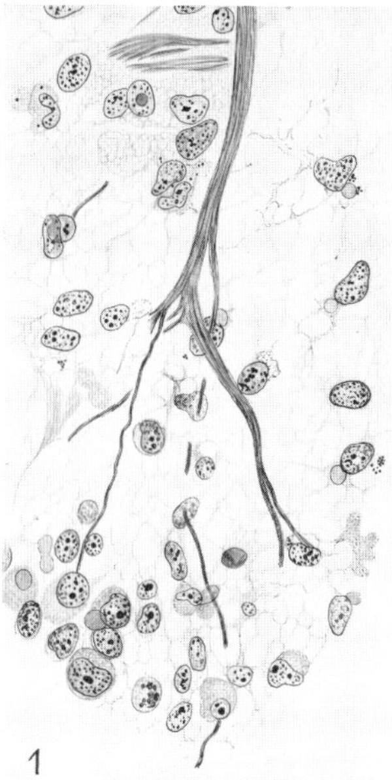

1

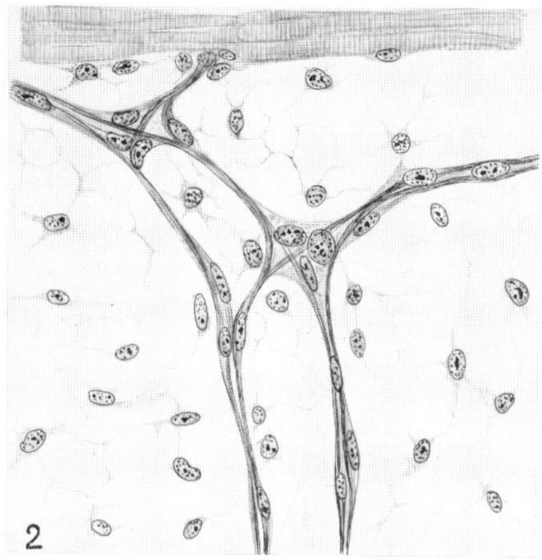

2

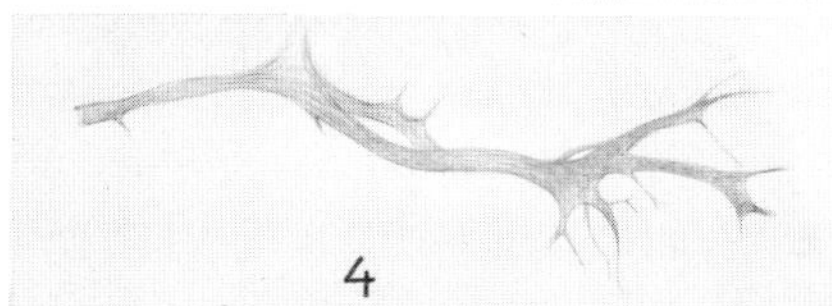

4

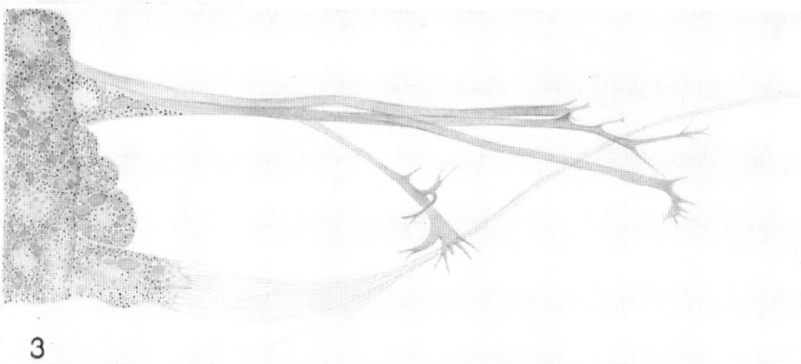

3

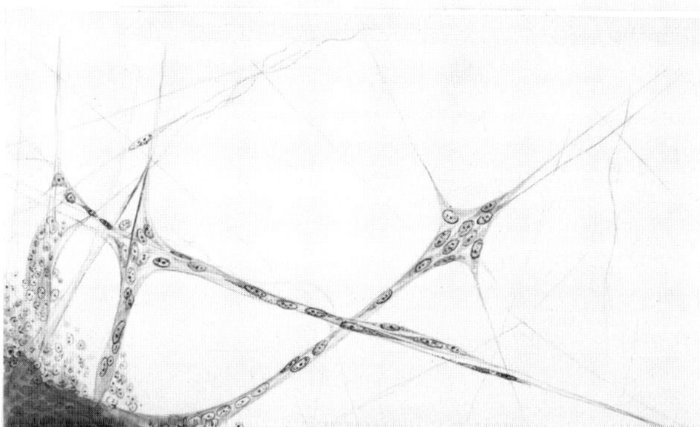

5

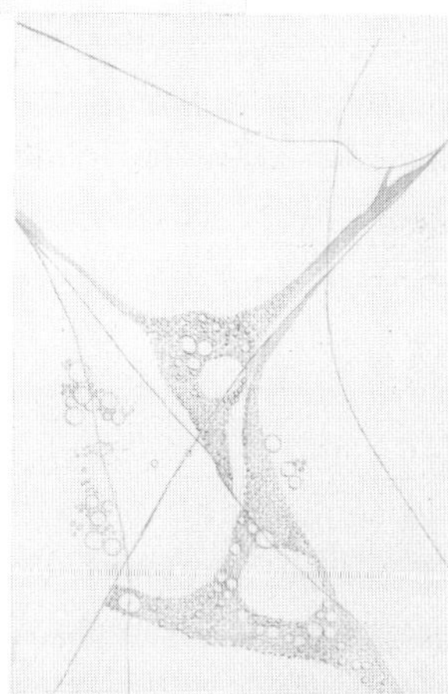

6

tions even wider than those here given. Speidel's observations on the living nerves in the tail fin of tadpoles also fall in line with the above observations. According to Cajal (1928, p. 230), the growth of the regenerating mammalian nerve in the peripheral stump averages from 83 to 125 μ per hour but may attain a rate of 170 μ, which is the maximum recorded.

The first nerve fibres to appear in a culture of embryonic nervous tissue usually grow into the medium from masses of cells, so that individual fibres cannot be traced to their particular cells of origin. However, there are numerous exceptional cases in which whole neurones are found isolated and visible in their entirety. The fibres usually run independently throughout their entire length but may form anastomoses with other fibres, which, however, are often resolved later. In cultures from the chick embryo, where the fibres are of finer texture than in the Amphibia, the formation of nerve nets is more frequent, and they seem often to be of a more permanent nature. Such nerve nets are particularly clear in the sympathetic fibres growing from ganglion cells in the intestinal wall of the chick (Lewis 1912), but they are also found in cultures of spinal cord and brain, where Levi (1917) observed that they were more frequent when the fibrin clot was not dense (Fig. 7). However, the anastomoses, even when they appear complete optically, frequently prove to be but temporary. On the other hand, Bauer (1932) finds that in cultures of the forebrain the formation of nerve nets is the usual occurrence and that the ganglion cells become united with each other by bridges. In the more highly differentiated cultures such structures as end feet (Held) and pericellular nets are formed. Various other observers have contributed to this field, but the observations are conflicting, and it is impossible at the present juncture to form a correct estimate of their significance.[3] Undoubtedly many of the discrepancies are due to differences in the source of the material, in its age and in the medium used for cultivation (Mihálik 1932). According to Esaki (1929), nerve fibres growing from spinal ganglia show numerous anastomoses, while those from the spinal cord or brain have few or none at all.

While the above experiments leave no doubt regarding the primary origin of the nerve fibre as the process of a single cell, we are obviously still confronted with the old and perplexing question, whether nervous connections are made by means of cytoplasmic continuity or by contact. This, in turn, is but part of the more general question of the occurrence of syncytia in the animal organism. Definite proof that the vertebrate nervous system in its primitive state consists of discrete cells was offered more than twenty years ago (Harrison 1910, p. 793). The relation of the sheath

3. For a full discussion of this subject *see* Levi (1934).

cells to the nerve fibre was also shown to be secondary, and attempts to classify it as syncytial, still frequently encountered, are mistaken. Even the penetration of nerve endings into the cytoplasm of epithelial cells and the fusion of nervous end plates with the sarcoplasm of muscle fibres, so clearly demonstrated by Boeke (1933), are but the final stages in the establishment of intimate nervous connections and in no way disprove the genetic and physiological independence of the nerve elements. Perhaps the most favourable evidence for the existence of syncytia has been given by tissue cultures, but both Lewis (1922) and Levi (1925), who have examined it critically, practically reject the supposed proofs. Levi points out that even in cases where actual fusion takes place, independent cell territories exist, and that in the supposed nerve nets each fibre is probably dependent upon a single cell.

The existence of neurofibrillae (Fig. 7) in nerves grown in culture media was first demonstrated by Burrows (1911) and then by Lewis in the sympathetic plexuses just mentioned. Levi has seen faintly fibrillar structures in living cultures, especially when the nerves are flattened against the coverslip, and it seems possible that they may change their form and arrangement according to conditions in the medium. It is questionable, however, if these are identical with the fibrillae which are revealed by the silver impregnation method.[4]

The physical properties of living nerve elements in tissue cultures, which have been studied by Levi (1926) and by Péterfi and Kapel (1928) by means of micromanipulation, are of much interest. According to the former the fibres may be readily stretched without tearing, after which they remain sinuous for a time before returning to their original length. When pressure is applied locally, a swelling arises from which pseudopodia are given off. Similar activities may also occur beyond the point of stimulation and also when the cell-body is touched (Fig. 8). If sharply pressed, the cytoplasm of the perikaryon suddenly becomes coarsely granular. Coalescence of two adjacent fibres may be produced by pressing them together and may persist for hours. Péterfi and Kapel (1928) find also that on pricking a young neuroblast with the microneedle the cytoplasm becomes granular. The cell becomes rounded by withdrawal of its processes, but soon reverse changes set in and, if a new axone is formed, it is always on the side of the cell opposite the point of stimulation.

The observations on the behaviour of fibres that have been cut off from their cells of origin are especially noteworthy. Ingebrigtsen (1913) had previously found that such fibres, like nerves within the body, undergo rapid degeneration, and that regen-

4. de Rényi (1929, *b*) has demonstrated, by means of micro-dissection, the presence of neurofibrillae in living axones of the lobster, but has been unable to do so in axones of the frog (1929, *a*).

eration takes place from their central ends. According to Levi, however, the activities of the distal portion are interrupted for a brief period only, after which the movement of pseudopodia at the peripheral end is resumed. The proximal end of the severed process also swells and sends out pseudopodia, so that the fibre may be stretched by pulling from both ends. This activity may continue for as long as twelve hours, but sooner or later it ceases, unless the two cut ends reunite. The fibre is then restored *ad integrum,* almost as if nothing unusual had taken place; fusions of this kind have been observed between fibres that did not originally belong together.

One of the most striking properties of the neuroblast, as seen from the foregoing, is the extreme sensitivity of its growing end to mechanical stimulation, and it is not improbable that this may be connected with the stereotropic responses known to occur in developing nerves (p. 133).

Plate 2

FIG. 7. Plexus of nerve fibres growing on the coverglass in a plasma medium from a culture of rhombencephalon of a 77-hour chick embryo, 29 hours after explantation. After Levi (1917). × 350.

FIG. 8. Cell from the mesencephalon of a 7-day chick, grown 50 hours in plasma culture. *a,* resting cell touched by micro-needle; *b,* effect of stimulation shown in formation of coarse granules and fine protoplasmic processes on the fibre. After Levi (1926).

FIG. 9. Nerve fibres from a piece of brain of a 4-day chick embryo, cultivated 72 hours together with muscle fibres and showing end organs. Gold chloride preparation. After Grigorjeff (1932).

FIG. 10. Colony of fibroblasts from the heart of a chick embryo growing in a plasma film in a triangular frame. The greatest growth is perpendicular to the sides of the frame. After Weiss (1929). × 7.

FIG. 11. A straight tract of spindle cells and nerve fibres growing between two spinal ganglia in an extremely thin plasma membrane. Ganglia from a 13-day chick embryo, 3 days in plasma. After Weiss (1934). × 37.

FIG. 12. From a culture of spinal ganglion of a 13-day chick, 3 days in plasma. Cells and fibres oriented in one direction, corresponding to the stretching of the clot. Photograph by Dr. Paul Weiss. × *ca.* 75.

FIG. 13. Same culture as in fig. 12, taken from a part of the clot not under tension. Cells and nerve fibres not definitely oriented. Photograph by Dr. Paul Weiss × *ca.* 75.

FIG. 14. Fragment of medulla (6-day embryo), grown for 2 days in diluted plasma surrounded by more concentrated plasma. A nerve plexus (*pl*) has developed along the boundary between the two media. After Weiss (1934). × 136.

FIG. 15. Formation of plexus by nerve fibres (from the medulla of a 6-day chick embryo) grown for 2 days next to the surface of the clot, the culture exhibiting slight liquefaction. After Weiss (1934). × 52.

7

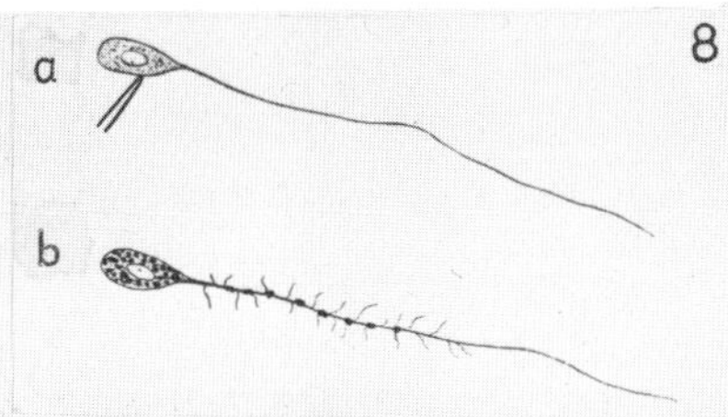
8
a
b

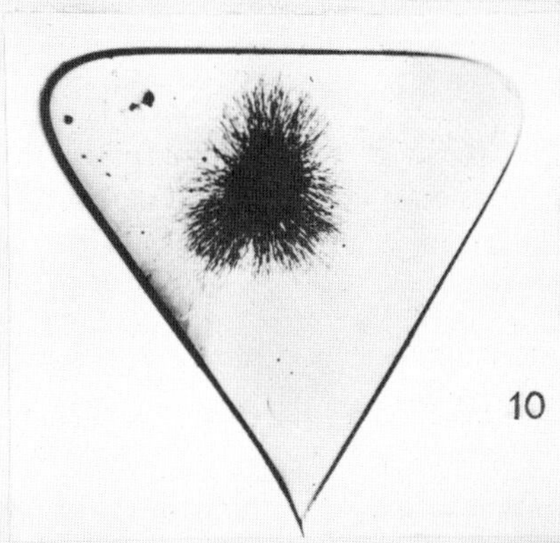
10

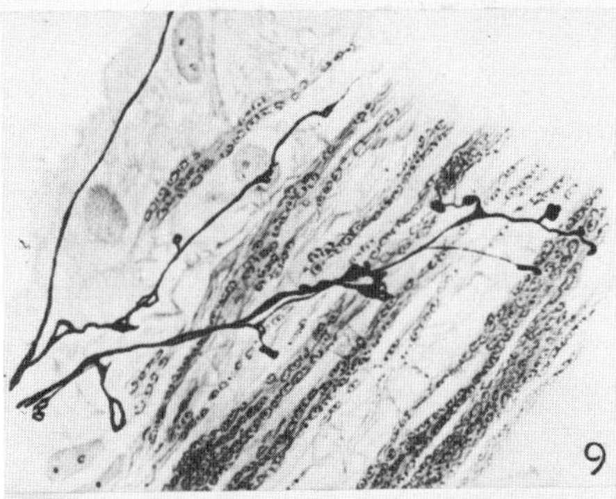
9

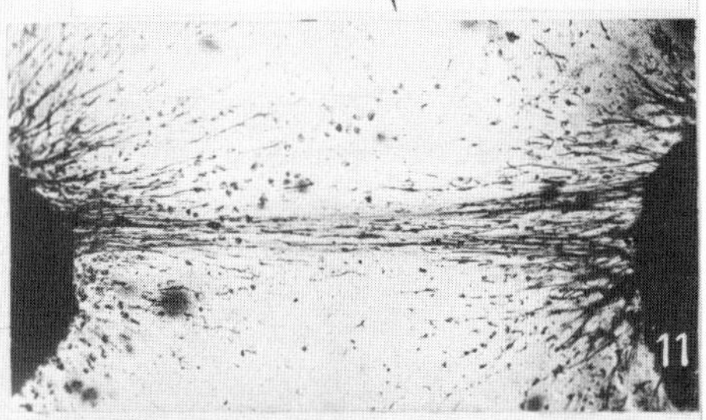
11

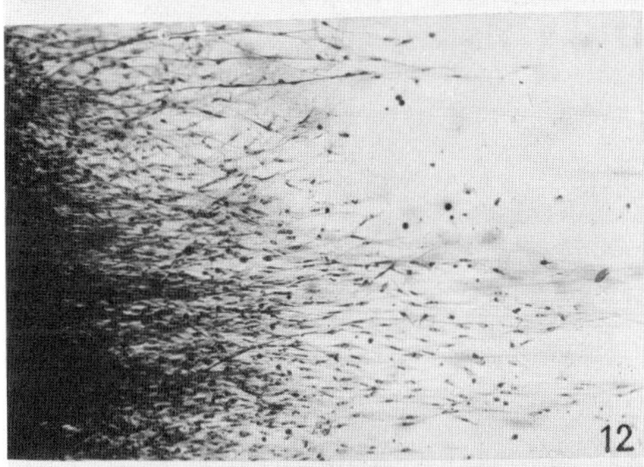
12

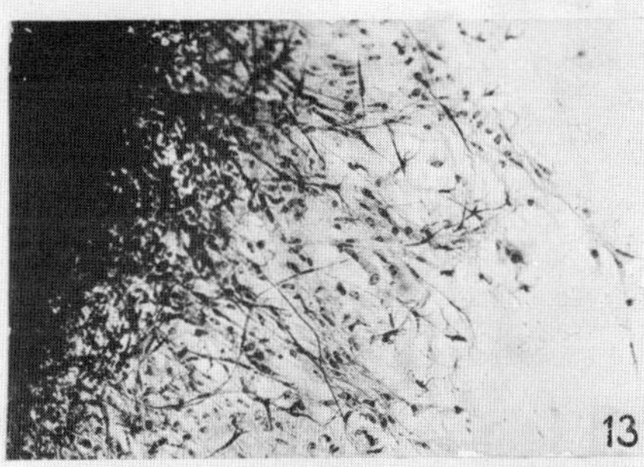
13

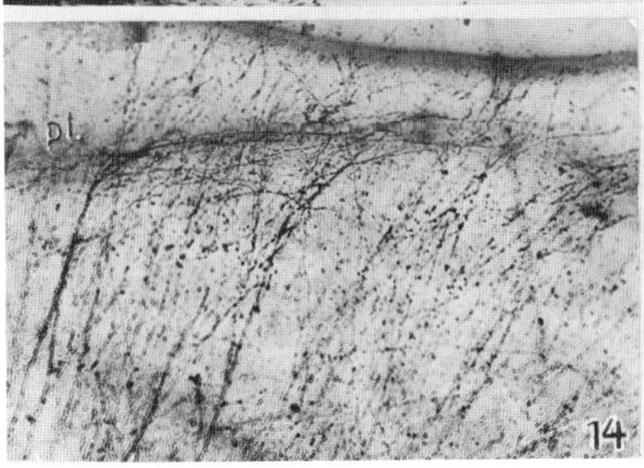
pl.
14

15

The Responses of Nerve Fibres to Conditions in the Medium

Before taking up the factors that direct the outgrowing nerve fibres within the organism, it will be of advantage to consider the responses made by nerve cells isolated *in vitro,* where the experimentally imposed conditions may be better controlled. Of the various factors assumed to influence the direction of outgrowth, the most discussed are localized chemical differences in the medium, electrical fields, and the configuration of the solids with which contact is made.

Chemotropism

The theory of chemotropism originated with Ramón y Cajal (1892) in his work on the retina. It was based primarily upon the postulated activity of the growth cones, which he had discovered at the ends of growing nerve fibres and which are identical with the amoeboid endings seen in tissue cultures. Later this was elaborated into a more general theory of neurotropism.[5] Cajal's conception of chemotropism is pictured vividly in the following sentence: "From the functional point of view the growth cone may be regarded as a sort of club or battering ram, endowed with exquisite chemical sensitivity, with rapid amoeboid movements, and with a certain impulsive force, thanks to which it is able to press forward and overcome obstacles met in its way, forcing cellular interstices until it arrives at its destination" (1899, pp. 544–5).

While it has seemed necessary to make some such assumption—and there are various observations upon developing and regenerating nerves that would be readily explained by it—no crucial proof of chemical sensitivity has ever been forthcoming. Obvious experimental methods of testing it are either the planting of other kinds of cells (muscle or epithelium) in the same culture with nervous tissue or the introduction into the medium of capillary tubes filled with known substances, according to Pfeffer's method, and then noting their influence upon the velocity, intensity and direction of outgrowth of the nerves.

The latter type of experiment, begun shortly after the introduction of tissue culture, did not yield definite results (Harrison 1912), and recent studies by Weiss

5. This term was first used by Forssman (1898, p. 96) in his celebrated work on nerve regeneration. For a general account of the theory *see* Cajal (1913, 1928), Tello (1923), and Mangold (1928).

(1934) have shown that in such experiments unsuspected mechanical strains are more likely to be the primary effective agent than the supposed chemical factors. Tubes filled with extracts of brain, muscle or liver fail to have any directive influence on outgrowing nerve fibres, as do pieces of degenerating peripheral nerves, when proper precautions are taken to avoid vectorial strains on the fibrin as it clots. Centanni and his pupils[6] have, it is true, described differential growth in such *colture affrontate,* but they were concerned with the growth of connective tissue and sheath cells from explanted pieces of peripheral nerves and not with nerve fibres themselves. Moreover, no assurance is given that the tubes and tissues introduced into the medium did not exert an influence on the arrangement of the fibrin net.

That the presence of other tissues in a culture may influence the growth and differentiation of nervous elements in other ways has been shown by Grigorjeff (1931, 1932). The vigour and manner of outgrowth of nerve fibres and the type of nerve ending formed are changed by the presence of mesenchyme. Nerve bundles growing over this tissue form fewer anastomoses, and their neurofibrillar structure is more fully differentiated; the nerve endings in contact with mesenchyme cells are more definite. When striated muscle is present, nerve endings within the sarcoplasm, similar to those found in the embryonic body, are formed (Fig. 9). Some of Grigorjeff's observations, such as the bending of a nerve fibre to reach a cell, or the convergence of a number of fibres upon a small group of cells, are interpreted as evidence for definite tropisms, possibly of a chemical nature, but here again there is no rigorous proof.

Galvanotropism

The possible effects of the electric current upon outgrowing nerve fibres have been the subject of much speculation and little experimentation.[7]

Ingvar (1920) reported that when very weak currents were passed through the medium into which small pieces of central nervous system of chick embryos had been

6. Cf. Bisceglie u. Juhász-Schäffer (1928), p. 160.
7. Already in 1892 Strasser proposed the hypothesis that the motor cells in the embryonic spinal cord send out their processes and move as a whole toward the muscle plate "in the direction of the greatest difference of potential of the electric charge" (p. 734). Hypotheses of this kind have been built up into the theory of neurobiotaxis of Kappers (1908, 1917, 1922, 1933) and the related theory of stimulogenous fibrillation of Bok (1915). It will not be possible to enter here into a discussion of these interesting theories, which, for the most part, are not based upon rigorous experimental proof, and, furthermore, are not in harmony with much of the available experimental evidence.

explanted, outgrowing cell processes were oriented almost entirely in the direction of the lines of force in the galvanic field. Those growing toward the anode showed morphological differences from those directed toward the cathode. Again, if a weak current was passed through the culture by means of a single conductor, the direction of outgrowth of fibres and cells was perpendicular to the conductor. These observations seemed to confirm the theories of neurobiotaxis and stimulogenous fibrillation.

Experiments by Weiss (1934), using different electrodes and a wide range of current density, have so far failed to confirm these findings, and Weiss has suggested that Ingvar's results may have been due to the oriented arrangement of the fibrin, brought about by the wick electrodes used. Moreover, there are serious theoretical difficulties involved in the simple interpretations given in the above observations. However intriguing the theory of galvanotropism may be in relation to the development of the nervous system, it must still be regarded as without substantial foundation.[8]

Péterfi and Williams (1933) have recently obtained effects of another kind on neuroblasts and axones from the central nervous system of chick embryos through the action of much stronger electric currents.[9] On stimulation the cell-body of the neuroblast becomes more highly refracting, then granules appear and the cytoplasm withdraws a little from the cathode. Similar granulations appear in axones, and the end bulbs become larger and more diffuse, but when the circuit is broken, there is an after-effect, the axone extending toward the anode. Whether reactions of this kind, in case of more prolonged stimulation, would have any lasting morphogenetic effect remains an open question.

Negative results were also obtained by Karssen and Sager (1934). Speidel's observations (1933) of nerves growing in opposite directions along a third fiber provides proof that electric factors under these circumstances can play no role in orienting the course of outgrowth of the nerve fibers. Marsh and Beams (1945, 1946), however, have obtained positive results with their experiments in subjecting nerves to electric currents. They used very much stronger currents than Ingvar

8. Huzella (1934) has recently reported positive results from growing heart fibroblasts in plasma cultures between the poles of an electro-magnet without material contact with the latter. The intensity of the field was alternately increased and diminished by rotating iron plates placed between the cultures and the magnet. Greater growth takes place in the direction of the magnetic poles, and the individual cells are stretched in the same direction. Whether this is a direct action upon the outgrowing cells or is effected through the fibrillar structure of the medium has not been determined.

9. The currents used by Ingvar were exceedingly weak, ranging from 2 to 4×10^{-9} amp. The effects noted by Péterfi and Williams followed the application of currents from 1000 to 14,000 times this strength (2 to 55×10^{-6} amp).

and point out in agreement with Weiss, that the results obtained by Ingvar must have been due to some other factor.

The current densities used by Marsh and Beams varied from 0.05 and 500 μ amp/mm^2. Tissue fragments of medulla oblongata of 7-day chick embryos were kept in heparinized chicken plasma clotted with embryo extract. They used large wedge-shaped chambers, 6 mm at the narrow end and 71 mm at the broad, giving great variety of current density. At suitable density, fibers grow out toward the cathode. It requires 170 μ amp/mm^2 density completely to inhibit growth toward the anode. At 80 μ amp/mm^2 no effect of current was observed. Fibers that start out perpendicular to the polar axis are deflected toward the cathode if the current is of proper strength. The fiber is also deflected if the direction of current is changed after fibers start growing and if the current is not applied until growth has started. Marsh and Beams point out that potential differences such as are here involved are of the same order of magnitude as some biolectric currents. The authors are conservative in their conclusions and accept, at least in part, the orienting effect of structure. [Lecture IV notes.]

Stereotropism

Inquiry into the possible mechanical factors that may act upon outgrowing nerve fibres has resulted in more definite and satisfactory conclusions. Such factors figured in the earlier accounts of the outgrowth theory and are implied in Cajal's description of the growth cones. They were rendered more probable by the first tissue culture experiments, and crucial evidence was obtained later from the study of the behaviour of cells and nerve fibres in media with different kinds of solid support. Loeb (1902, 1907) had already concluded from earlier observations that stereotropic stimuli play an important part in the movement of epithelial cells. Experiments, in which mechanical and chemical factors in the medium were varied independently, definitely showed that no cellular growth or migration occurs except when the cells make contact with a solid body, such as the cover slip, the fibrin net, or fibres (spider web) immersed in fluid (Fig. 6), or with the surface film of the medium (Harrison, 1914).[10]

10. Carrel and Burrows (1911), using cotton fibres and silk in tissue cultures, also called attention to the necessity of solid support for outgrowing cells. Matsumoto (1918) added much further evidence for stereotropism in a study of corneal epithelium in relation to the surfaces over which it grows. More recently Huzella (1931; with Lengyel, 1932) has demonstrated by very ingenious methods the adaptation of growing fibroblasts and other cells to the supporting reticulum. The latter may be either natural (frozen sections of various organs) or artificially prepared by coagulation *in vitro* under special mechanical conditions by the method first introduced by Nageotte (1927, 1931).

The reactions are characteristic for each form of support, and, in the case of spider web, cells often surround the fibres as sheath cells cover a nerve (Fig. 5).

The foregoing tests were made mainly with fibroblasts and epithelial cells, but evidence was obtained that nerve fibres react similarly. Furthermore, the relation of outgrowing nerve fibres in the embryo to solid masses of cells and to the interstitial network bears out the interpretation that they too react to solids and are oriented by the configuration of the latter. It was to be expected, therefore, that further light would be thrown upon this question by the study of the relations of developing nerve fibres to the fibrin net, formed under various mechanical conditions.

Experiments to this end were carried out in the Zoological Laboratory at Yale University by Weiss (1934). Earlier work by the same author (1929) had shown that fibroblasts from the chick embryo respond to the structure of the fibrin clot in which they are cultivated. When a small mass of these cells is explanted to a film of plasma stretched across a frame of suitable size and of the shape of an equilateral triangle or a rectangle, the greater elastic tension lies in the direction of the shortest lines between the mass and the frames. This orients the micellae of the colloidal medium during coagulation and, in these sectors, leads to a parallel arrangement of the fibrillae in the laminated membrane of which the clot is composed, whereas elsewhere, in the absence of a sufficiently strong directive factor, a more diffuse net is formed.[11] The growth and migration of the fibroblasts is greatest in the sectors which have oriented fibrillae and least in the diffuse sectors (Fig. 10). In the former the cells are spindle-shaped and radially oriented (Fig. 12), while in the latter they tend to be stellate or multipolar (Fig. 13). The more intense growth is due largely to the greater availability of fluid in the sectors with parallel fibrillae, and there is a reciprocal effect, since the local dehydration of the membrane by the rapidly growing cells increases the tension and leads to accentuation of its structural characters.[12]

In ordinary cultures of nervous tissue the outgrowing fibres are not definitely oriented. If, however, either before or after implanting a piece of nervous tissue from a chick embryo in a drop of plasma, the drop is stroked with a needle or a brush in a given direction, the fibrin threads, as they form on the surface, assume a parallel arrangement. The growing nerve fibres follow this same general direction, though

11. According to Weiss (1929), the fibrillar arrangement in the individual laminae of the film is submicroscopic. However, in Baitsell's (1917) experiments, made under different conditions, a parallel arrangement of fibrillae was just as clearly visible under the microscope when the fibrin clot was subjected to tension as was the diffuse net in regions where tension was not applied.

12. According to Huzella (1929), the fibroblasts, as they move along over the fibrin, form a gelatinous secretion which condenses into the argyrophile reticulum, superimposed upon the fibrin net and ultimately replacing it.

they often appear wavy instead of straight. Such fibres branch very much less than those growing without definite orientation, as already found by various observers in regenerating nerves. There are fewer mechanical obstacles in the fibrin which is clotted in bands.

Two or more pieces of tissue placed not far apart in a single plasma culture lead to the formation of parallel fibrin bundles running between them (Fig. 11). In ordinary coverslip preparations this occurs only when the explanted pieces project above the surface. It takes place also in films stretched on glass frames. Under these conditions heart fibroblasts tend to grow and migrate along the fibrin bundles from one piece of tissue to the other, and this has led to the erroneous assumption of chemical or physical attractive forces between the cultures. Similarly, connective tissue and sheath cells grow out from explanted spinal ganglia, and nerve fibres follow in the same definitely oriented path (Figs. 12 and 13). When, however, pieces of the central nervous system are taken, there is little or no cell proliferation and the fibres that grow out are not definitely oriented. In such cases the dehydrating action of the proliferating cells and the consequent increase in tension along the lines between the pieces of tissue are lacking. On this account the structure of the clot is not sufficiently differentiated vectorially to influence the direction of outgrowth of the axones.

Plexus formation may be brought about by implanting embryonic nervous tissue in a plasma medium and surrounding the first clot by a similar medium of different density. When the outgrowing nerve fibres reach the border zone between the two media, they are in part deflected in the direction of the boundary (Fig. 14), and are thus brought into contact with other fibres, only to bend again and run through the outer zone of the medium. Similar effects are produced by the partial liquefaction of the clot, which often occurs in the zone immediately surrounding the grafted tissue (Fig. 15).

The foregoing descriptions show some of the ways in which the configuration of the medium may affect the direction of growth and the arrangement of cells and nerve fibres. They show at the same time some of the possible complications of relatively simple conditions and the great difficulty of separating chemical, electrical, and mechanical factors even in such a simple environment. The evidence for the direct action of chemical and electrical factors is meagre and not rigorously proved. While the evidence for the influence of the structural arrangement of the medium, both visible and ultramicroscopic, rests on a firmer foundation, it is not certain whether such factors act directly or through the mediation of induced conditions of some other nature.[13]

13. For an able discussion of these questions *see* Weiss (1933).

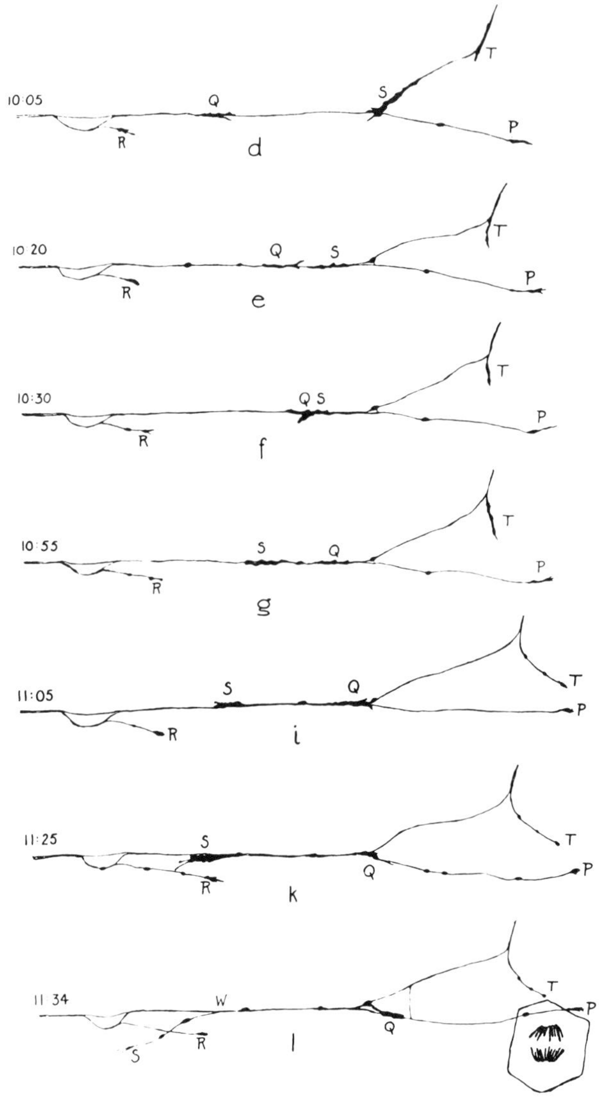

FIG. 16. Growth of two nerve fibres (*Q* and *S*) in opposite directions along a third fibre. The fibres meet in (*f*) and pass. Time intervals given on the left. Observations on the regenerating tail fin of a larva of *Hyla crucifier*. After Speidel (1933).

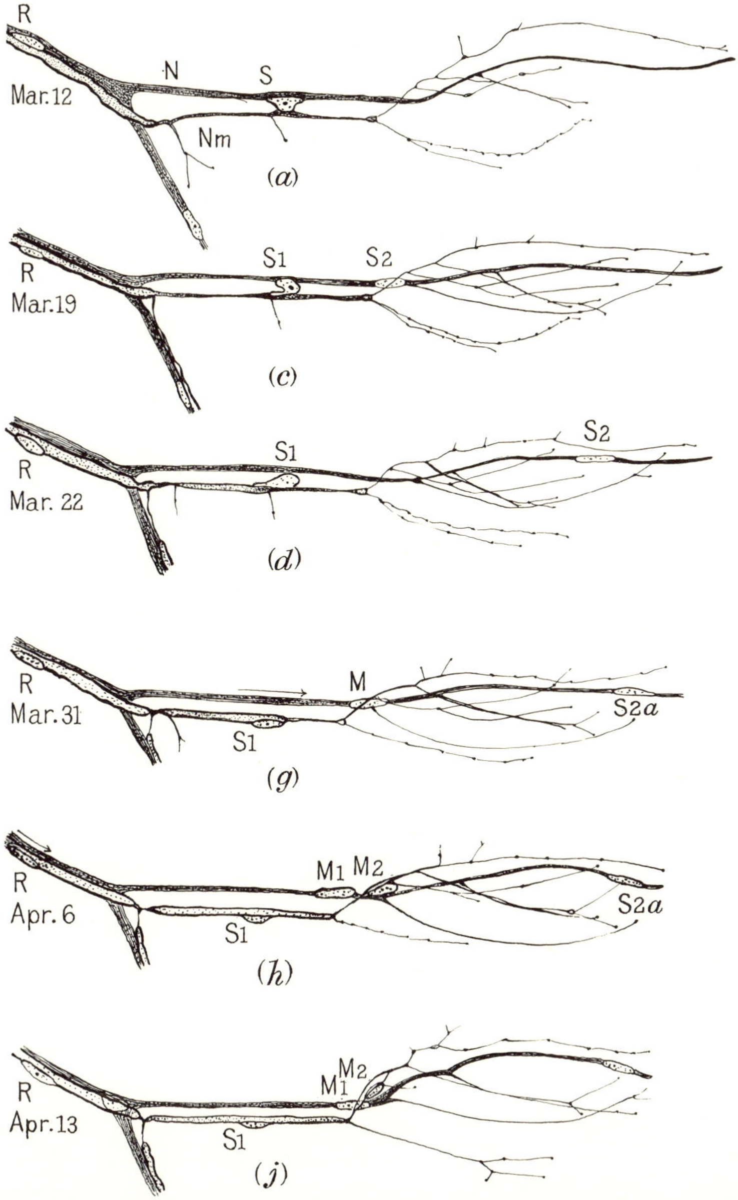

FIG. 17. Myelinization of a sensory nerve fibre in the tail fin of a larva of *Hyla crucifer* observed in the living animal. Two nerve fibres, one the continuation of a myelinated fibre (myelin emergent, *Nm*) and the other *N* of a non-myelinated fibre. The sheath cell (*S*) touches both in (*a*), then divides, one (*S2*) remaining on *N* and the other (*S1*) gradually withdrawing to *Nm* (*c* and *d*). The latter then adds another myelin segment in relation to *S1* (*d–h*). A second cell (*M*) moves out upon the fibre *N* and divides into *M1* and *M2*; the latter moves over to *Nm* and another myelin segment is laid down. Observations on a normal living tadpole. After Speidel (1933).

Nerve Paths and Connections within the Organism

Observations on Living Material

The study of structural relations in the embryo at the time when the first nerve fibres are growing out suggests that factors similar to those outlined above play a role in determining their course. Special attention should be called to some of these, such as the tendency of the fibres to grow in a straight line unless deflected by some obstacle, their predilection for surfaces of solid structures and grooves between them, their guidance by the finer structure of the interstitial tissue. It is important, however, to verify inferences from observations upon preserved material by the study of actual living fibres within the organism. Up to the present this has been accomplished in but one object, the tail fin of the amphibian larva.

The nerves in the fin fold have engaged the attention of many investigators since the time of Schwann (1839) and Kölliker (1846). In living material the independent movements of sheath cells along the fibres were observed some years ago by Harrison (1904, 1924). Now the notable observations by Speidel (1932, 1933) have given us detailed information regarding the outgrowing sensory nerve fibres themselves and the behaviour of the sheath cells, including their relation to the process of myelinization.

Growth cones at the ends of nerve fibres behave in much the same way as in tissue cultures, and there can be no question but that they are the active agent in drawing out the axone. Their advance through the mesenchyme of the fin fold is a halting one. At times it may reach 1 μ per minute, but it is usually much slower and sometimes almost ceases. Temporary obstructions may call forth the production of varicosities, giant growth cones or actual branching, and a prolonged encounter with an insuperable obstacle may be followed by retraction of the growth cone or even by its autotomy. After the pioneer fibres are laid down they are followed by others which glide along them, thus indicating a stereotropic response. Anastomoses are frequently formed between neighbouring nerves, and a fibre may double back on a nerve bundle and run centripetally, so that recurrent fibres may meet and pass others growing in the opposite direction (Fig. 16). This last observation is of special importance in excluding electrical or chemical gradients as possible factors in directing the movement.

In the formation of the myelin sheath, axone and sheath cells react together, and the latter behave in a characteristic way. The myelin is laid down section by section, and in a given segment it first appears on a level with the nucleus of the sheath cell. The process extends along one fibre of a bundle, while the others may remain myelin free (Fig. 17), but frequently the myelinization is discontinuous, in which case a myelin segment is added later to fill the gap. Sheath cells frequently move from non-medullated fibres to myelin-emergent fibres but very rarely in the opposite direction.

The nerve fibres of the tail may be cut either singly or in small bundles and their subsequent behaviour observed (Harrison 1908; Williams 1930; Speidel 1933). Degenerative changes, which occur rapidly in the peripheral stump, involve neurilemma, myelin sheath, and axone. Regeneration takes place from the central stump through the activity of new growth cones, the regenerating fibres frequently following the degenerating peripheral tract but rather more often breaking new paths. The velocity of this secondary outgrowth, like that of the primary, varies within wide limits but the mean is considerably higher when the new nerve is growing in the old sheath (Williams).[14]

The question of healing of the individual fibres *per primam* (Harrison 1908) has been answered negatively by Williams, who finds that the rapidity of outgrowth is sufficient to explain the phenomena observed. Speidel's observation of recurrent fibres in the fin nerves might also account for the appearance of primary healing when such a nerve is cut, for in these cases both ends are in part central stumps. In spite of these explanations, the possibility of primary healing must be admitted in view of Levi's (1926) observations on isolated fibres in tissue cultures.

Polarity of Neuroblasts and Conditions First Encountered by Outgrowing Fibres

By virtue of the protoplasmic movement which takes place at their ends, the nerve fibres extend to all parts of the embryo. They must move in paths which conform to the topography of the primitive organs, for the distribution of nerves in any species of animal is remarkably constant, the variations being of a minor nature.

At the time when the first peripheral fibres (pathfinders) are laid down, the absolute distances they have to grow are very short and are of the same order of magnitude as the length of isolated neurones actually observed in tissue cultures. Their initial course is usually straight, and the configuration of the surrounding organs is, on the whole, sufficient to account for the deflections observed. The sensory spinal nerves

14. In the spinal nerves of *Hyla* the mean velocity is $7 \cdot 1 \pm 0 \cdot 77$ μ per hour in the old sheath and $2 \cdot 9 \pm 0 \cdot 25$ μ outside, while in the lateral line nerve the corresponding figures are $9 \cdot 9 \pm 1 \cdot 36$ μ and $3 \cdot 1 \pm 0 \cdot 22$ μ respectively.

from the dorsal or Rohon-Beard cells run at first laterally, some branches enter the myotomes as proprioceptors (Coghill 1929) and the others, on reaching the skin, are deflected dorsally or ventrally and to some extent headward or tailward to form anastomoses with branches from neighbouring segments. Motor nerves reach the myotomes almost immediately on leaving the spinal cord, some entering and others bending dorsally or ventrally. Motor anastomoses are apparently not found except in the limb plexuses, and here the peripheral distribution in relation to segmental origin varies considerably. Sensory fibres from the spinal ganglia, which arise later than those from the dorsal cells, run at first with the motor nerves, and fibres of the dorsal rami extend into the fin fold. The olfactory nerves have but a short distance to cover between the nasal placode and the brain, and the optic fibres use the optic stalk as a pathway. The ganglia of the lateral line and the ear develop in close association with the sensory epithelia; in the case of the former the first axones develop before the main placode starts on its long journey from head to tail. This is the most remarkable case of passive stretching known in the development of nerves.

Within the central nervous system the conditions are different in so far as the outgrowing fibres have to pass through a more solid matrix. In the salmon embryo, where the early relations of the neuroblasts are shown clearly, there is evidence that the nerve processes bore their way through the supporting cells[15] (Fig. 18, *gc*). At first rounded and lying in the marginal veil (Fig. 19 A, *e*), the neuroblasts transform themselves either into bipolar cells (Figs. 19 A, *f* and 19 B, *g* and *h*) with processes running up and down the cord, or into unipolar cells with a single process running at right angles to the longitudinal fibres (Fig. 18, *com*). The latter are either commissural cells, connecting the two sides of the cord, or motor neurones (Fig. 18, *m*), which break through the external limiting membrane and reach the muscles. The primary peripheral sensory fibres arise directly from cells of the dorsal column (Fig. 18, *s*), or by branching from the fibres of the dorsal longitudinal bundles (Fig. 19 C). Coghill (1913, 1914) has shown that in the urodele, *Amblystoma,* both motor and sensory peripheral fibres arise as collaterals from funicular fibres (Fig. 20). Dendrites and other collaterals develop later in time to effect the better co-ordinated movements of the swimming stage.[16] According to Coghill, differentiation of neuro-

15. This was pointed out by Harrison (1901) and is confirmatory of the view expressed previously by von Lenhossék. It has also been accepted by Coghill (1929, p. 54) on the basis of observations on *Amblystoma.* Weiss's observations on nerves in tissue cultures (1934, p. 433) are interesting in this connection.

16. This most important work of Coghill on the correlation of structure and function in the developing nervous system has been brought together in his University College Lectures, published in book form, 1929.

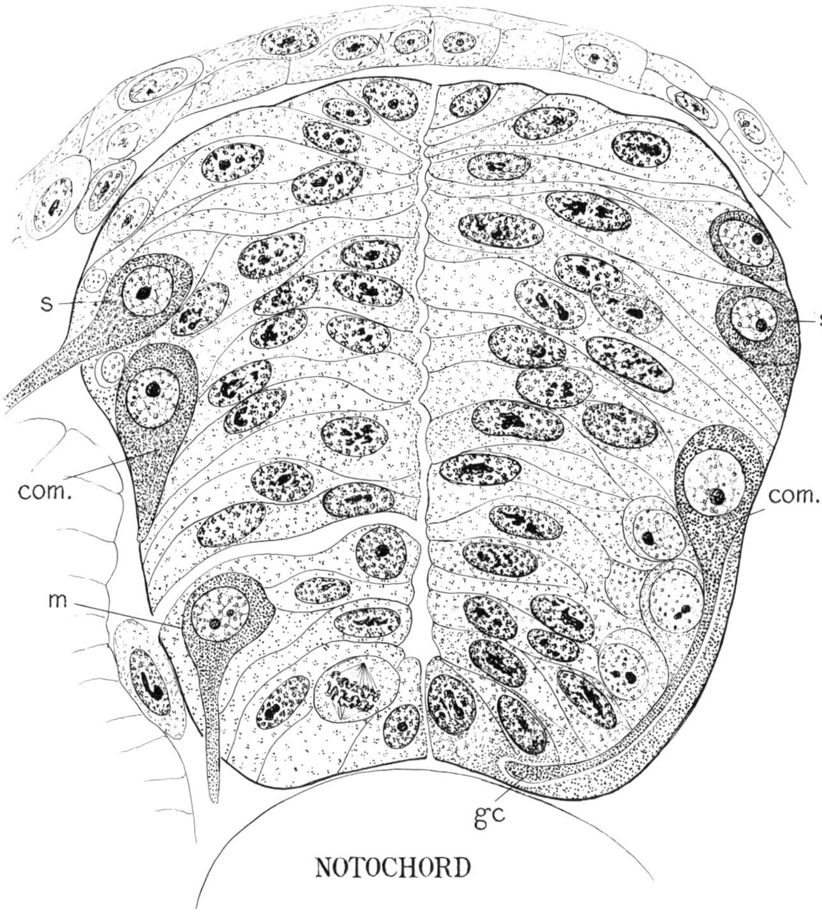

FIG. 18. Composite cross-section of the spinal cord of a salmon embryo with about 29–33 somites. On the right are two Rohon-Beard sensory cells (*s*) and a large commissural cell (*com*) with process running nearly to the mid line, ending in a growth cone (*gc*). On the left are a sensory cell (*s*) with peripheral fibre, a commissural cell (*com*), the axone of which is cut off, and a motor cell (*m*) with axone emerging from the cord. Modified from Harrison (1901). × 932.

blasts begins in localized centres and other fibres tend to grow towards these active foci.

The outgrowth of a given neuroblast is not necessarily dependent upon the specific qualities of its surroundings, as is shown by tissue cultures and by experiments in which the surroundings of growing nerves are radically altered. In the former the

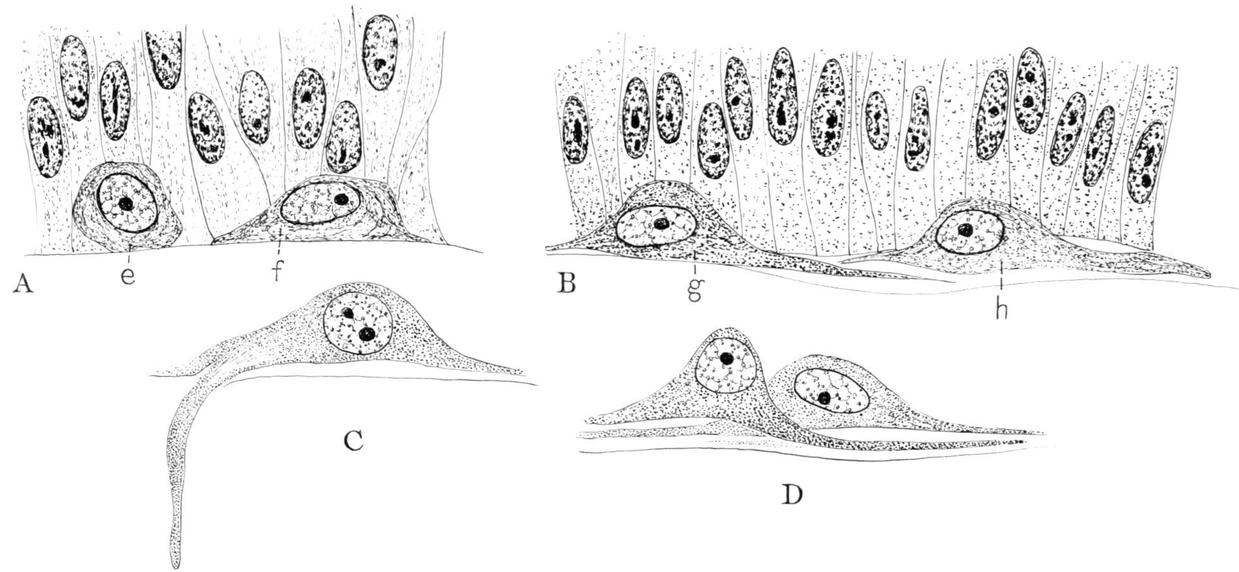

FIG. 19. Cells shown in oblique longitudinal sections of the spinal cord of the salmon embryo. A, from an embryo with 24 somites; two neuroblasts (sensory) embedded in the ependyma, one (*e*) polygonal and the other (*f*) showing beginning of longitudinal processes. B, from an embryo with 31 somites; two similar cells, with well-differentiated processes. C, single dorsal sensory cell (Rohon-Beard) with ascending and descending processes and peripheral fibre running off from the latter; from an embryo with 34 somites. D, two sensory cells from the dorsal column of an embryo with 33 somites. After Harrison (1901). × 932.

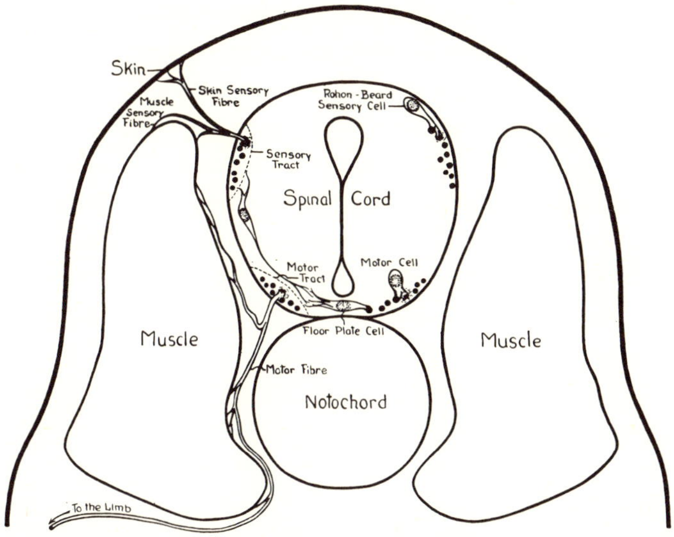

FIG. 20. Diagrammatic cross-section of the spinal cord of an *Amblystoma* embryo showing the relation of peripheral fibres to skin and muscle and certain of the primitive connecting neurones within the cord. After Coghill (1929).

nerve fibres grow radially from the explanted tissue into the medium; in the embryo they tend to preserve their original direction of growth. For example, when the spinal cord is removed in a frog embryo with neural folds just closed, longitudinal fibres, proceeding from the medulla, run in the same direction through the mesenchyme, a tissue of very different texture, that fills the space normally occupied by the excised cord (Harrison 1904, 1910). The olfactory nerves likewise grow into the mesenchyme after removal of the brain or after transplantation of the nasal pit (Lewis 1907), and many cases have been observed in which nerve fibres grow for long distances in strange surroundings from pieces of embryonic brain or spinal cord transplanted to other regions of the body (Harrison 1906, 1910, *a;* Lewis 1907). Hoadley (1925) has made similar observations on nerves from grafts of the midbrain of the chick growing on the chorio-allantoic membrane.

The foregoing observations indicate that the primary direction of outgrowth of the axone is already determined in the neuroblast, but what determines it is unknown. Most of the explanations that have been given must be relegated to the limbo of pure hypothesis. The earliest visible polarization of the cells in the medullary tube lies in the direction, external limiting membrane, central canal, which corresponds to the surface-depth axis of the medullary plate. Visible differences between the external and internal zones of the cells may be detected in the prenervous stage when the plate is forming, but there is nothing to prove that these apparent differences are due to anything more than the respective relations of the two surfaces, the one exposed to the external world and the other adjacent to the deeper layers of the embryo.[17] When the axones begin to grow out from the neuroblasts, they are found in almost all cases to extend from the side of the cell adjacent to the external limiting membrane (Mall 1893),[18] but some (primary sensory neurones) grow longitudinally along the membrane, while others (peripheral motor nerves) grow directly toward the membrane and perforate it, and again others (commissural cells) are deflected by the membrane to run dorsoventrally or transversely. The result of this is that, in spite of the uniform polarity of the cells, the axones seem to run in all directions (Figs. 18 and 19).

There is also a pre-nervous differentiation of the neural plate with reference to the antero-posterior and the medio-lateral axes, and this is controlled by the material underlying it, as shown by Spemann's (1918) rotation experiment with the upper hemisphere of the early gastrula. When the first longitudinal fibres develop in the medullary cord, the direction of their outgrowth takes place in accordance with the antero-posterior polarity; sensory neuroblasts in the dorsal part send out their axones toward the head and motor neuroblasts in the ventral part send theirs in the opposite direction.[19] In the case of the former, however, it is not possible at first to distinguish, histologically, between dendrite and axone.

The polarity of the medullary cord, when first established, is labile, and even after the neuroblasts have undergone a certain degree of differentiation, their functional polarity may be changed, as Hooker (1917) has shown by reversing segments of the spinal cord in the frog embryo. In such experiments the neuroblasts within the graft develop at first normally in accordance with their original orientation, and there is

17. Recent experiments by Luther (1934), in which ectoderm of amphibian embryos was transplanted with its inner surface toward the outside, indicate that the surface-interior polarity is not fixed until the neurula stages, and cast serious doubt upon the assumption of a fixed polarization, in this direction, of the elements of the neural plate. In *Triton* the indifferent condition lasts until the neural folds close; in the axolotl, until shortly after the formation of the plate.

18. Mall in his study of the histogenesis of the retina (p. 417, footnote) states: "The general indication is that the pole (of the neuroblast) which gives rise to the nerve fibre is the original basal end of the cell." He suggests that cells for which this rule does not hold, for example the cortex cells in the mammalia, have undergone a rotation from their primitive position. Cajal (1908, p. 100) has described in the medulla of the chick certain *neuroblastes intervertis,* in which the axone first grows toward the ventricle only to bend later 180° to run back through the brain substance toward its outer layer.

19. Coghill (1929, pp. 55–58), in agreement with Child (1921) and Herrick (1924), refers the polarity of the neuroblasts to the metabolic gradient of the embryo. It seems to me more likely that the two are but separate manifestations of a more fundamental polarity of the protoplasmic units.

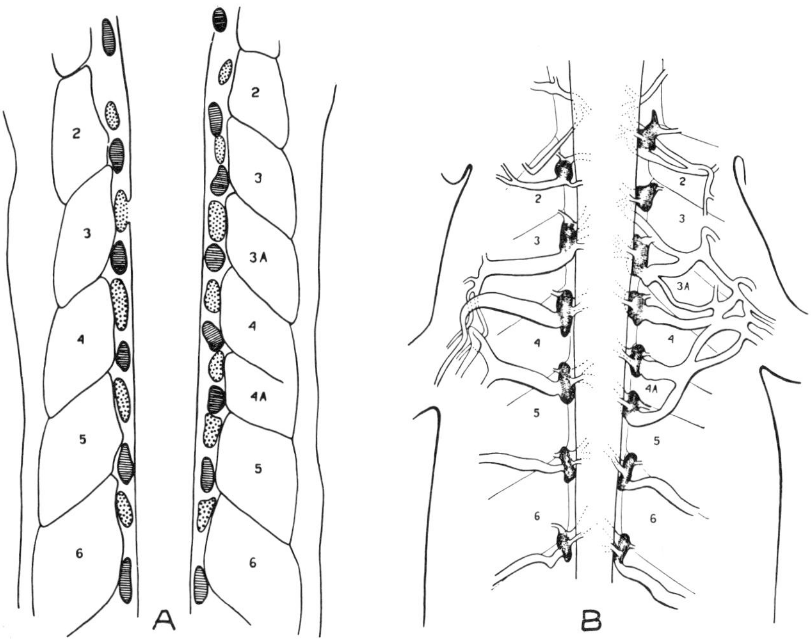

FIG. 21. A, frontal section through an *Amblystoma* larva with two additional somites (3*A* and 4*A*) on the right side which have developed from grafted unsegmented mesoderm; corresponding to these are two additional vertebral arches (striped) and spinal ganglia (stippled). B, reconstruction of spinal nerves in the same case. Five segmental nerves supply the fore limb on the side of operation and only three on the normal side. After Detwiler (1934). × 16.

evidence of some resistance to fibres growing into the reversed segment or from it into the normal cord, where two like poles are apposed. Muscular movements, controlled by the reversed piece, are at first poorly co-ordinated with those of head and tail, but function improves later and may become normal, indicating that the primary morphological polarity of the neurones has been overcome functionally, and that the fibres in the longitudinal bundles which originate in the reversed graft conduct in a direction opposite to normal. After similar operations on younger *Amblystoma* embryos, the reactions are normal from the beginning (Detwiler 1923).

When a short segment of the spinal cord of an *Amblystoma* embryo is excised and replaced transverse to the body axis, the graft maintains its original polarity, and the anterior part of the nervous system re-establishes continuity through the growth of descending fibres which enter the transverse graft near its original anterior end and pass through it to the posterior part of the cord (Wieman 1922, 1925). Only then do the ascending sensory tracts from the posterior part of the cord grow into the graft. The fibre tracts change their direction of growth both on entering the graft and on leaving it.

After rotation of a segment of the spinal cord 90° or 135° about its longitudinal axis, the outgrowing fibre tracts adjust themselves to the changed conditions by undergoing torsion sufficient to bring them into line with the proper sectors of the rotated piece. After rotation of 180° the adjustment is not so perfect, though evidence of dorso-ventral decussation of fasciculi has been found (Hooker 1930).

Peripheral Factors, Especially Those Concerned with the Innervation of Limbs

The segmental arrangement of the spinal nerves is dependent upon the muscle plates, as first shown by Lehmann (1927). However, when the myotomes are completely excised in a region, the spinal ganglia, according to Detwiler (1932, 1934), do not fail altogether to develop but are irregular, and the nerve roots also, though to a less degree. When shorter segments from further tailward are transplanted to the region of the anterior limb, more than the normal number may be inserted into a given space. The number of nerves then tends to correspond to the number of muscle segments (Fig. 21 A), but not always exactly, since the ganglia and the vertebral arches are sometimes less in number than the myotomes. In the above cases the limbs

---

Plate 3

FIG. 22. Fully-grown larva of *Amblystoma punctatum,* with large grafted right limb from *A. tigrinum,* 75 days after operation. Exp. NE19. × 1·75.

FIGS. 23 and 24. Reciprocal pair in which the right anterior limbs were interchanged; preserved 18 days after operation. Exp. NE133. × 4·5.

FIG. 23. *Amblystoma punctatum.* Grafted limb (*gr*) much further advanced than the normal (*n*).

FIG. 24. *A. tigrinum.* Grafted limb (*gr*) much less developed than the normal (*n*).

FIG. 25. Larva of *A. punctatum* with large grafted right eye from *A. tigrinum,* 11 days after operation. (*Cf.* figs. 31 and 32.) Exp. Oc20. × 1·75.

FIG. 26. Larva of *A. punctatum* with two eyes grafted on the side, the anterior one replacing the fore limb. *Cf.* fig. 29 for the arrangement of the spinal nerves in a similar specimen. After Detwiler and Van Dyke (1934). × 3·7.

FIG. 27. Young toad (*Bombinator*) about completing metamorphosis with fore leg grafted to side of head. After Braus (1908).

FIG. 28. Young toad (*Bombinator*), with fore leg in place of hind leg on the left side. After Braus (1908).

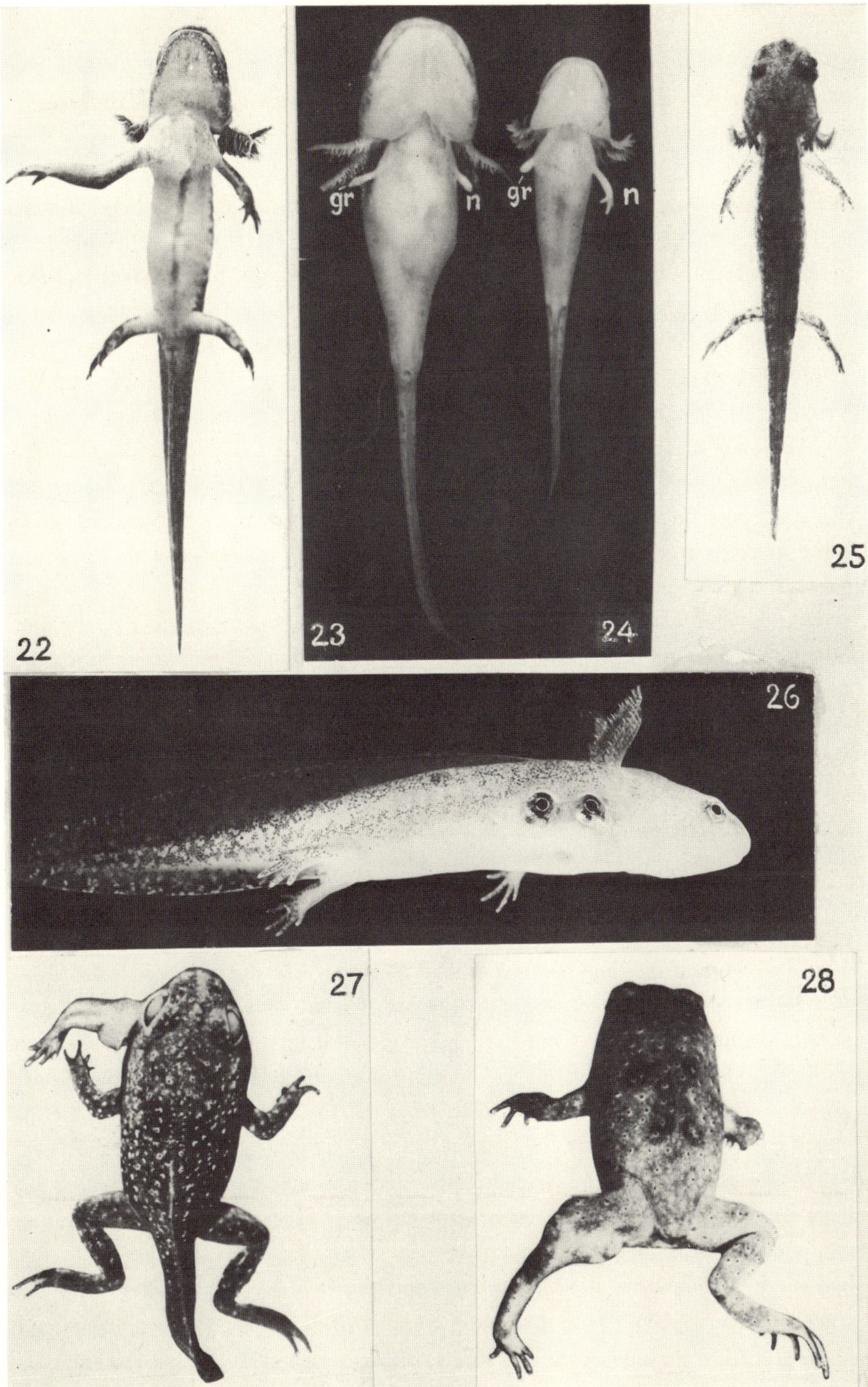
gr
n
gr
n
22
23
24
25
26
27
28

are innervated by all of the nerves of the region (Fig. 21 B), and function properly, no matter if the nerves are irregular or are more numerous than in normal cases.

Experiments in transplanting embryonic limbs, begun by Braus (1904, 1905), have thrown much light on the problems of nerve plexuses and peripheral distribution (Figs. 27 and 28). Two fundamental facts were discovered in the earlier experiments in this field: transplanted limbs tend to receive their innervation from the nerves belonging to the region of implantation; the intrinsic arrangement of the nerves within the limb is normal. Interpreted in the light of the outgrowth theory, these facts offer an explanation of the segmental variation in nerve plexuses, which, as Fürbringer long ago pointed out, may be coupled with constancy of peripheral distribution. They show that there are two independently varying factors affecting the innervation of the limb. The first, which determines the source of nerve supply, is the position and extent of the limb bud at the time the nerves enter. The second is the mode of segregation and growth of the individual structures of the limb; this fixes the mode of intrinsic distribution (Harrison 1907, p. 277).

That the problem of the plexus is not so simple as this was shown later by the more extensive studies of Detwiler (1920, 1922), in which limb buds were implanted systematically from one to five segments caudal to their normal position (in extreme cases opposite the eighth, ninth, and tenth somites), or from one to two segments anterior to this in the region of the gills. If the position of the limb were the only factor involved, the nerves entering should correspond exactly to the region of implantation, but this is by no means so. Limbs moved one or two segments tailward receive the normal brachial nerves, and it is only when the graft is placed from three to five segments back that there is a shift in the nerves entering it. There is also a tendency to a greater spread in the nerve supply in these cases, four nerves frequently entering instead of three. If the limb bud is displaced anteriorly into the gill region, the plexus may become slightly shifted in that direction—there are only two spinal nerves anterior to the brachial plexus—but here again the shift in the plexus is less than the displacement of the graft. When the normal limb bud is left intact or allowed to regenerate, and the graft is placed so near that the brachial nerves may reach either bud, the nerves tend to divide and go to both.

The foregoing points to the conclusion that the rapidly proliferating limb bud gives rise to conditions that direct the growing nerve fibres, and that the nerves normal to the limb are thereby affected more than others. That the attraction to the limb bud is not specific is indicated by further experiments of Detwiler (1928) and Detwiler and Van Dyke (1934) with optic vesicles or nasal placodes, instead of limb buds, grafted to the side of the body (Fig. 26). In such cases the spinal nerves also

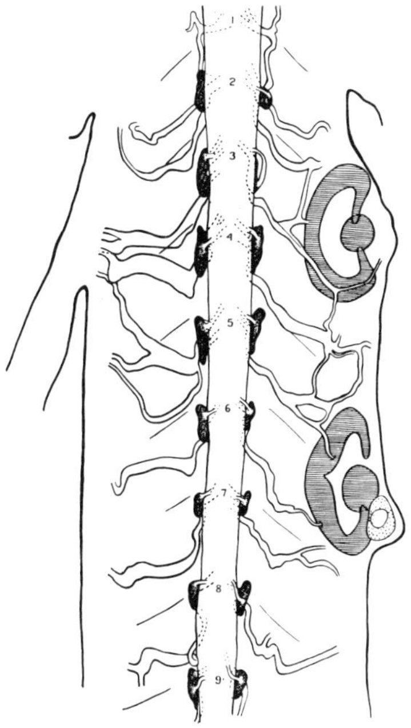

FIG. 29. Reconstruction showing relation of spinal nerves to two grafted eyes, fig. 26. Branches of the fifth and sixth nerves penetrate the sclerotic cartilage. After Detwiler and Van Dyke (1934). × 16.

tend to run toward the graft (Fig. 29), with which, however, they usually form no nervous connection.[20]

In explanation of these observations, Weiss (1934, p. 430) has suggested that the rapidly growing limb bud or sense organ may alter the density and arrangement of surrounding interstitial structures with which the nerves come into contact when growing out, and that this may facilitate their growth in the direction of the graft (p. 134).

The grafted limb does not function normally except when fibres from at least one of the segmental nerves of the brachial region contribue to its innervation (Detwiler 1925; Detwiler and Carpenter 1929), even though it be well provided with other nerves. Co-ordinate movements require a direct connection with the normal centres in the spinal cord, which, however, need not be nearly so extensive as it normally is. When two or more limbs receive nerves from the brachial plexus, they exhibit syn-

2c. Wieman and Nussmann (1929) and Nussmann (1931) made similar experiments but concerned themselves with the question of the actual innervation of the graft rather than with direction of growth.

chronized movements, homologous muscles contracting simultaneously in each member. The explanation of this is too controversial at present to be entered upon here.

That the intrinsic distribution of nerves within the limb is due to factors inherent in the limb itself, follows from the fact that a normal arrangement of nerves is found in limbs which have developed under a great variety of circumstances. Grafted to the head (Fig. 27), a limb receives its supply from one or more cranial nerves (Braus 1905; Detwiler 1930; Nicholas 1929, 1933); a fore limb put in place of a hind limb (Fig. 28), becomes innervated by the lumbo-sacral plexus. Furthermore, as Hamburger (1928, 1929) has shown, the normal arrangement occurs, irrespective of the quantity or the quality of the nerves entering it, as well as of their origin (p. 152).

The following consideratons may facilitate an understanding of this relation. The limb bud in the Amphibia is small and a mere nodule of mesenchyme when the first nerves reach it close to the blood vessels that precede them. At least some of the primary connections may be assumed to be made then, although the definitive nerve endings, according to observations upon embryos of birds and mammals (Tello 1917; Cajal 1919), are not developed until later. Subsequently the elongation of the bud takes place, and the individual muscles and cartilages segregate out of the original blastema. As this proceeds, the nerves are drawn out and shifted with the individual segregates and thus assume a normal arrangement within the appendage, whatever their origin may be. [All of the first fibres to develop which have only short distances to go are termed pioneer fibres. After they have laid down the paths, other fibres which follow them run very close much as they run along a glass rod in the culture medium. With the growth of the animal these newly differentiated fibres have much longer distances to cover but they are guided by the pioneers which are already there. The nerve is further drawn out by the towing action of its terminal tissues. However, as Piatt (1940, 1949) has shown by an ingenious experiment, the later fibres are not dependent entirely on the pioneers but can still find their way along the blood vessels and fascia between the muscles and under the skin. (Lecture IV notes.)]

The adjustment between ingrowing nerves and end organs, while exact as to final results, may accommodate itself within a wide range of time and space, as shown by the following examples.

When by an appropriate experiment a limb bud is made to undergo its early development in the absence of nerves and then is grafted to a normal embryo, the nerves of the latter form their connections in the limb later than they usually do, and yet they assume a normal arrangement (Harrison 1907).[21] [cf. Piatt; R.G.H.] Be-

21. Braus (1905), who originated this experiment, obtained different results and interpreted his findings in accordance with the protoplasmic bridge theory.

tween the two species, *Amblystoma punctatum* and *A. tigrinum,* there is a marked difference with respect to the relative time of growth and differentiation of the fore limb. When a tigrinum limb bud is grafted to a punctatum embryo, its relative growth is so slow at first that the limb is still but an elongated bud (Fig. 24, *gr*), when the normal limb (*n*) of the host shows elbow joint and two long digits and has nerve trunks running to the hand. Yet, when the grafted limb finally reaches this condition, the nerves, which have grown in from the host, show normal distribution. In the reciprocal experiment the punctatum limb on the tigrinum host is precociously developed (Fig. 23, *gr*), and the nerves are normally distributed days before the proper time. Sections taken at suitable stages show nerves reaching at least to the forearm of the grafted limb, while on the normal side they run merely to the base of the bud which is but a nodule of mesenchyme.

The adjustment of nerves to a different spatial scale is shown in regenerated limbs. In the salamander, *Eurycea,* Weiss and Walker (1934) found that the nerves in regenerated limbs correspond in their arrangement precisely to the nerves of the primary limb, although the regenerated appendage develops from a mesenchymal blastema that is much larger than that of the embryonic limb bud. The nerve fibres are more numerous in the regenerated limb, but the individual units are of the same order of magnitude in both cases.

Attempts to obtain limbs without nerves have led to the discovery of some very interesting facts regarding the paths taken by fibres that do succeed in reaching their destination. To prevent their entrance into the hind limb in embryos of the frog or toad, a longitudinal horizontal incision is made through the posterior part of the trunk just below the notochord (Harrison 1904; Hamburger 1928, 1929). An object, best a thin plate of mica, is then thrust through the slit in order to keep the wound from healing completely (Fig. 30 A). Even so it is difficult to prevent all nerves from reaching the limb, and out of seventeen cases investigated by Hamburger only four had the limbs on one or both sides entirely without nerves.

The nerveless limbs do not interest us particularly here. They are normally formed with joints and toes, and the tissues are differentiated, but they are markedly atrophic and functionless. The partly innervated limbs, on the other hand, are very important for our present purpose.

To reach the limbs, nerves may take the most circuitous courses. They may form a plexus across the mid line above the wound and curve as much as 180° around either the posterior or the anterior border to reach the limb (Fig. 30). When one-half of the medullary cord in the region of the hind limb is excised, although a considerable amount of regeneration usually takes place across the median plane, there

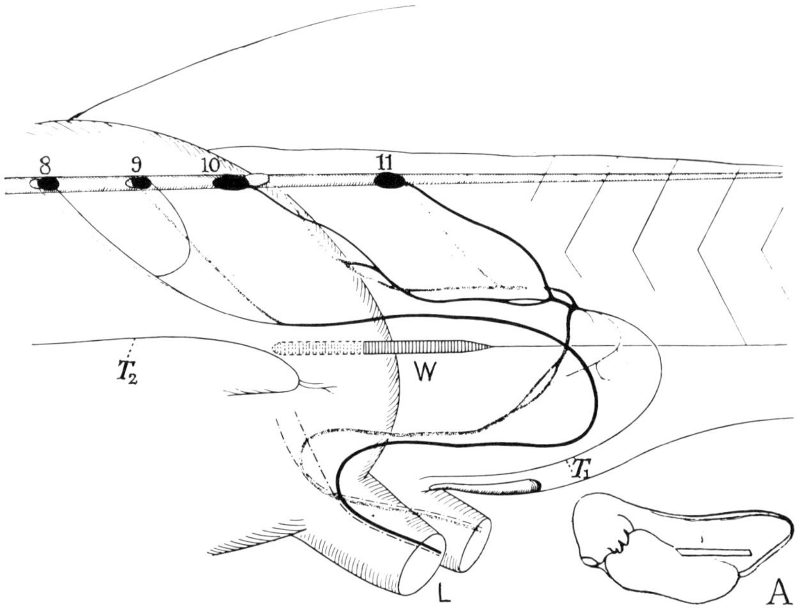

FIG. 30. Innervation of hind limbs of a frog (*R. temporaria*), in which the nerves were prevented from entering the limbs directly by the insertion of a mica plate (*m*) between notochord and limb rudiment of the embryo, shown in the small figure (A). The nerves 8–11 run around the posterior border of the scar (*W*), completely reversing their direction to reach the base of the limbs (*L*). After Hamburger (1929).

are some cases in which no nerves arise on the side of the operation. In some of these, nerves from the opposite side cross the mid line from the intact side and thereby reach the opposite limb, a course that no nerves normally ever take. Whether these facts are to be taken as showing that regions without nerves, particularly the growing limb bud, exert a positive directive influence upon developing nerve fibres, or whether they are better explained by assuming that the configuration of the interstitial tissue is such as to guide the nerves in their peculiar course around the wound, cannot be decided from the evidence at hand.

When the motor areas are eliminated by cutting out a median strip of the medullary plate, there is a tendency for the cord to reorganize itself and form some motor elements. However, one case was obtained by Hamburger in which only sensory nerves were present on both sides, with the exception of a small motor component on the right. While only two small branches to muscles, probably proprioceptors, were found, the plexus was normal and nerves were normally distributed in both legs, all of the sensory branches being present. Partial motor innervation, without sensory nerves, was found in one case, as a result of incomplete restitution of the motor part of the cord after excision of one lateral half. The motor nerves were distributed normally as far as they went. Both sciatic and crural nerves were present and the former could be followed to below the knee.[22]

Although sensory and motor nerves normally run together, the foregoing experiments show that either may find their way alone. In so doing the sensory nerves disregard possible short cuts to the skin, and follow the usual paths between the muscles.

Hamburger finds that the cases in which the nerve trunks accompany the main arteries have much more complete innervation. When the nerves run alone, they tend to scatter in the proximal part of the limb, innervating only single muscles or restricted areas of skin. While there is no rigorous experimental proof that the nerves are actually attracted by the arteries, the blood vessels do precede the nerves in the developing limb bud (Tello 1917). It has been suggested by Weiss (1934, p. 433) that the close contact between the two may be explained by the assumption that outgrowing nerves and vessels respond to a common agent, the ultrastructure of the interstitial ground substance.

One of the most baffling questions in the development of peripheral nerves is the selectivity of the fibres in establishing their proper terminations—motor neurones with muscle fibres and sensory neurones with the epithelium of the skin or mucous membranes or with muscle spindles. In all of the spinal nerves motor and sensory fibres join beyond the ganglia and run together, yet without confusion. The growing fibres connect with the proper end organs, where they form characteristic arborizations. There is even a strict selectivity among sensory nerves, as pointed out by Cajal (1919), in connection with the nerves of the tongue, where the trigeminus forms general sensory endings and the facial and the glossopharyngeal run to the taste buds. It seems necessary to assume some specific reaction between each kind of end organ and its nerve (Harrison 1910), and Cajal (1919) and Tello (1923) have both pointed out that this could scarcely be of other than a chemical nature. However, there is no experimental evidence to prove the latter, and the experiments of Boeke (1913, 1914, 1917) in cross-suturing sensory and motor nerves render a satisfactory explanation on this basis very difficult. These experiments, which concern especially the lingual and the hypoglossal nerves in the hedgehog, show that regenerating fibres may be forced to form anomalous connections. Motor hypoglossal fibres, conducted through the peripheral stump to the mucous membrane, form there typical sensory

22. By excising the dorsal half of the cord in the limb region of *Amblystoma,* Detwiler and Van Dyke (1934) have also obtained limbs with efferent nerves only. Their mode of action is of great interest, but as yet the exact distribution of the intrinsic branches has not been followed.

end organs and may even stimulate the rehabilitation of taste buds. Similarly, regenerating sensory lingualis fibres, passing through the peripheral hypoglossal trunk, reach the muscles and form typical motor end plates. While the latter are almost exactly like normal end plates, the sensory end arborizations formed by the hypoglossal are frequently atypical and may show some influence of the quality of the regenerating nerve. Another more recent experiment by Boeke and Dijkstra (1932), in which skin was interchanged between leg and bill in the duck, shows likewise that nerve fibres are non-specific. The corpuscles of Grandry and of Herbst, found normally only in the bill, first degenerate after transplantation, and then regenerate, becoming innervated by nerves of the leg. In skin from the leg grafted on the bill, no such sense organs are formed in spite of the presence of nerves normally innervating them.

The heterogeneous mass of facts referred to in this section cannot at the present time be brought under one rubric. However, many of them may be interpreted in the light of the stereotropic responses demonstrated by the tissue culture experiments. Most difficult to explain is the formation of terminal connections, and here, notwithstanding the inconsistencies that have been recorded, one can scarcely but agree with those who postulate a response to some sort of chemical stimulus.

Responses of the Nerve Centres to Changed Conditions at the Periphery and in Other Parts of the Central Nervous System

After the first fibres are laid down, the nervous system passes into another phase of development, in which functional activity of some sort begins to play a role.

It has long been known that when the end organ or the peripheral end of a neurone is destroyed and fails to regenerate, atrophy of the nerve cell follows. In the embryo an analogous process may occur, but here it is rather a failure to develop than an actual atrophy of nerve units already differentiated.

Responses of Sensory Centres

In addition to sporadic observations on defective monsters, there is now much experimental evidence of hypoplastic development of nerve centres following peripheral injury. Early removal of the eye in amphibian embryos is followed by a marked reduction in the walls of the mid-brain on the opposite side, as shown by Steinitz (1906) and Dürken (1913). A similar reduction in the olfactory centres occurs as a consequence of removal of the nasal pit, and in the absence of the nose the fore-brain does not regenerate after excision (Burr 1916). In the chick embryo, after removal of the wing rudiment, the spinal ganglia of the region are much reduced (Shorey 1909). [Hypoplasia after limb extirpation has been shown by Barron in the sheep. (Lecture IV notes.)]

The positive counterpart to this, the hypertrophy or hyperplasia of nerve centres in response to peripheral overloading or to increased stimuli from within, was first discovered by Detwiler (1920, 1923) in connection with the spinal ganglia innervating supernumerary grafted limbs. A limb bud, implanted on the flanks in *Amblystoma,* is supplied by nerves that ordinarily have a more limited distribution. Their ganglia become larger than normal and show an actual increase in cell number amounting to about 40%. The corresponding hypoplasia of the ganglia of the brachial plexus, after removal and non-regeneration of the limb bud, is about 50%.

Experiments of the same general kind, but varied much in detail, have all confirmed these results. Limbs grafted to the head bring about hyperplasia of the cranial ganglia innervating them (Detwiler 1930). Heteroplastically transplanted limbs that grow larger than the normal (Fig. 22), have intrinsic nerves approximately proportional to their size, and the spinal ganglia supplying them usually show hyperplasia (Schwind 1931), although the results here do not seem to be quite constant. Spinal ganglia give evidence of hyperplastic response to supernumerary limbs, grafted in the tail bud stage, as early as twelve days after operation, which corresponds closely to the time at which function begins in the limb,[23] and the capacity of the ganglia to respond to either augmentation or reduction of the peripheral field innervated by them persists until after metamorphosis (Carpenter 1932, 1933).

Further cases of sensory hyperplasia have been obtained after transplantation of other organs, and, in fact, this phenomenon may be said to be of general occurrence after peripheral overloading. When a second nasal pit is grafted alongside the normal one and its olfactory nerve penetrates the telencephalon, hyperplasia of the latter ensues (Burr 1930), and an analogous effect is obtained after implantation of a supernumerary eye (Pasquini 1927).

Perhaps the most nearly normal way to bring about overloading is by the method of

23. Unpublished work by Muschenheim referred to by Detwiler (1933, p. 289).

heteroplastic grafting just referred to. The eye lends itself well to such experiments because of the ease and precision with which it may be measured in the living animal. The eye of *Amblystoma tigrinum* grafted upon *A. punctatum* (Fig. 25), may attain twice the linear measurements of the normal eye of the latter species, the average ratio between the two eyes in such experiments being about 1 : 1·8 at the time of metamorphosis (Harrison 1929). The retina of the large eye thus has from three to four times the area of the smaller one in the same animal, and although the number of cells in the ganglionic layer is less per unit area in the larger eye, the total number is much greater. Consequently, the optic nerve, which is derived directly from these cells, is larger (Fig. 31). If it connects normally with the brain and crosses to the opposite side, that side of the mid-brain undergoes hyperplasia and bulges markedly (Fig. 32). Twitty (1932) has found that the number of cells in the grey substance may exceed the normal by more than 20%. Similar experiments with nasal pits (Burr 1930) and with the ear (Richardson 1932) give comparable results, though the amount of hyperplasia is apparently less.

In order to induce hyperplasia, it is not necessary that the peripheral overloading be of a nature to produce normal function. After implantation of a supernumerary nasal pit close to the normal one, the olfactory nerve of the former, instead of joining the latter, may run independently and enter the *pars dorsalis thalami,* in which an

---

Plate 4

FIG. 31. Cross-section through the diencephalon of a larva of *A. punctatum,* 41·5 mm long, showing the inequality of the two optic nerves, the one on the right (left of figure) coming from a large grafted axolotl eye. Preserved 102 days after operation. Exp. Oc23. × 35.

FIG. 32. Cross-section through the midbrain of similar specimen with an *A. tigrinum* eye. The wall of the left side (right of figure), which receives the larger optic nerve, is thicker and more bulged than that of the right. Specimen preserved 90 days after operation. Exp. Oc10. × 35.

FIG. 33. Cross-section of spinal cord in the region of the fifteenth ganglion of a chick embryo, 8 days after beginning of incubation, and 5 days after the removal of the right wing rudiment. The right side of the cord (left of figure) is smaller. The hypoplasia is most marked in the lateral part of the ventral horn, and is less in the dorsal horn. There is no significant hypoplasia of the medial portions. After Hamburger (1934). × 113.

FIG. 34. Section through spinal cord of an *Amblystoma* larva from which the seventh, eighth, and ninth somites had been entirely removed on the right side in the tail bud stage. Cord essentially symmetrical and no significant difference between the two sides in the number of cells. After Detwiler (1929). × 260.

FIG. 35. Another section from the same individual.

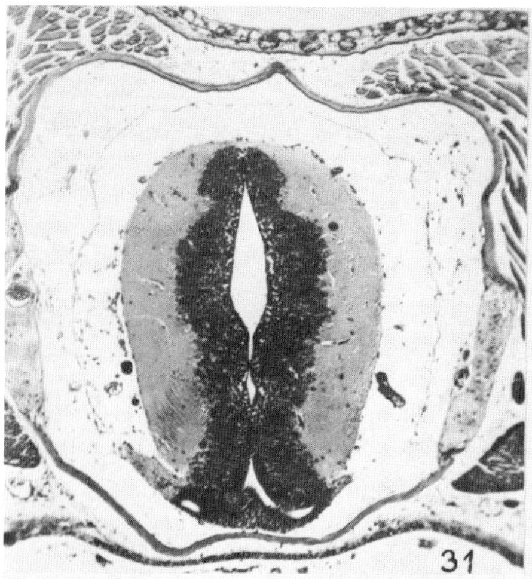

31

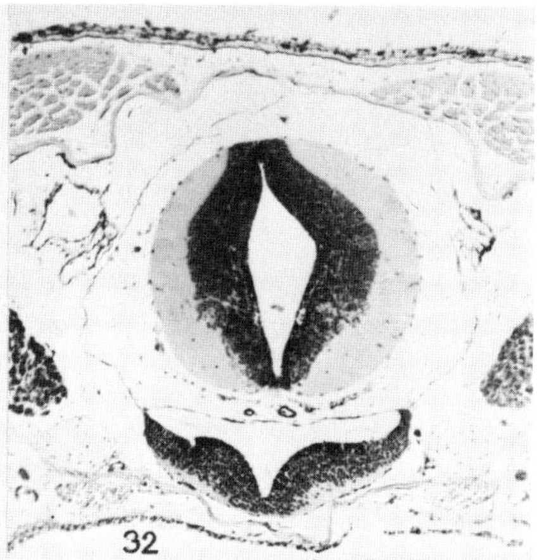

32

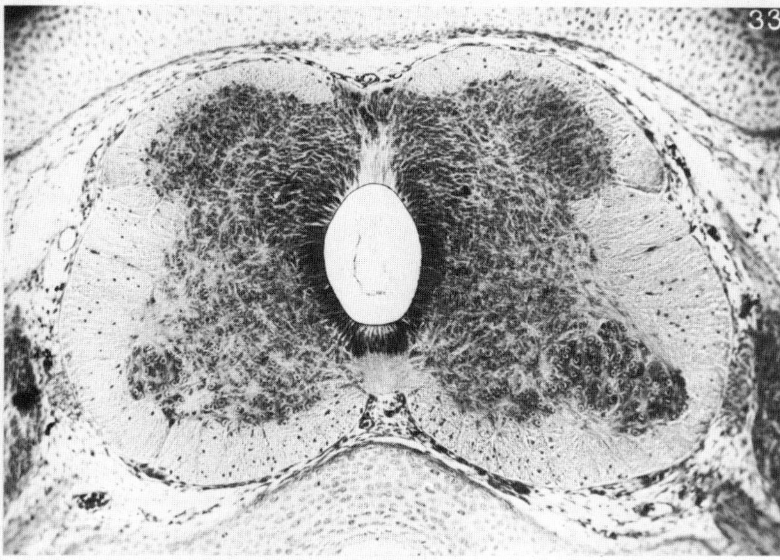

33

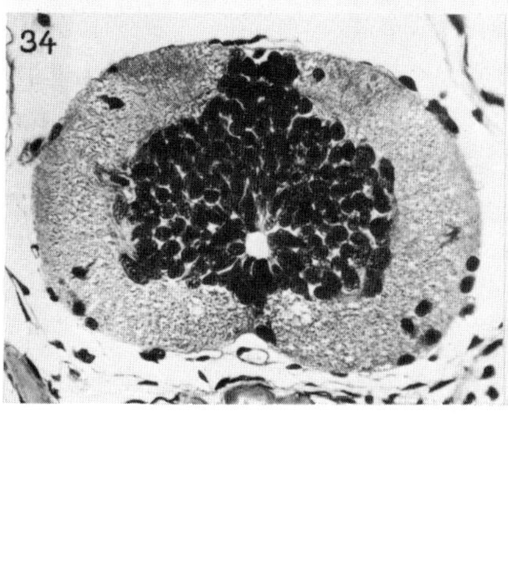

34

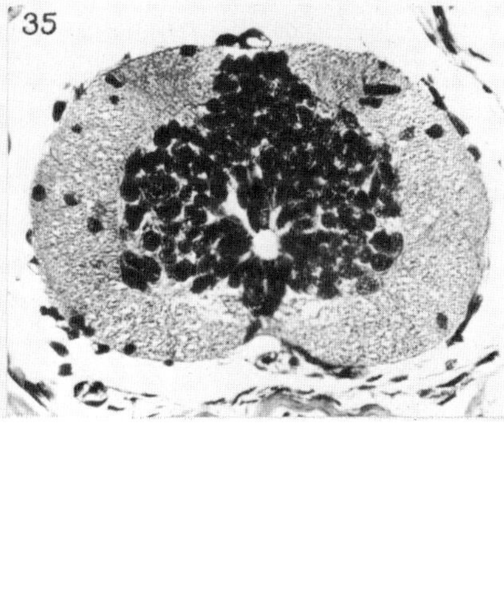

35

increase in the number of cells amounting to about 20% may then be observed (Burr 1924, 1930). Sometimes olfactory fibres may join the ophthalmic nerve; if so its ganglion becomes larger. Analogous results were obtained by May and Detwiler (1925) after transplantation of the eye and the nasal pit to the region of the ear. The optic nerve then runs directly into the acoustic region of the medulla or into the vagus ganglion with a resultant hyperplasia that may reach 20% in the former and 50% in the latter.[24] That function in the ordinary sense is not concerned with this reaction follows also from experiments by Burr (1920) in which it was shown that even when the nasal pit is buried under the skin, so that stimuli cannot reach it from the outside, hypoplasia does not occur in the telencephalon, provided the olfactory fibres enter it.

These findings accord well with the results of earlier experiments in which frog embryos were reared in chloretone between the time of closure of the neural folds and complete resorption of the yolk, a period of about seven days, during which normally the primary nerve paths are all laid down and various coordinated movements initiated (Harrison 1904). Embryos so treated remain motionless throughout and do not respond to tactile stimuli. Nevertheless, when finally placed in pure water at the expiration of the period, they exhibit perfect coordination in their movements as soon as the anaesthetic is eliminated; this may not require more than fifteen minutes, while normally the movements develop by gradual stages, correlated with the degree of differentiation of the nervous connections (Coghill).

The above facts all indicate that the factors influencing the development of coordinating mechanisms are not nervous stimuli of the usual kind. Thus even in these relatively late stages, the developmental functions remain distinct from the common physiological processes.

That the aberrant olfactory nerves, mentioned above, usually enter the brain at a particular place in the thalamus, is correlated, according to Burr (1932), with the presence of a centre of active cellular proliferation at that point when the nerve approaches it. This is a reciprocal relation, for the ingrowing fibres, in turn, stimulate the cells of the centre to increased activity, leading to the observed hyperplasia. Something of this kind may take place in normal development, for, according to Coghill (1924), the distribution of dividing cells in the central nervous system of the embryo is unequal, with a series of zones of more intense activity alternating with zones of less, and Burr (1932) finds that the active regions correspond to the points of entry of the sensory roots of the cranial nerves.

24. *See also* May (1927).

Responses of Motor Centres

The foregoing observations and experiments, which have to do with afferent centres, are remarkably consistent. With regard to the motor centres the evidence is conflicting.

Shorey (1909) reported that when the limb bud or myotomes are excised, both motor and sensory nerves which normally run to them are defective, with a corresponding reduction in the ventral horns of the grey matter of the spinal cord as well as in the spinal ganglia. This was held to apply to urodele and anuran amphibians (*Amblystoma* and *Bufo*) and to the chick. With regard to the latter, Shorey's results have recently been confirmed by Hamburger (1934), who finds that after extirpation of the wing rudiment the hypoplasia of the lateral motor group of ventral horn cells ranges from 28 to 61% (Fig. 33). May (1933) has found essentially the same in the anuran, *Discoglossus,* and maintains also that overloading by implantation of a supernumerary limb is followed by hyperplasia of the motor centres.

On the other hand, Detwiler's earlier and much diversified experiments upon *Amblystoma* have yielded invariably negative results. There is no reduction in the number of cells in the ventral (motor) sector of the spinal cord after extirpation of myotomes or limb buds and no hyperplasia after implantation of supernumerary appendages (Figs. 34 and 35). Weiss (1931) likewise failed to find motor hyperplasia in an adult frog with supernumerary hind legs. There is, however, a small reduction in the total mass of the motor region of the cord, accompanying defects at the periphery, which is due to the smaller size of the motor cells, and this is greater when the defects (ablation of limbs) are bilateral (Detwiler and Lewis 1925).

The stimuli to the motor cells from within the central nervous system constitute in the urodeles the principal factor regulating the number of these elements. This has been shown by Detwiler (1923–1933), by means of a series of experiments in which portions of the hind brain and spinal cord were removed and transplanted in various ways. Substitution of a piece of spinal cord, comprising the seventh, eighth, and ninth segments, for the brachial region (third, fourth, and fifth), which normally contains a larger number of motor neurones, is followed by hyperplasia of the motor cells of the graft to the extent that the normal number for the brachial region is attained. This means a hyperplasia of about 80%, and it takes place even after excision of the fore limb, which receives its innervation from that region, showing that it is independent of peripheral loading. If the corresponding segments are left in place and a limb is transplanted so that it becomes innervated by them, the hyperplasia does not occur. A similar effect is obtained after reversal of the brachial region of the cord, whereby the fifth segment, which normally contains fewer motor cells than the third, is placed, with polarity reversed, in the position of the latter and vice versa. The number of motor cells found then in each segment in the mature larva is the number proper to its new position. Independent positive evidence of the influence of centres higher up in the central nervous system has been obtained by substituting for the first five segments of the spinal cord a piece which included the posterior part of the medulla oblongata and the first two spinal segments. An appreciable hyperplasia occurred both in the two transplanted spinal segments and in the first two intact segments (sixth and seventh) posterior to the brachial region. This can be ascribed only to the influence of the increased number of neurones in the two medullae combined, which give rise to a larger than normal number of descending fibres (bulbo-spinal tract) in the cord.

If, on the other hand, the medulla is extirpated, there is a reduction in the number of cells in the first three spinal segments, which is the case only to a much less degree when the medulla is left intact and the mesencephalon removed (Nicholas 1929, 1930). Also in the chick embryo isolation of the lumbar cord from the regions above and below is followed by hypoplasia (Williams 1931).

At present there is no ready explanation of the conflicting observations regarding the response of motor centres to changes in conditions at the periphery. The most probable one is that they are due to real differences in the conditions obtaining in the species used. In the chick the motor nuclei are compact and well marked off from other centres (Fig. 33), while in the urodele amphibians the motor cells are widely dispersed and without definite limits. Inasmuch as hyperplasia follows stimulation which does not pass over the neurones that are themselves affected but over neighbouring ones (p. 162), Hamburger (1934) suggests that they may be too scattered to be influenced in the urodeles while readily affected in the chick. The anuran amphibians occupy an intermediate position. Whatever the ultimate solution of the discrepancies may be, it is well estabished that the motor centres are markedly influenced by central impulses, i.e. by stimuli received from within.

That still other factors may play a part in regulating the proliferation of neuroblasts is shown by the experiments of Severinghaus (1930) in grafting segments of the spinal cord, together with notorchord and myotomes, to the flank of another individual without direct connection with the central nervous system of the latter. In such cases cell proliferation runs wild, and in the dorsal (sensory) half of the cord the number of cells may reach three times the normal, as if some inhibitory factor had been removed.

Summary and Conclusion

Proof that a kind of functional regulation is active in determining the final relations of the elements in the nervous system must not be taken to imply that influences of this nature are the only factors involved or even the most important ones, for the material in which this regulation takes place has already acquired in advance most of its essential qualities. The protoplasmic characteristics of the species are already laid down in the egg and, being shared by the material of the presumptive brain and spinal cord, render the nervous system of one species different from that of any other. This prospective nervous material is, moreover, subject to the influence of the "organizer" underlying it in the gastrula, and in the absence of this substratum its potential qualities are not realized.

In the period of the medullary plate, which is less understood for lack of appropriate experiments, considerable plasticity still remains, the form of the central nervous system gradually becoming fixed under the continued action of the underlying layer. By the time of closure of the medullary folds its main topographic features are established, although the differentiation of nerve cells has not yet begun. Then follows a period of differentiation, during which the neuroblasts give rise to the first fibre tracts within the central nervous system and to the first peripheral nerves. The original direction of outgrowth seems to be governed by the polarity of the neuroblasts, i.e. it depends primarily upon qualities inherent in the latter at that time, although the mode of establishment of this polarity is still conjectural. The outgrowing fibres may divide or be swerved from their original course by the configuration of the embryonic organs and the membranes surrounding them, and are guided by the micellar structure of the interstitial ground substance. Beginning at this time and continuing for a considerable period thereafter, there is an active localized proliferation and differentiation of neuroblasts by which the principal nerve centres are segregated and their relations to the nerve tracts established. This period, as we have just seen, is characterized by a delicate adjustment of cell number to stimuli received from other regions of the embryo, both from the periphery and from the central nervous system itself.

In the last period influences affecting the neurones are probably in part growth regulating factors of a general nature, while others are of a kind approaching specific nervous stimuli but not yet identical with them. Even the mode of reaction to changes in the peripheral sensory field can hardly be exactly the same in all cases.

When a sensory area like the eye or nose is decreased by excision or augmented either by doubling or by means of heteroplastic grafting, the operation itself changes the prospective number of primary sensory neurones and consequently brings about an actual decrease or increase in the number of axones entering the central nervous system. It is not difficult to imagine how the altered number of nerve fibres, with their terminal arborizations, might influence cell division and differentiation in the centres with which they connect, although even here it is not a strictly nervous function but something more of the nature of trophic stimulus that produces the result (Burr 1920). How the spinal ganglion cells become similarly affected by an alteration in the peripheral area supplied by them is another question, since here the primary neurones are not changed directly by the operation. One would naturally think that the outgrowing sensory fibres might accommodate themselves to the peripheral disturbance by an adjustment in the frequency of branching of their axones, as the motor fibres undoubtedly do. That this direct response at the periphery may occur to some extent is not disproved by any of the experiments, but the hypoplasia or hyperplasia of the ganglia, which is the more conspicuous response, is difficult to understand, for it is not the cells with processes reaching to the periphery that respond in this way, but the undifferentiated cells far off in the ganglia. The former must therefore carry impulses to the centre that stimulate neighbouring cells to send additional axones to the periphery.

With the motor neurones, the reaction to stimuli reaching them from within the central nervous system is again not so difficult to comprehend and is probably of the same nature as the reaction of nerve centres to changes in the primary sensory neurones. Response of motor cells to peripheral influences must take place in urodeles largely by means of axonal division, as is shown by the difference in size between the proximal and distal portions of the nerves. On the other hand the central responses to such influences, reported by Shorey, Hamburger, and May in the anuran *Amphibia* and in birds, must be mediated in the same complicated way as those of the spinal ganglia.

The experiments last referred to deal almost entirely with the middle stages of development, in which regulatory responses of structure are obviously related to changes experimentally produced in other parts of the organism. As development proceeds, such responses become more and more difficult to follow, and yet many structural changes must take place in older animals, corresponding to the functional processes known as conditioning of reflexes, habit formation, and learning. Here again the traditional conflict arises between the treatment of the organism as a whole and the analytical method of approach by means of anatomical and physiological experiment with constituent parts. In Lashley's experiments on cerebral localization in the rat, the importance of a specific topographic arrangement of neurones seems to fade in comparison with that of some kind of representation of totality in every region. [On the other hand the experiments of Sperry in crossing nerves and tendons (1941–1945), Stone's work in inverting eyes (1944, 1947) and Weiss' work (1937) in interchanging limbs from right to left, all show that the animal cannot adjust to such interference, whereas in man adjustment seems possible, at least to some extent. (Lecture IV notes.)] Nevertheless, the adjustments made to meet the loss of a large amount of cortex, which require considerable time to be effected, must have a structural basis and must result in a new configuration of the remaining material. Thus at the last stage of development of the organism we are still confronted, as at all previous stages, with this most significant but perplexing problem of biology: how an organism can be made up of independently working parts, and, at the same time, have the whole represented in each part.

In the new experiments that will be required to throw light on this field, embryology will need the assistance of the most delicate methods of modern physiology. In return it will be able to contribute materially to the advancement of physiology, whenever the latter is ready to make use of the methods which experimental embryology has developed. By ablation and transplantation of parts of the embryo it is possible to fashion almost any kind of nervous system desired and subsequently to study its function without the disturbing effects of trauma. This is one of the most promising lines of investigation leading off from the field covered in the present lecture.

References

References not listed below may be found in one of the four following papers:

Detwiler, S. R. 1933. *Biol. Rev., 8:* 269.

Harrison, R. G. 1910. *J. Exp. Zool., 9:* 787.

———. 1924. *J. Comp. Neurol., 37:* 123.

Mangold, O. 1928. *Ergebn. Biol., 3:* 151.

Baitsell, G. A. 1917. *Am. J. Physiol., 44:* 109.

Bauer, K. 1932. *Zeit. mikr.-anat. Forsch., 28:* 47.

Bethe, A., and E. Fischer. 1931. *Handb. norm. pathol. Physiol., 15:* 1045. Berlin, Springer.

Bisceglie, V., and A. Juhász-Schäffer. 1928. *Die Gewebezüchtung in vitro.* Berlin, Springer.

Boeke, J. 1933. *Zeit. mikr.-anat. Forsch., 33:* 23, 233.

Boeke, J., and C. Dijkstra. 1932. *Proc. Akad. Sci. Amst., 35:* 1114.

Burr, H. S. 1932. *J. Comp. Neurol., 56:* 347.

Cajal, S. R. y 1908. *Anat. Anz., 32:* 1. Reprinted in *Études sur la Neurogenèse de quelques Vertébrés,* Madrid, 1929, pp. 81–125.

Cajal, S. R. y 1919. Trab. Lab. Invest. Biol. Univ. Madr., *17:* 181. Reprinted in *Études sur la Neurogenèse de quelques Vertébrés,* Madrid, 1929, pp. 297–348.

———. 1928. *Degeneration and Regeneration of the Nervous System,* trans. from the Spanish edition (1913) and ed. by R. M. May, vol. 1. Oxford Univ. Press.

Carrel, A., and M. T. Burrows. 1911. *J. Exp. Med., 14:* 244.

Coghill, G. E. 1929. *Anatomy and the Problem of Behavior.* Cambridge Univ. Press.

Detwiler, S. R. 1932. *J. Comp. Neurol., 56:* 465.

———. 1934. *J. Exp. Zool., 67:* 395.

Detwiler, S. R., and R. W. Lewis. 1925. *J. Comp. Neurol., 39:* 291.

Detwiler, S. R., and R. H. Van Dyke. 1934. *J. Exp. Zool., 68:* 321.

Esaki, S. 1929. *C. R. Assoc. Anat.,* 24me Réun., 13 pp.

Grigorjeff, L. M. 1931. *Arch. exp. Zellforsch., 11:* 483.

———. 1932. *Arch. exp. Zellforsch., 13:* 195.

Hamburger, V. 1934. *J. Exp. Zool., 68:* 449.

Harrison, R. G. 1914. *J. Exp. Zool., 17:* 521.

Herrick, C. J. 1924. *Neurological Foundations of Animal Behavior.* New York, Holt.

Hooker, D. 1930. *J. Exp. Zool., 55:* 23.

Huzella, T. 1929. *Roux' Arch., 116:* 430.

———. 1931. *C. R. Soc. Biol., Paris, 106:* 179.

———. 1934. *Arch. exp. Zellforsch., 15:* 250.

Huzella, T., and J. Lengyel. 1932. *Arch. exp. Zellforsch., 13:* 141.

Kappers, C. U. A. 1908. *Folia neuro-biol., 1:* 157.

———. 1933. *Psychiat. neurol. Bl. Amst., Jaarg., 1933,* p. 579.

Lashley, K. S. 1926. *J. Comp. Neurol., 41:* 1.

———. 1929. *Brain Mechanisms and Intelligence.* Univ. of Chicago Press.

Levi, G. 1925. *Arch. exp. Zellforsch., 1:* 1.

———. 1926. *Arch. exp. Zellforsch., 2:* 244.

———. 1934. *Ergeb. Anat. Entw., 31:* 125.

Lewis, W. H. 1907. *Am. J. Anat., 6:* 461.

———. 1922. *Anat. Rec., 23:* 177.

Luther, W. 1934. *Roux' Arch., 131:* 532.

Mall, F. P. 1893. *J. Morph., 8:* 415.

Matsumoto, S. 1918. *J. Exp. Zool., 26:* 545.

May, R. M. 1933. *Bull. Biol., 67:* 327; Pl. 13–14.

Mihálik, P. 1932. *Anat. Rec., 54:* 149–56.

Mossa, S. 1927. *Arch. exp. Zellforsch., 4:* 447; Pl. 2.

Nageotte, J. 1927. *C. R. Soc. Biol., Paris, 96:* 172, 464, 828, 1268; *97:* 559.

———. 1931. *Ann. Anat. path. méd.-chir., 8:* 1; Pl. 1.

Nicholas, J. S. 1933. *J. Comp. Neurol., 57:* 253.

Olivo, O. M. 1927. *Arch. exp. Zellforsch., 4:* 43.

Péterfi, T., and O. Kapel. 1928. *Arch. exp. Zellforsch., 5:* 341; Pl. 29.

Péterfi, T., and S. C. Williams. 1933. *Arch. exp. Zellforsch., 12:* 210.

de Renyi, G. S. 1929a. *J. Comp. Neurol., 47:* 405.

———. 1929b. *J. Comp. Neurol., 48:* 441.

Richardson, D. 1932. *J. Exp. Zool., 63:* 413.

Speidel, C. C. 1933. *Am. J. Anat., 52:* 1.

Steinitz, E. 1906. *Roux' Arch., 20:* 537.

Van Campenhout, E. 1930. *J. Exp. Zool., 56:* 295.
Weiss, P. 1929. *Roux' Arch., 116:* 438.
———. 1931. *Pflüger's Arch., 228:* 486.
———. 1933. *Am. Naturalist, 67:* 322.
———. 1934. *J. Exp. Zool., 68:* 393.
Weiss, and R. Walker. 1934. *Proc. Soc. Exp. Biol. Med., 31:* 810.

5

Relations of Symmetry in the Developing Embryo

Samuel Butler[1] answered the old riddle of the egg and the hen by reminding us "that a hen is only an egg's way of making another egg." This remarkable anticipation of the Weismann theory, by directing attention away from the more notorious relation that an egg is only a hen's way of making another hen, goes at once to the root of one of the most perplexing biological problems.

It was Weismann (1883, 1885) who first gave currency to this view in his theory of the continuity of the germ plasm,[2] according to which the only direct continuity from generation to generation is through the germ cells, which are set aside for the function of reproduction. The mysterious cycle of eggs and hens is thus resolved into a row of parallel lines, each representing the development of a single egg into a hen and all springing from a base line, the succession of germ cells.

In most organisms the process is complicated by the sexual mode of reproduction,

Revision of a paper read at the Harvard Tercentenary Conference, September 9, 1936, and published provisionally without figures or references in the "Collecting Net," Vol. 11, December, 1936.—R.G.H.

Reprinted from *Transactions of The Connecticut Academy of Arts and Sciences, 36* (July, 1945), 277–330.

1. Life and Habit, p. 134. Butler's discussion is both provocative and amusing, but there is no indication that he had any profound understanding of the subject. In the latter part of the book he reveals himself as an out-and-out Lamarckian.

2. See Weismann's essays (1892a). Similar ideas had been expressed previously by Owen (1849, p. 5) in his essay on parthenogenesis, Jäger (1876; 1878, p. 196; 1879, p. 378) and Nussbaum (1880). Nussbaum, who based his theory on the study of the origin of sex cells in vertebrates, used the more appropriate expression "continuity of the germ cells." For the historical and controversial bearings of the question, see Weismann (1892b) and Nussbaum (1893).

in which the cooperation of two parent cells, male and female, is required to initiate development and produce a complete new organism. However, to the biologist this complication has proved to be a rare advantage, for it has afforded an unparalleled means of experimental analysis, by which it has been possible to penetrate deeply into the nature of the structure and arrangement of the germinal material in the cell nucleus. Thus modern genetics has been raised to the plane of an exact science.

This phase of the problem of reproduction deals with the mixing and reassortment of qualities from generation to generation. Its working hypotheses have been clear; its principles are relatively simple and capable of precise formulation. The other principal phase, which is concerned with the development of the individual, has only begun to yield to experimental analysis. Embryology, long mainly a descriptive science, has gone its way independently of genetics without a master hypothesis like the theory of the gene to guide it.

Embryonic development consists of a series of events which are usually dated from the time of maturation and fertilization, but which should, more appropriately, be carried back to include the growth period of the egg. The egg, a single cell, is quite as specific as the mature organism in that it develops only into an adult of its own species. Environmental factors may produce modifications, they may even change sex, but they do not change the species.

There is abundant evidence that each species differs chemically from every other. Experiments in heteroplastic grafting and in species hybridization, the serological study of blood relationship, and the chemical and crystallographic study of similar proteins and carbohydrates from different species all point unequivocally to this conclusion. We must, therefore, assume some sort of chemical configuration as characteristic of the protoplasm of each species of organism.

Because of the visible complexities which arise in the embryo, development is often described as a change from the simple to the complex. This is, however, illusory, for the egg has implicitly all the characters of the species. Egg and mature organism are but different phases of the same entity. While the changes which we observe involve localized chemical and physical transformations in the original substance of the egg, it must be assumed nevertheless that a residuum of this material preserves throughout life the essentials of its finer specific configuration.

Cells have a dual organization in their make-up of nucleus and cytoplasm. The exact relations between these two constituents are obscure, and, as pointed out by East (1934), they present problems of the type which are difficult to approach experimentally. Nevertheless, there are certain relevant facts that may help toward an understanding.

1. Neither nucleus nor cytoplasm can function indefinitely without the other, as shown by numerous experiments on unicellular forms. The most striking case of cytoplasmic survival and activity is that found by Hämmerling (1932, 1934) in Acetabularia, a large unicellular alga, which is capable of living for months and of regenerating complicated structures without the presence of the nucleus.

2. When nucleus and cytoplasm of different origin are associated together, sooner or later the nucleus assumes control of the characters which develop. However, it may require at least a whole generation before this control becomes effective (p. 171).

3. Nucleus and cytoplasm must not be of too diverse origin if they are to function together. Even in combinations of different species of the same genus disturbances usually arise, which are less marked in ordinary hybrids than in those arising from non-nucleated egg fragments fertilized by foreign sperm (merogonic hybrids), where all of the nuclear material is from one species and all of the cytoplasm from the other. The nature of these disturbances is not understood but they may be checked by the cellular environment, as shown by the experiments of Baltzer (1930), Hadorn (1930, 1932, 1934) and Baltzer and de Roche (1936).

4. Practically all visible differentiations undergone by the developing embryo take place in the cytoplasm. It is mainly with respect to this constituent that the various types of tissue cells differ.

5. Changes in the nucleus in consequence of cytoplasmic influence have been observed only rarely and then only in somatic cells in contrast with germ cells, as in the so-called diminution phenomenon described by Boveri (1887, 1892, 1899, 1910) in Ascaris.

Since the nucleus can act only with cytoplasm of its own or of closely related species, and since all differentiations are carried out in the cytoplasm, the latter is not to be regarded as neutral or plastic material which can be immediately changed by the nucleus. There must be a finer cytoplasmic configuration which fits with the nuclear configuration.

The nuclear material has a definite arrangement, and even changes in the order of this arrangement without addition or subtraction of material may produce changes in the organism. It is precisely divided at each mitosis and is evenly distributed to the daughter cells. There is no fundamental nuclear differentiation except at the maturation divisions, when paired chromosomes separate. On the other hand the cytoplasm may be very unevenly divided both quantitatively and qualitatively, since there are already visible regional differences in the egg before segmentation and even before maturation. It would be false, however, to conclude that cytoplasmic differentiations are necessarily coincident with visible materials accumulated in the several regions of the egg, for in many cases the distribution of such substances may be changed by centrifuging without interfering with normal development.[3] Whatever the basis for these regional differences, the cleavage process produces an embryo made up of cells in which the nuclei are alike but the cytoplasms different. If, then, in what follows emphasis is laid upon the cytoplasmic changes occurring in differentiation, it is because of their direct involvement in the process and not that they are assumed to be independent of the activities of the nucleus.

Methods most generally used in experimental embryology involve removal or incomplete separation of parts of the embryo or their rearrangement by compression, transplantation or centrifuging. They were devised for the study of what has been known as the "determination problem," which concerns itself with the temporal and spatial restriction of potencies of the several parts of the embryo as development proceeds, and with the interaction between parts in differentiation. They do not disclose the actual nature of the physical and chemical changes underlying differentiation.

From the large mass of work done by these methods one important generalization stands out. In practically all eggs a part is capable of giving rise to more than it would if left in its normal surroundings, and in many cases a part may give rise to the whole. This applies to egg fragments when the nucleus is left intact, to fertilized non-nucleated fragments, to whole blastomeres separated in cleavage stages, and in many cases also to circumscript areas in much later stages of development. The power of regulation decreases rapidly with development, especially after fertilization, but eggs of different species differ immensely in this respect and in some cases the regulatory power is very limited.

Even when there are regional differences in the embryo, the prospective fate of cells may be changed by subjecting them to different surroundings either within the embryo itself or in the medium. In the segmenting sea urchin egg, which Hörstadius (1935, 1936b) has submitted to a most exhaustive study, the cells are arranged in tiers perpendicular to the axis of the egg. Each plays a definite rôle in normal development, and, when separated, each gives rise to a certain type of defective embryo. Recombined in abnormal ways, the resulting embryos may be either abnormal or normal, depending upon the balance between materials from the two hemispheres of the egg. The experiments point to the existence of two gradients running in opposite directions from pole to pole, the potential intensity of action of the cell tiers being a function of the distance from the poles. The original polarity of the material is

3. First clearly shown by Lyon (1907). Morgan (1927), who has contributed much to this field, gives an excellent review of it in his "Experimental Embryology," pp. 493–535.

either preserved or else modified in a definite way by the conditions experimentally imposed.

Other eggs behave similarly. In the amphibia not only are early segmentation and gastrulation stages capable of regulation, but also later embryos, where particularly the ectodermal areas have been shown each to have diversified prospective potencies, only one of which is normally realized. This quality of "wholeness" in the parts of the organism, particularly the embryo, has led to much speculation and even to a system of philosophy. It is the capital problem of embryology to find the physico-chemical basis for it.

Since each nucleus carries the whole genom of the individual, one might be inclined at first sight to attribute manifestations of "wholeness" solely to it. However, this would be insufficient, and we must seek also in the cytoplasm, which effects differentiation, a basis for the characters of the organism. This is assumed to be some kind of a "repeat" configuration, in each unit of which the qualities of the whole, including its symmetry and polarity, are in some way implied. We have to explain, then, how such a system can shake out of itself the macroscopic figure, composed of localized organs and tissues, characteristic of the adult organism.

Models which resemble a complex system in certain respects may be useful in leading to a better understanding of it. Crystal models have long been employed in morphology,[4] although their appropriateness has not gone without challenge. A rigid and angular crystal seems very different from the living organism. In recent years, however, the concept crystalline has been much extended, and the sharp distinction between crystalloid and colloid, even that between solid and liquid, has broken down. X-ray analysis shows that many substances, formerly regarded as colloidal, have crystalline arrangement. Also, a paracrystalline state, in which the mobility of liquids is combined with a definite configuration, is now recognized.

In the crystalline state atoms or atom groups are arranged in a space lattice, which is a point system such that the arrangement of points around any single point is the same as that around any other. This internal structure determines the fundamental characteristics of the crystal. Its external form, while related to its lattice pattern, is not necessarily an enlarged copy of it, being dependent also upon external factors.

Nearly all animals, like crystals, have characteristic external form and symmetry, assumed here to be related to the molecular configuration of their substance. The most common type is the bilateral symmetry found in representatives of many different phyla. This is a very low degree of symmetry, but in many coelenterates and particu-

4. Cf. Przibram (1906) for references to earlier work, also Przibram (1926) and Rinne (1931).

larly in the radiolarians a higher order of symmetry prevails. In many animals, however, the superficial bilateral form merely masks a more fundamental asymmetry, which comes to fullest expression in the gastropods and is quite definite internally in the vertebrates. In fact, all organisms are probably asymmetric in some degree.

In asymmetric species the members usually show this quality in a constant sense, but occasionally an individual may arise which is asymmetric in the opposite sense or the mirror image of the normal. Left handedness constitutes a partial inversion of this kind, and, more rarely, in man and other vertebrates complete *situs inversus* of the viscera may occur. In any one species of snail the shell is usually coiled in the same direction, some species being dextral and some sinistral, but here again individuals may arise which are oppositely coiled.

Asymmetric crystals also occur as rights and lefts, as do the molecules of many chemical compounds, particularly those with asymmetric carbon atoms. The protein molecule, which contains many asymmetric carbon atoms, is also asymmetric. When such compounds are formed, the right and left enantiomorphs are always present in equal quantities except when another asymmetric molecule is involved in the reaction. Organisms frequently react with or form one enantiomorph to the exclusion of the other, as in the case of enzymes, and this quality has consequently been thought by some to be one of the most important distinctions between living and non-living matter. However, since the protein molecules entering into the composition of living matter are themselves asymmetric, an unbalanced synthesis of asymmetric compounds on the part of living organisms is no more than should be expected. What needs to be explained, then, is not the asymmetric syntheses accomplished by organisms, but the fact that the amino acids of which proteins are composed are all of the laevo-form. But this is an historical question harking back to the origin of life.

Embryological observations of Crampton (1894), Kofoid (1895) and Conklin (1897, 1903) [Fig. V–1] have shown that the direction of coiling of the snail may be traced back to the early segmentation stages of the egg, the spiral cleavage being opposite in direction in right and left handed forms. Genetic experiments of Boycott, Diver, Garstang and Turner (1930) made with the pond snail, *Limnaea peregra,* in which inversely coiled individuals were crossed with the normal, show that there is a cytoplasmic basis for the inverse symmetry, which is not changed immediately by nuclear influence, for in crossing eggs from sinistral snails by sperm from the dominant dextral form no effect is shown in the F_1 offspring, all being sinistral. In the following generation, however, the dominant gene functions and all of the offspring are dextral. Mendelian segregation in the ratio 3 dextral to 1 sinistral then occurs in the F_3 generation, but by broods instead of by individuals. Presumably during the

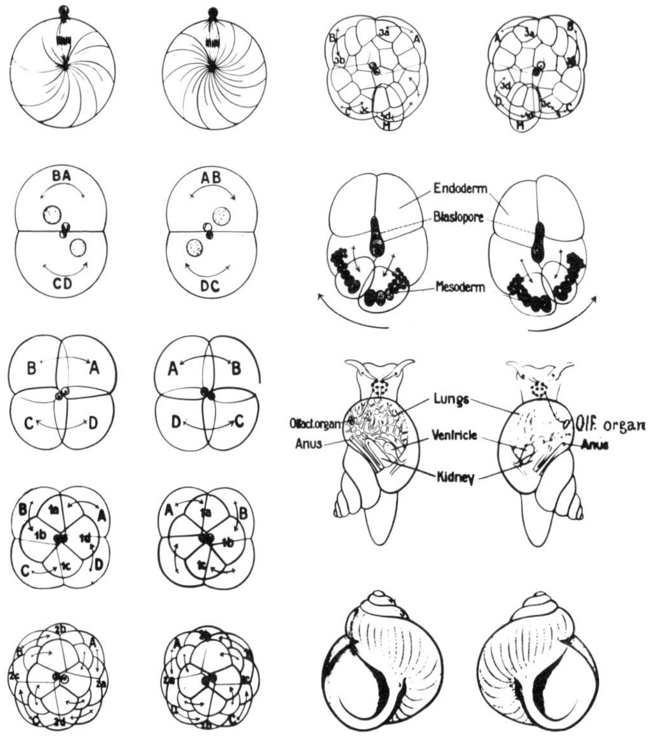

FIG. V–1. Inverse symmetry in the development of snails. After Conklin, 1903.

growth period of the egg, the dominant gene does bring about the inversion of the cytoplasmic configuration. An off-hand explanation of the phenomenon would be that the gene concerned is asymmetric and either catalyzes at different rates the formation of two enantiamorphs in the cytoplasm, or only one to the exclusion of the other. However, molecular inversion is not easy to bring about chemically, and with all of the evidence on the other side it cannot be assumed that inverse organisms are composed of molecularly inverse material (p. 199).

In bilateral organisms there are many paired organs, such as the limbs and the ear, which are asymmetric in themselves but are the mirror images of the corresponding organs on the opposite side. Reversal of the asymmetry of such structures can be readily induced experimentally in amphibian embryos of appropriate age (Harrison 1917, 1921, 1936).

It will not be possible here to muster the large array of facts bearing upon either of the two main topics under discussion, the problems of localization and of sym-

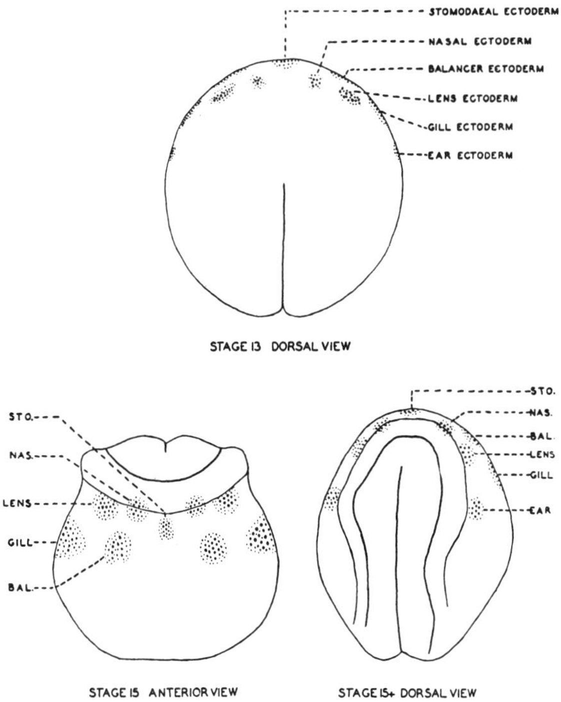

FIG. 1. The relative position of the presumptive ectodermal organ rudiments of the head in early neurula stages of *Amblystoma punctatum*. After Carpenter (1937).

metry. The earlier stages of development will not be considered in detail, but instead, attention is invited to the consideration of a single organ, the internal ear of the amphibian embryo, which both illustrates the principle of localization and shows significant relations of symmetry.

The differentiation of the ectodermal areas of the embryo is activated by the underlying living material known as the organizer. The neural plate develops only when in contact with this underlayer, or with other organizing material that may be substituted for it. Lying outside of the neural plate, particularly in the head, there are areas destined in normal development to become sense organs and a variety of other structures (Fig. 1).

There is another system of organ rudiments: pronephros, limb, muscle plates, etc., which are formed in the mesoderm, i.e. directly out of organizer material; and still another in the endoderm, part of which is also included in the organizer. The cells in each of these systems are assumed to be at first potentially alike and capable of differentiating in various directions (Fig. 13), the localization of their several potencies

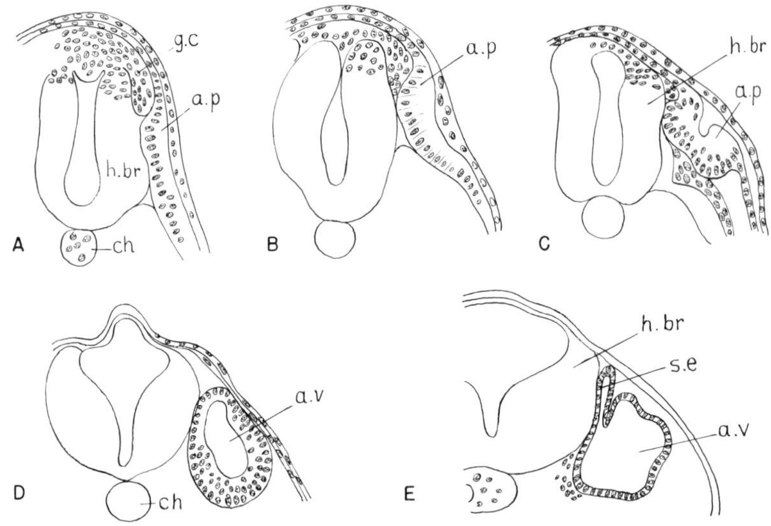

FIG. 2. Development of the ear vesicle from the ectodermal placode, shown in transverse sections. After Norris (1892). a.p., ear placode; a.v., ear vesicle; ch., notochord; n.c., neural crest; h.br., hind-brain; s.e., saccus endolymphaticus.

gradually becoming more and more restricted as development proceeds. There is a difference in the manner in which this takes place in the different layers. In the mesoderm and probably also in the endoderm it is largely an autogenous process, whereas in the ectoderm, as just stated, it is dependent upon the underlying layer.

In the case of the ear we are concerned particularly with the ectodermal localizations. What is it, for instance, that makes this organ form at a particular spot and the nose, the lens, the balancer and other structures at other spots definitely related topographically to the embryonic axis and to each other? This is but part of the general problem of localization, one, however, that lends itself readily to experimental analysis.

Before proceeding with this it will be appropriate to recall briefly the morphological changes that take place as the ear develops. First, there is a gradual marking out of the auditory placode or disk from the general ectoderm (Fig. 2A), as described in *Amblystoma* by Norris (1892), Kaan (1926) and Yntema (1933). The inner layer of this area of ectoderm sinks in, making a pit (Fig. 2B, C), which closes over to form a vesicle (Fig. 2D). The latter swells rapidly, its walls becoming generally thinner, with certain areas differentiating into sensory epithelium. Accompanying this the walls of the vesicle fold and become constricted in places (Fig. 2E). The gross changes are, roughly speaking, the transformation of a circular plate into a sphere, followed by unequal growth with consequent local constriction and bulging of its walls to form the semicircular canals and other chambers of the labyrinth.

It is possible from data given by Norris and Kaan to project approximately the definitive parts of the ear upon the circular plate. The endolymphatic duct and sac and the sensory areas of the saccule arise from the dorsal half of the disk, mainly from the dorsoposterior quadrant. The ventral half gives rise to the semicircular canals and the utricle.

The physico-chemical factors which underlie the transformation of the disk of ectoderm cells into the complicated labyrinth are unknown. The first changes, which consist in thickening followed by pitting in and folding of the original plate, involve an alteration of the shape of the cells similar to that occurring in the neural plate and the nasal pit, probably as a result of molecular rearrangement.[5] A certain degree of turgor in the otic vesicle seems to be necessary for this process, but if exceeded, as occurs after some operations, particularly in those cases in which the ductus and saccus endolymphaticus fail to develop, the resulting ear may swell and remain vesicular.

In considering the factors concerned with the localization and subsequent development of the ear, there are several possibilities that suggest themselves immediately. Its differentiation might be due either to intrinsic conditions already established in the ectoderm out of which it develops or to factors outside of this layer. The latter might be located in the underlying primary organizer or in a secondary inductor in the central nervous system, with which the ear placode comes into contact as the neural folds rise and close. Transplantation experiments with embryos in the stage when the medullary plate is barely outlined (Stages 13–14) show that from that time on all of these factors affect the end result and that the position at which the ear develops is also of consequence. In other words, the presumptive ear ectoderm is already predisposed to form an ear, probably through the early action of the underlying mesoderm, although the whole non-medullary ectoderm is still pluripotent with respect to the various structures that ordinarily develop from it. A certain spot on the underlying mesoderm can still initiate the formation of an ear from strange ectoderm in the absence of the medullary cord. On the other hand, a certain region in the medullary cord, centered at the level of Mauthner's cell, which is still not visibly

5. It was maintained by Glaser (1914) that the thickening of the neural plate is accompanied by the taking up of water. However, Brown, Hamburger and Schmitt (1941) conclude on the basis of density studies that there are but slight changes in volume or water content in the neural tissue during early development. They ascribe the changes involved in the formation of the neural tube to molecular surface action.

differentiated, also has the power to induce an ear vesicle when brought into contact with ectoderm from any region. However, the ears produced in this way by one agency alone, or even by two acting together or in sequence, are far from perfect, and the degree of development that they attain varies with the combination of factors involved. The best-developed cases are those obtained after simple removal of the underlying mesoderm, in which the predisposed ectoderm is acted upon by the neural cord, reinforced possibly by the small amount of mesoderm that moves in to fill the gap due to the operation. The most rudimentary are those in which the mesoderm from the ear region acts alone upon strange ectoderm in a position where no ear is normally formed.[6]

A brief description of some of the experiments will indicate how these conclusions have been reached. The inductive action of the hind-brain may readily be shown by transplanting it to other regions of the body, where it frequently induces a small and imperfect ear vesicle, usually with ductus endolymphaticus, from ectoderm at the

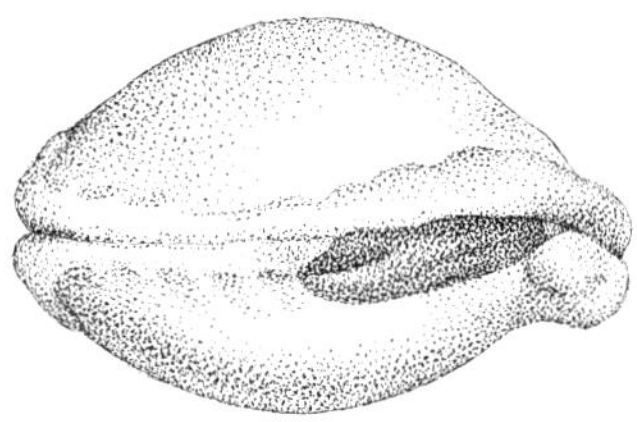

FIG. 3. Dorsal view of embryo with right hind-brain replaced by ventral ectoderm (darkly stippled area), from a donor stained in Nile blue, one day after operation. Exp. EMP40. × 10.

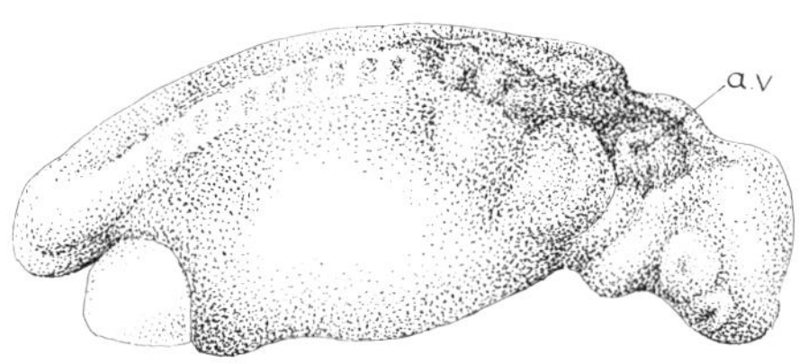

FIG. 4. Lateral view of same embryo two days later; the ear vesicle (a.v.), lying underneath the graft, showed in life as a blue spot. Exp. EMP40. × 10.

6. The inductive power of the mesoderm by itself is at this stage very weak and apparently less than that of the hind-brain. This is to correct the statement in the "Collecting Net" (p. 221) made before all of the present evidence was available.

location of the implant. The effect of the elimination of the hind-brain may be tested by removing it either unilaterally or bilaterally, after which the wound may be left or else covered by ectoderm from another part of the body (Fig. 3). After the unilateral operation, however, the missing half is reconstituted by the remaining portion (Fig. 4) and may act as an inductor if the ectoderm is brought into contact with it, though usually the reconstituted medulla lies below the surface, with mesenchyme between it and the induced ear (Fig. 5), in which case the mesoderm functions as inductor. Both ear and lateral line organs may be induced in this way. The bilateral

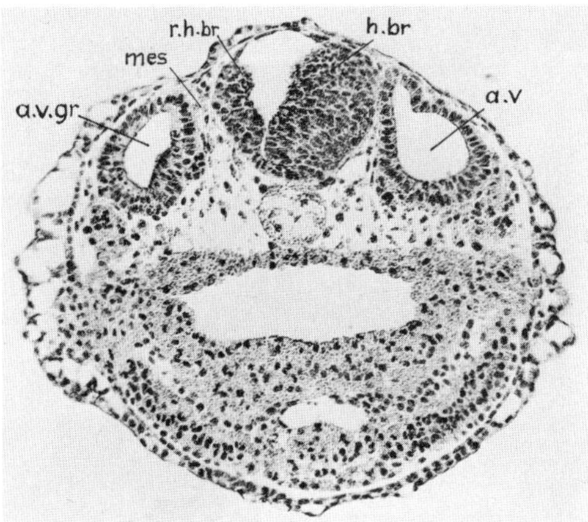

FIG. 5. Section of similar embryo five days after operation. The normal ear vesicle (a.v.) of the left side is on the right side of the figure; a.v.gr., induced ear vesicle from graft not so far advanced as the normal; h.br., hind-brain unoperated side; r.h.br., reconstituted hind-brain of operated side; mes., mesenchyme between ear and reconstituted hind-brain. Exp. EMP41. × 60.

operation yields a crucial experiment, for then there is no reconstituted hind-brain to introduce an element of doubt, and the underlying mesoderm is the only possible inductor present. When not only the prospective hind-brain but also the ear ectoderm is removed on both sides and the wound is covered by abdominal ectoderm, one ear vesicle develops on each side. When, however, the ear ectoderm is left intact and only the hind-brain portion of the neural plate is replaced by abdominal ectoderm, then two ear vesicles develop on each side (Fig. 6), one (a.v.gr.) induced in the graft through the action of the underlying mesoderm and the other (a.v.) from the ear ectoderm of the host by self-differentiation, possibly reinforced by the continued action of the underlayer. How definitely the area of ear-inducing mesoderm is circum-

TABLE 1. Results after Varying the Factors of the Complex That Influences Ear Development

No.	Exp.	Factors involved: Ect.	N.Pl.	Mes.	Locus	Resulting ear
1.	Norm. contr. Ear region	+	+	+	+	Normal
2.	IMA, IMMP h-tr. mes.	+	+	+	–	Complete or with slight defects; usually smaller
3.	IAxM tr-h AxM	+	+	–	+	Smaller, slightly vesicular or 2 separate vesicles
4.	IMP, IMMP tr-h, no mes.	+	–	+	+	Vesicular, thin sensory areas. Sac. end. usually absent, no duct
5.	EAV PNAV No mes.	–	+	+	+	Vesicular, sometimes nearly normal or defective, sac. end. usually absent
6.	IMA, IMMP h-tr, no mes.	+	+	–	–	Incomplete, irregular, smaller. Sac. end. present, duct usually
7.	AVPN AVML Mes.	+	–	+	–	Either absent or small vesicles without sac. end., usually without sensory areas
8.	IMP, IMMP h-tr. mes.	–	+	+	–	Small vesicles, often absent. Sac end. present, duct usually
9.	IMP, IMMP tr-h. mes.	+	–	–	+	Variable, usually small without sac. end. One case nearly all sensory areas
10.	PNAV with mes.	–	+	–	+	Small, without partitions, sometimes single sensory area. Sac. end. absent
11.	IMA, IMMP tr-h. no mes.	–	–	+	+	Defective vesicles. Some sensory areas. Sac. end. usually absent
12.	AVML, AVPN No mes.	+	–	–	–	Absent altogether, or small vesicles without sensory areas
13.	IMP, IMMP h-tr. no mes.	–	+	–	–	Small, occasionally absent. Some sensory areas. Sac. end. present
14.	IAxM h-tr	–	–	+	–	Usually no ear; some cases small round vesicles. No sac. end
15.	IMA, IMMP tr-h, mes.	–	–	–	+	Variable—small, lacking sensory areas. No sac. end. Host tissue involved
16.	Norm. contr. Non-auditory	–	–	–	–	No ear developed

Ectoderm of ear region (AV); neural plate (N.Pl) hindbrain level (HB); mesodermal substratum (Mes); normal ear locus. Ectodermal interchange with and without underlying mesoderm, head (h) and trunk (tr) levels: *IMP,* HB and SC (spinal cord, unilateral); *IMA,* HB with AV and SC with PN (pronephric region); *IMMP,* HB (bilateral) with and without AV and SC (bilateral) with and without PN; *AVPN, PNAV,* ear and pronephric region; *AVML,* ear and ventral (tr) midline ectoderm; *EAV,* AV removed; *AxM, IaxM,* interchange of mesodermal substratum alone (see Fig. 7). Harrison, unpublished.

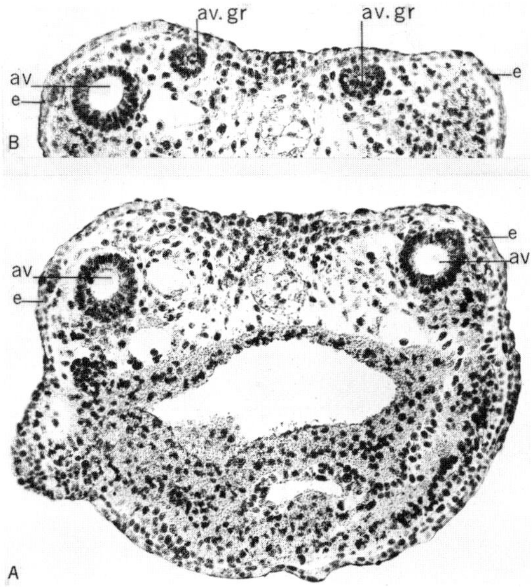

FIG. 6. Two sections through the ear region of an embryo from which the hind-brain has been completely removed (bilateral operation). A, showing the two ear vesicles (a.v.) derived directly from the ear placode of the embryo; the pigment is of the normal brown color; e, boundary of graft. B, dorsal part of similar section, 60 μ further forward, showing, along with one of the autogenous vesicles, two smaller ear vesicles (a.v.gr.) derived from the grafted ventral ectoderm through the inductive action of the mesoderm; the pigment in the small vesicles is deep blue in the sections from the Nile blue sulphate with which the donor embryo was stained. Exp. EMP165. × 60.

scribed is difficult to determine, since the direct experiment of transplanting the mesoderm alone usually fails and the few ears so induced are exceedingly small and imperfect.

In a second series of experiments, in which each factor enumerated above was varied both singly and in all combinations, the possible effect of a reconstituted hind-brain was eliminated by interchanging other parts of the central nervous system with the hind-brain (Fig. 7) instead of leaving a defect in the neural plate on one or both sides. The results confirm the conclusions reached from the first series.[7] [They are summarized in Table 1.]

7. These observations have been published only briefly and in part (Harrison 1935, 1938). The question of the induction of ears directly by the action of the underlying mesoderm, which is part of the primary organizer, has been previously raised by Holtfreter (1933, p. 645), and Dalcq (1933) postulates an induction center for the ear in this layer. However, in the large number of induced embryos, described by various observers, ears are always associated with a central nervous system. Szepsenwol (1933) has found in omphalocephalic chick embryos independent ear vesicles, the induction

From the above experiments the conclusion may be drawn that the differentiation of the ear is not a sudden action but one that is long drawn out and involves several different factors acting in succession.[8] Only imperfect ears arise when the factors work singly or even in combination, unless the whole complex is brought into play in an orderly manner.

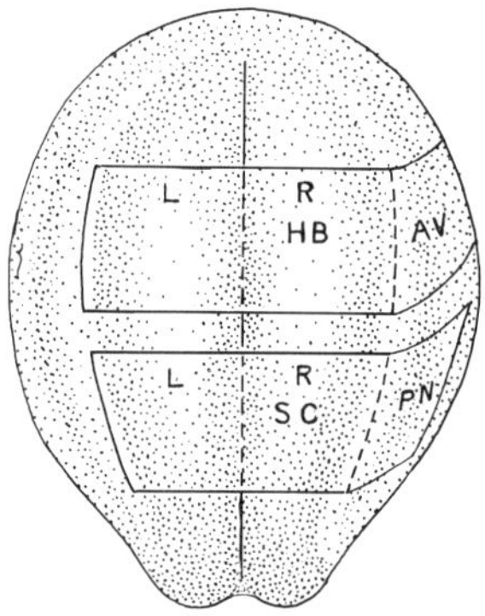

FIG. 7. Scheme of operation in the second series of experiments, in which tissues from the hind-brain level (HB) were interchanged with those from the spinal cord level (SC). The diagram shows an embryo (Stage 14) from the dorsal side. The solid lines outline two transverse strips of tissue corresponding to the regions just mentioned. Whole strips may be interchanged or any part of them, either with or without mesoderm. The broken lines indicate the boundaries within the strips, dividing the right and left halves of the hind-brain, and spinal cord from each other, as well as the right half of the central nervous system from the adjacent ectoderm. AV, ectoderm of ear region; PN, ectoderm of pronephric region.

The title of this lecture is misleading, for we live in an asymmetric world and we ourselves are asymmetric. We like to think of a symmetrical thing as an aesthetic ideal and the bodies of many organisms approach symmetry, at least in external form, even if they do not achieve it throughout their organization. For this residual asymmetry there is a fundamental reason.

Etymologically, symmetry means commensurability but it has come to have the general connotation of order, regularity, arrangement. Geometrically symmetry is

of which he ascribes to the action of the acoustic ganglion, which he considers to be derived from the neural crest. Since, however, the acoustic ganglion is itself derived from the ear placode, some other inducing agent must be sought. Ear vesicles develop normally in the absence of neural crest and there is no evidence that the latter has any effect on the development of the ear.

8. See Lehmann (1929, 1933), according to whose terminology the ear would be a "combinative Einheitsleistung."

defined by certain operations of rotation, reflection, inversion, and translation (helicoidal symmetry) which need not be gone into here.

Organisms, like crystals, which are the best known symmetrical objects found in nature, may be referred for description, as we have already seen, to three axes and the width of the angles between each pair, the degrees of symmetry vary greatly—from the triclinic system, where axes and angles are all unequal, to the regular or cubic, where they are all equal.

In crystallography, interest has shifted from external form to the internal arrangement of atoms and molecules. This arrangement is a regular one and forms the kind of pattern in space known as a space-lattice, which is a point system such that the arrangement of points around any single point is the same as that around any other.

X-ray analysis shows that many substances formerly considered colloid have a crystalline arrangement. This includes such organic structures as hair, horn, muscle, tendon, etc. Also, a paracrystalline state in which there is mobility of liquids in one or two directions combined with a definite configuration is now recognized. Such arrangements are found in the tissues of living organisms. Bilateral symmetry (reflection in a median plane) is the commonest form in animals but it is never perfect. There are many other types of symmetry in animals and plants but some of them, as for instance five-fold radial symmetry, cannot be referred to any kind of a space lattice.

Organisms, as just stated, give evidence from X-ray analysis, of being made up of substances (proteins, cellulose, lipids, etc.) with repeat patterns. There is much indirect evidence for this type of organization from regeneration and experimental embryology. Further evidence will be given in the experiments to be described today. [Lecture V notes.]

Preceding and to some extent accompanying the changes just described are certain more recondite changes of orientation and polarization which involve the microstructure of the material composing the ear rudiment. They are revealed by the effect on the asymmetry of the ear produced by rotating the ear plate when grafting.

In considering these experiments it is convenient to refer the ear placode to three arbitrarily chosen axes, one of which runs through its center perpendicular to the surface. The other two lie on the surface and are designated the antero-posterior and the dorso-ventral axes respectively.[9] A square plate of ectoderm concentric with

9. The surface of the ear placode does not run strictly dorso-ventrally but this designation of the axis seems to be the most appropriate one practically and corresponds to its direction as seen when operating.

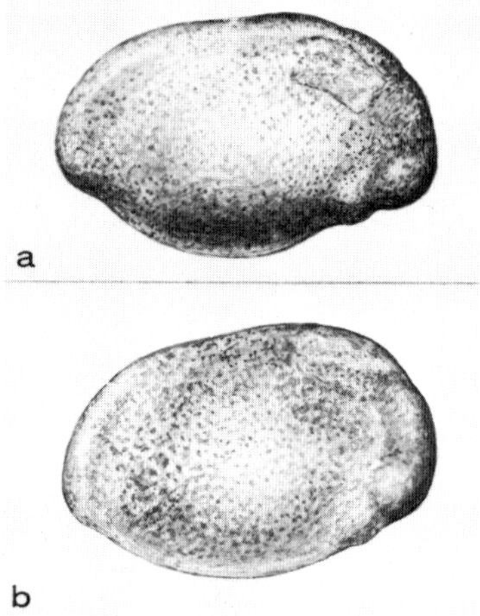

FIGS. V–2 and V–3. Transplantation of ear ectoderm (*A. punctatum,* stage 15) between unstained (a, a′) and vitally stained (b, b′) embryos. Fig. 2 shows ear region at stage 20; Fig. 3 at stage 35. Harrison, unpublished.

the ear placode is lifted and replaced in some particular orientation by a similar piece of ectoderm from another embryo, one of the two having been stained *intra vitam* in Nile blue sulphate [Figs. V–2 and V–3]. With rotations of 0° or 180° there are four possible ways of placing the grafts (Fig. 8). The plate from the same side may be grafted upside down or right side up, thus reversing both the axes at the same time or neither of them. It may be taken from the opposite side, in which case, ac-

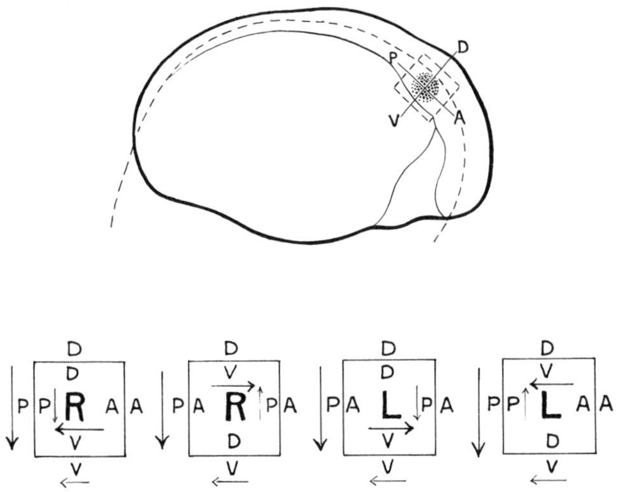

FIG. 8. Scheme of operations to test the effect of changed orientation. A, Profile of an embryo in the stage of monaxial polarization of the ear placode (Stage 21) is shown above. The curved broken line represents the embryonic axis; the stippled area, the position of the ear placode not yet differentiated; the square area (broken lines), the graft. The four squares below show the position of the axes in the four orientations, the operations being represented as on the right side of the embryo. R, graft from the right side; L, graft from the left side; A, anterior; P, posterior; D, dorsal; V, ventral. The letters inside the squares give the orientation of the graft; those outside, the direction of the cardinal points of the embryo. The arrows denote the intensity and direction of the vectorial factors acting to determine the asymmetry of the ear in the second period of axial polarization; thus the first and fourth arrangements give normal ears, the second and third disharmonic ears (left on the right side).

cording to its orientation, either the dorso-ventral or the antero-posterior axis is reversed, the remaining one not. The conclusions are based upon a study of nearly eight hundred individual cases.

The ectoderm of the ear region and surrounding area is at first isotropic about the axis perpendicular to the surface. The inner surface is differentiated from the outer, but, according to Luther (1934), in stages slightly younger than those used here a graft may even be placed inside out without changing the course of development.[10] Just as the neural folds are closing a change supervenes. The ear plate, which is still not visibly differentiated from the surrounding ectoderm, becomes polarized with respect to its antero-posterior axis.

10. Luther does not mention the ear specifically but does refer to lens, stomodeum and cilia.

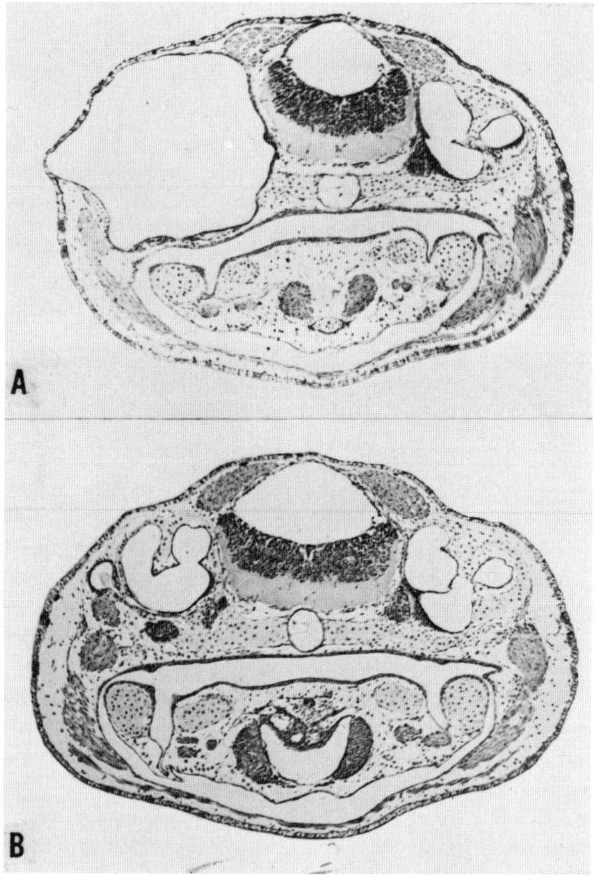

FIG. V–4. A. Photomicrograph through the ear region of a larva (*A. punctatum*); large vesicular ear on the right side, a.p. and d.v. axes reversed at stage 24 (viewed on the left).

B. Well-developed reversed ear lying on its side after inversion of the a.p. axis of the ear placode at stage 29, right side of embryo. Harrison, unpublished.

In the isotropic period all orientations yield normal ears, with few exceptions.[11] In the period of monaxial polarization, which lasts for some hours after closure of the neural folds, two of the orientations—grafts which are oriented normally and those from the opposite side with antero-posterior axis normal but dorso-ventral inverted —yield ears with proper asymmetry, while after the other two orientations ears de-

11. The exceptions indicate that even at this early period there is a polarization of the antero-posterior axis which must, however, be readily reversible. Both Yntema (1939) and Hall (1939) have found additional evidence that this is the case.

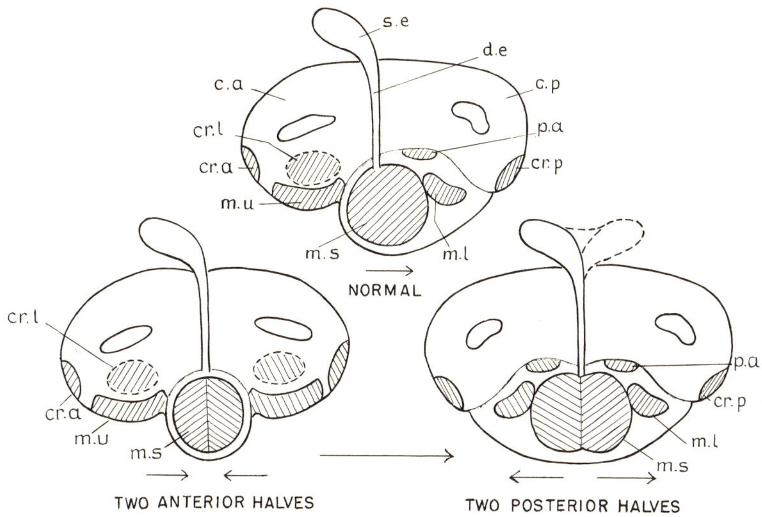

FIG. 9. Normal right labyrinth and the two types of enantiomorphic twins, medial surface. The large arrow points toward the posterior end of the embryo; the small arrows point posteriorly with respect to the labyrinth or half labyrinth to which they refer. c.a., anterior semicircular canal; c.p., posterior canal; cr.a. and cr.p., the corresponding sensory cristae; cr.l., lateral crista, projected from the ventrolateral wall but not visible in this view; d.e., ductus endolymphaticus; m.l., macula of lagena; m.s., macula of sacculus; m.u., macula of utriculus; p.a., papilla amphibiorum; s.e., saccus endolymphaticus.

velop which are disharmonic in the sense that they are of the side opposite to that on which they are placed, e.g. left ear on the right side (Fig. 8). This occurs irrespective of the side of origin of the graft. A little later conditions again change, indicating a transition from an indifferent dorso-ventral axis to one which is definitely polarized. When this axis is normally placed, a graft from the right side always develops into a right, and one from the left into a left, but when the graft is placed with dorso-ventral inversion, a marked disturbance arises in many cases. The vesicle swells excessively, sometimes with almost complete suppression of partitions and sensory areas [Fig. V–4A]. Finally, when the vesicle is about to be constricted off from the ectoderm, a stage is reached in which the ear when transplanted develops essentially as placed.[12] [Fig. V–4B.]

During the isotropic period and the period of monaxial polarization, the ear rudiment is an equipotential system,[13] by which is meant that any part of the rudiment can give rise to any part of the ear. This is shown by the following experiments: a

12. In this connection see experiments by Spemann (1906a, 1906b, 1910), Streeter (1907, 1914), Ogawa (1921).

13. Driesch (1899, p. 72; 1905, p. 679; 1929, p. 88); Harrison (1921, p. 87).

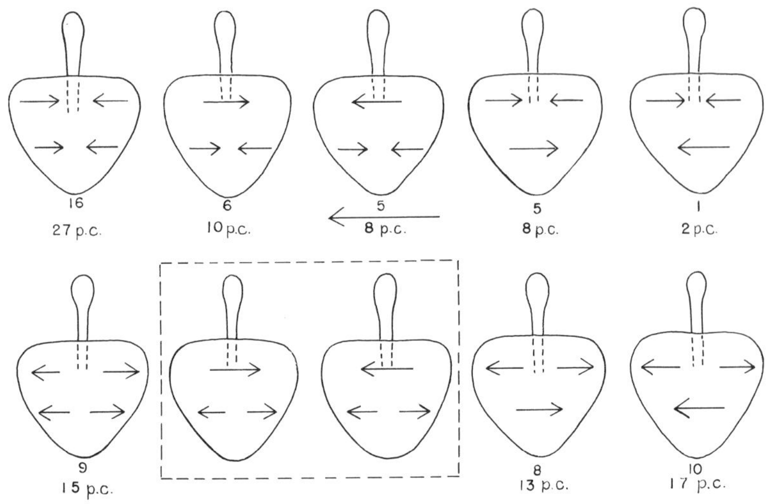

FIG. 10. Types of twinned ears found in the experiments; upper row, two anterior portions; lower row, two posterior. The large arrow outside the figures points toward the posterior end of the embryo. Within each figure the arrows point posteriorly with respect to the parts in which they lie; the upper arrows concern the saccule and utricle; the lower arrows the semicircular canals. Total number of cases, 60. Number in each class and percentage of total number given under each figure.

fractional part of the plate may give rise to a whole ear; two ear plates may be fused together to form a single one; and a right ear may develop from a graft taken from the left side. This forces the conclusion that the vectorial properties of the organ are represented in the ultimate structural units of the rudiment, and not by a mosaic of coarser grain.[14]

Grafts made during the first critical stage, i.e. when the antero-posterior axis is becoming polarized, frequently result in enantiomorphic twin ears (Fig. 9). These are always mirrored across a transverse plane and may consist of either two anterior or two posterior halves. Partial twinning, involving either the semicircular canals, which develop from the ventral half of the ear plate, or the saccule, which develops from the dorsal half, also occurs. There are ten possible classes of these twins, of which eight have been actually observed (Fig. 10). In these experiments twinning never takes place across a longitudinal plane, i.e. on the dorso-ventral axis. Such twins can be produced imperfectly, however, by grafting a second hind-brain on the outer or lower side of the ear placode at a sufficiently early stage. The latter is then

14. Driesch (1899, p. 44; 1908).

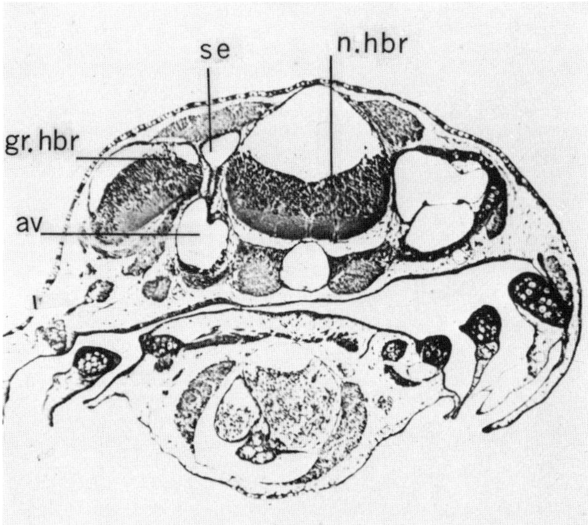

FIG. 11. Photomicrograph of section through the ear region of an embryo with a supernumerary hind-brain grafted on the ventro-lateral side of the ear plate. a.v., ear vesicle, developed between the normal hind-brain (n.h.br.) and the graft (gr.h.br.); s.e., saccus endolymphaticus, developed under the inductive action of the two hind-brains. ×33.

under the influence of two hind-brains, acting in opposite directions. The ears which develop under these circumstances are irregular and badly distorted, since it is difficult to gauge the proper amount of ectoderm to leave between the two hind-brains. A number of cases with double endolymphatic apparatus have, however, been obtained (Fig. 11).

Prior to the experiments with the ear, I had made a large number of experiments with the forelimbs of the salamander, which gave results comparable to those of the ear experiments. The limb is quite a different system from the ear (see Normal Stages 21–46 and Fig. V–5). It consists of a thickening of the mesoderm of the body wall which is covered by ectoderm—the outer (epidermal) layer which gives rise to the ear. The mesoderm is the more important constituent. There are other differences between the ear and limb, however, since the latter has not yet been found in an isotropic condition. The earliest stages investigated show an antero-posterior polarization. At this time and for some time later, the parts of the limb bud, as the rudiment is called, are equipotential, as shown by tests similar to those made with the ear.

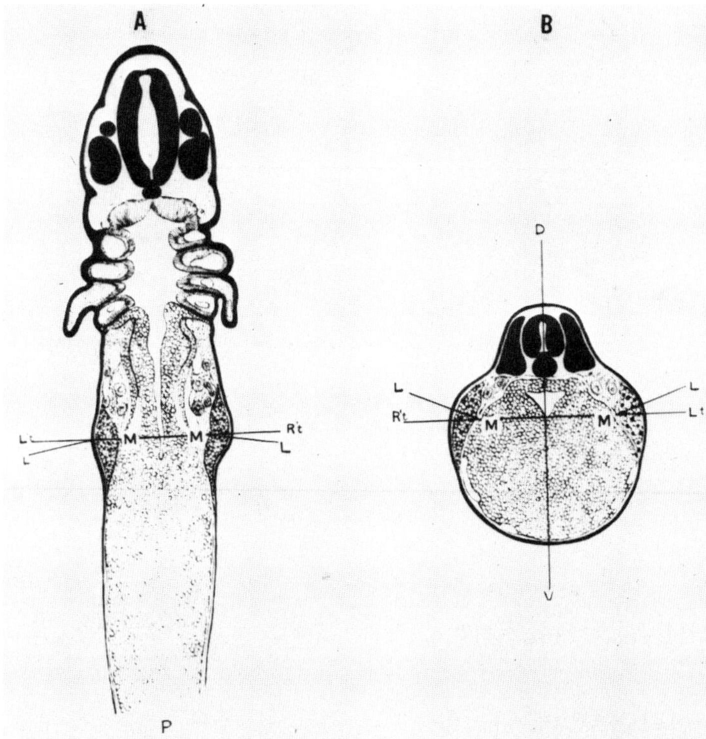

FIG. V–5. A. Frontal section through the forelimb of an Amblystoma embryo showing the position of the limb bud in the body wall. B. The limb bud in cross section of a similar embryo (stage 36). Axial relations of the limb buds are indicated. After Harrison, 1925.

The salient events in the development of a limb are the following: thickening of the body wall in a circumscript area below the third, fourth, and fifth body segments, the budding of this rudiment and its pointing in a posterodorsal direction, bifurcation at the end forming the first two digits; then the elbow joint is formed, the third and fourth digits (ulnar) are produced on the upper border of the hand by budding. This is all completed by the time the yolk in the intestine is absorbed and the animal begins to feed. The limb is then fully functional.

The operations which are similar to those with the ear are described in the diagram shown in Fig. V–6. The results are complicated, however, by the frequency of reduplication which occurs usually by budding instead of by splitting of the original rudiment as in the ear (Fig. V–7). It is thus possible to distinguish the primary from secondary, tertiary, and even quaternary buds. The cases must

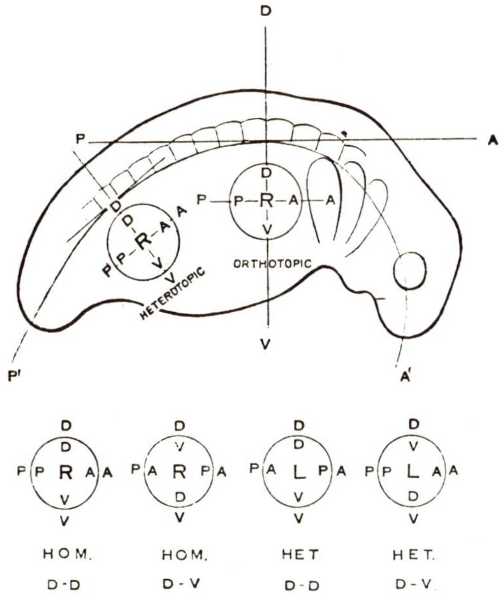

FIG. V–6. Diagram delineating an Amblystoma embryo, stage 29. The inner circles represent the limb bud, in the normal (orthotopic) and the abnormal (heterotopic) location. The four circles below represent the four different ways in which the limb buds may be oriented with reference to the cardinal points of the embryo; the letters (D, dorsal; V, ventral; A, anterior; and P, posterior) within the circles designate the original cardinal points of the transplanted limb, those outside the corresponding points of the embryo. The operations are represented as on the right side. R, right limb bud; L, left limb bud; hom., homopleural; het., heteropleural.

The geometrical axes used in orienting the graft are respectively the tangent and the perpendicular to the curved axis where the plane of the dorsoventral axis of the limb bud cuts it. After Harrison, 1925.

be observed frequently in order to be sure which is which, since otherwise accurate interpretation may become impossible. When this is done, it is found that the same rules obtain as in the ear, in its second stage of polarization, i.e., (1) when the antero-posterior axis is reversed in grafting, the limb that develops is disharmonic; that is, of the opposite side, no matter whether originally derived from the same or the opposite side (Fig. V–8); (2) when the antero-posterior axis is placed in normal orientation at the time of grafting, the limb that develops is proper to the side on which it is placed, no matter whether derived from that side or from the opposite side; (3) when double limbs arise, the original member, the one first to begin its development, follows either rule one or two, depending on the orienta-

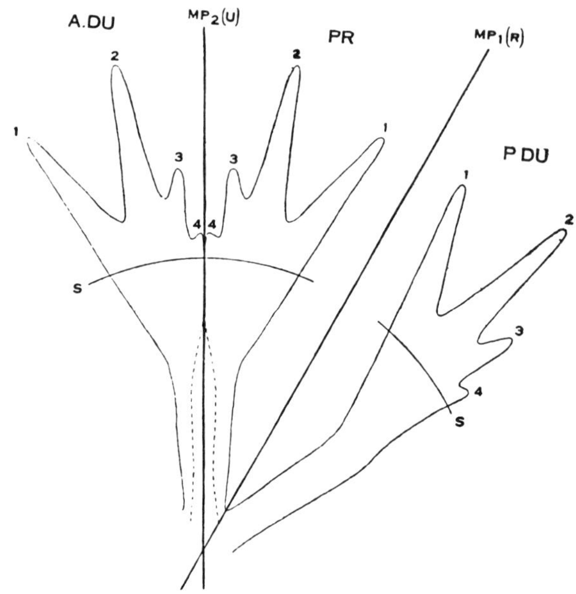

FIG. V–7. Diagram showing mode of reduplication. PR, primary limb; P, DU posterior reduplicating member; A, DU, anterior reduplicating member; MP_1 (R), primary (radial) mirror plane; MP_2 (U), secondary (ulnar) mirror plane; 1 to 4, first to fourth digits, respectively. Dotted lines show the outlines of limbs as they would have been had there been no coalescence. After Harrison, 1921.

tion of the graft, while the secondary member is the mirror image of the first (Fig. V–7). Tertiary members mirror the primary or the secondary limbs, according as they are nearer to one limb or the other.

The behavior of some of the reduplicating limbs is tricky, particularly when the primary starts to develop into a disharmonic appendage. The reduplicating member may get an early start and grow much more vigorously so that the primary member becomes a mere stump or rudimentary appendage on the secondary. The effect of this is regulatory and in certain cases the primary member may disappear entirely and the animal comes out of its experience with a perfectly normal limb. Other cases may be regulated, where possible, by rotation; both of these adjustments are adaptive.

Double and triple limbs, such as the reduplicated limbs produced in these experiments, occur not infrequently in nature; Bateson collected many instances in insects, crustaceans, and vertebrates. They are probably produced by regeneration after injury.

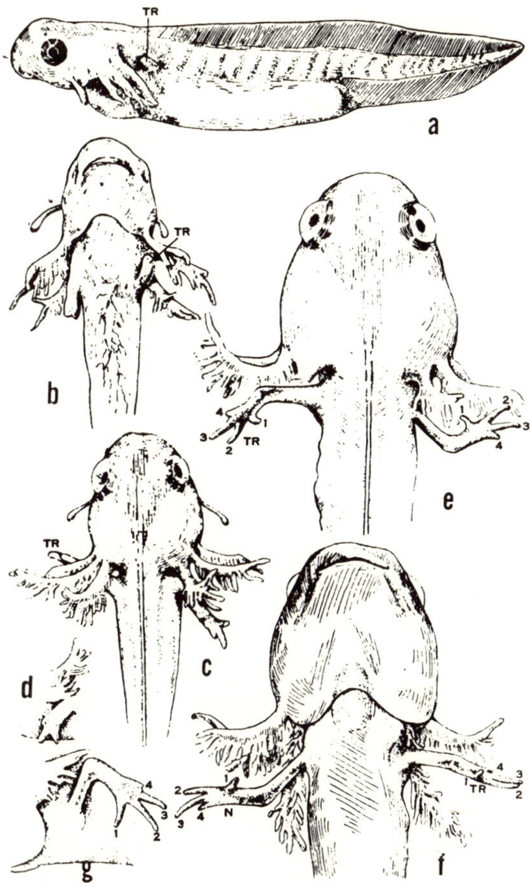

FIG. V–8. Orthotopic transplantation; left limb to left side inverted (hom. dv.), resulting in a normal right limb on the operated side. TR, transplanted limb.

a. Lateral view, five days after operation.
b. Ventral view, ten days after operation.
c. Dorsal view, sixteen days after operation; transplanted limb covered by gills.
d. Lateral view of limb, sixteen days after operation.
e. Dorsal view, twenty-three days after operation.
f. Ventral view, twenty-three days after operation.
g. Lateral view, twenty-three days after operation. After Harrison, 1921.

Nicholas (1922, 1924) showed that it is not the organism as a whole which influences the posture of the limb but the tissue immediately surrounding it. This is proved by a double operation in which a ring of tissue is placed around the grafted limb bud or oriented in different ways. Swett (1928–1940) has confirmed this and shown that it is the tissue dorsal to the limb that is active in producing this result. He has also shown that the dorso-ventral axis becomes polarized at Stage 34–35, just after the first movements of the body begin and the medio-lateral axis is fixed at about Stage 37.

The significance of all this detail, which may seem trivial, is that it gives crucial evidence that these two very different systems, ear and forelimb are made up of units arranged in a repeat pattern which gradually become more and more definitely oriented. [Lecture V notes.]

The facts just assembled and many others that might be added necessitate the assumption that the processes of development leading up to the final form of the organism are but the expression of the forces involved in the molecular configuration of the egg protoplasm. Yet it seems idle to speculate about this when students of the living cell cannot agree even as to the physical state of protoplasm—whether it is a fluid or is more of the nature of a lyophilic colloidal gel. Microdissectionists and most observers of the streaming of the protoplasm of the plant cell describe the living substance as an emulsion with a liquid dispersion phase. On the other hand it has been observed to exhibit elastic properties,[15] and observations by Harvey and Marsland (1932) on the amoeba subjected to high centrifugal force indicate that visible particles are moved through the protoplasm in a jerky way, as if they were bumping against solid obstacles. Again, Moore's (1933, 1935) experiments in forcing myxomycete plasmodium through filters of various degrees of fineness show that even this organism, which is about the most amorphous imaginable, is dependent for its existence upon the structural integrity of its protoplasm. It has been shown that many unsegmented eggs, when subjected to strong centrifrugal force, still develop normally even though visible particles may be quite definitely stratified in an abnormal way. Particularly striking are the experiments of Conklin (1917) in centrifuging the egg of Crepidula. The maturation spindle may be stretched from its anchorage near the pole, so that very large polar bodies are given off. On release from the centrifugal force the protoplasm returns to its original form and development proceeds normally. It seems quite impossible to explain such phenomena as occurring in a purely fluid system. Biochemists are recognizing more and more the importance of molecular configuration and orientation in the organization of protoplasm.[16] The

15. Seifriz (1936, p. 263), Scarth (1927).
16. Hardy (1934), Peters (1929), Jordan Lloyd (1932), Needham (1936) and others.
Although the avenue of approach is different, the views expressed in the present paper fall in line with the micellar theory of Nägeli (1858–1884, edited by Frey, 1928). After years of disfavor this theory has again come into good standing, supported by much new evidence on the physico-chem-

work on the chemical side may now be woven in with the results of X-ray analysis and other newer methods of study to give a picture, however imperfect it may be, of the finer structure of living matter.

Living protoplasm is a complex system, consisting of water, electrolytes, proteins, and other organic compounds—enzymes, carbohydrates, and lipoids. There are many who think that proteins, as such, do not exist in protoplasm and that in studying them we are only dealing with dead material. But organic chemistry leans heavily on the assumption that the molecular structure of substances is revealed by the character of their decomposition products. Who will say that we are not justified in applying this same principle to the study of living matter? Proteins are certainly the most specific constituents of protoplasm and the ones most likely to account for the innumerable varieties or species in which the living substance occurs, although specific carbohydrates and lipoids have been shown to occur and are probably of wide distribution.[17] A consideration of the nature of the protein molecule should therefore prove instructive.

Proteins, though of wide diversity, have certain fundamental features in common. As determined by their decomposition products and chemical synthesis, as far as this has been accomplished, they are composed of chains of *l*-α-amino acids joined in peptide linkage, each segment or link consisting of a nitrogen and two carbon atoms. These constitute the stem or backbone of the repeat design. One of the carbons has ketone oxygen linked with it, the remainder of the carboxyl group of the amino acid. The other carbon has attached to it, as a sort of appendage or side chain, the amino acid residue. Many of these segments are added together end to end in the pro-

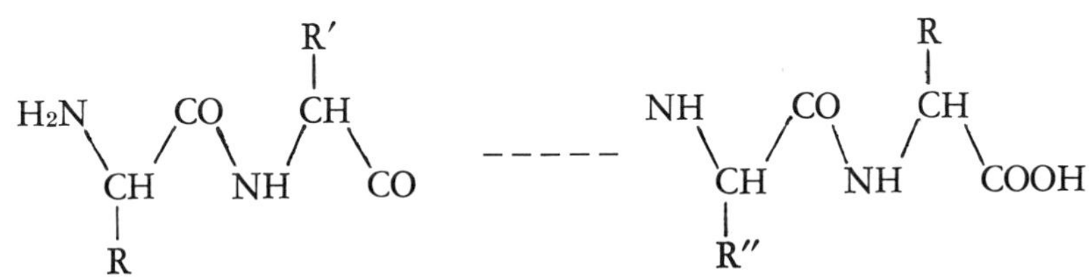

ical basis of organic form. Recently a number of general articles on this subject have appeared, of which reference is made to the following: Koltzoff (1928), Frey-Wyssling (1938), Schmidt (1939), Bernal (1940), Freundlich (1940), Sponsler (1940), Schmitt (1941), Moyer (1942), Sponsler and Bath (1942); also to various papers in Cold Spring Harbor Symposia on Quantitative Biology, Vol. VI, 1938.

17. For the carbohydrates this was first shown in the case of the polysaccharide capsules of the different types of Pneumococcus by Heidelberger and Avery (1923, 1924), Avery and Heidelberger (1923) and Heidelberger, Goebel and Avery (1925). See Landsteiner (1936).

tein molecule, into the composition of which a number of different amino acids usually enter, so that the side chains are of different kinds. The chain may be extended almost indefinitely, according to the number of amino acid molecules which are linked together.

While the protein molecule is fundamentally a long chain, Svedberg's (1934) determinations of molecular weights and dimensions of a variety of proteins by means of the ultracentrifuge show it to be globular or oval or, in some cases, rod-like (myosin) with molecular weights in multiples or sometimes in submultiples of 34,500. The shortening of the wool fiber and the assumption of globular form by many protein molecules might be explained in part by the tendency of the simple peptide chain, when not subject to tension, to close up into a series of hexagonal rings, derivatives of diketopiperazine.[18]

A further transformation by a compact three dimensional folding would lead to an ellipsoidal or a globular form. Wrinch (1936, 1937a, 1937b), making use of a hexagonal folding of this type, has described a system of "cyclols" which would include the arrangements both in monomolecular films and in the three dimensional "globular" proteins.[19]

Certain proteins such as keratin, collagen and myosin tend to assume fibrous form in the organism, and it may now be regarded as certain that this is due to the orientation of their elongated molecules end to end in parallel arrangement. According to Edsall and von Muralt (1930) myosin, the globulin of muscle, shows ordinarily no orientation when extracted in aqueous solution, but if forced through a capillary tube or subjected to shearing stresses between cylindrical walls one of which is rotating, its particles become oriented, as shown by the birefringence of the solution (p. 204). In a similar way the orientation of the fibroin molecules in the silk fiber may be accounted for.

X-ray diffraction pictures of the above protein structures reveal a definite spacing of the molecular groups that accord well with their molecular structure as determined by chemical methods. Astbury and collaborators (1930–1937), who have made many important contributions to the subject, have shown that the spacing is differ-

18. Originally suggested by Frank in 1933 in discussing a lecture by Astbury. See Frank (1936), Astbury (1936b, p. 781). According to Jordan Lloyd and Shore (1938, p. 261), however, the balance of evidence is against diketopiperazine rings as units in the protein molecule. Other patterns of folding, as for instance by the establishment of hydrogen bonds, have been considered by Wrinch and Jordan Lloyd (1936), Mirsky and Pauling (1936), and by Astbury (1940).

19. Wrinch has elaborated this theory in numerous subsequent publications and has received strong support from Langmuir.

ent in the keratin of the wool fiber according as it is unstretched (α-keratin) or stretched (β-keratin).

The dipole character of the protein molecule[20] must also be taken into account. At one end of the amino acid chain there is an uncombined carboxyl or acid group, while at the other there is an amino group which is alkaline. Moreover, many of the amino acids themselves have an unequal number of basic and acid groups, giving basic or acid side chains in the proteins into which they enter. As electrolytes proteins are therefore amphoteric, dissociating as acids or bases according to the hydrogen ion concentration of the medium. At the isoelectric point they form double or *Zwitterions*. Thus, according to Jordan Lloyd (1932, p. 258), the protein molecule "has a backbone, which is chemically stable, and limbs which have possibilities of chemical activity . . . the backbones of the protein molecules are not highly solvated but the limbs and the head and tail may be, in fact generally are, in the circumstances we are considering here. This ingenious arrangement gives physical stability with chemical activity, which is what is wanted in protoplasm." Indeed, one might say with Pangloss that in this best of all possible worlds the protein molecule was designed just to form living protoplasm.

In the cell such molecules would tend to become oriented in a definite way, particularly at interfaces. Under certain conditions they might either form fibers with parallel arrangement or spread out in films, though usually, it may be assumed, they would keep a more compact or folded form, in which they would likewise tend to assume a definite paracrystalline configuration, involving a large amount of water of crystallization. Such a structure would be very tenuous and might under certain conditions go over into the fluid state, only to resume its structural form when intracellular conditions again change.

Considering protein molecules with reference to the developing egg, it is necessary to go back to the early growth period of the oöcyte to relate their form and arrangement to the configuration of the future embryo, particularly to the qualities of polarity, superficial bilaterality, and fundamental asymmetry. In one polar field of the egg, such molecules may be supposed to be anchored with one end at the surface of the cell and around the other pole to be fixed by the opposite end. They would run in nearly parallel rows from one polar field to the other all oriented in approximately the same direction (Fig. 12), or they might have a less regular arrangement in the intermediate zone. The figure reduces the arrangement to very simple terms and leaves out many complicating features, e.g. the cortical layer of cytoplasm in which a mo-

20. Cohn (1935).

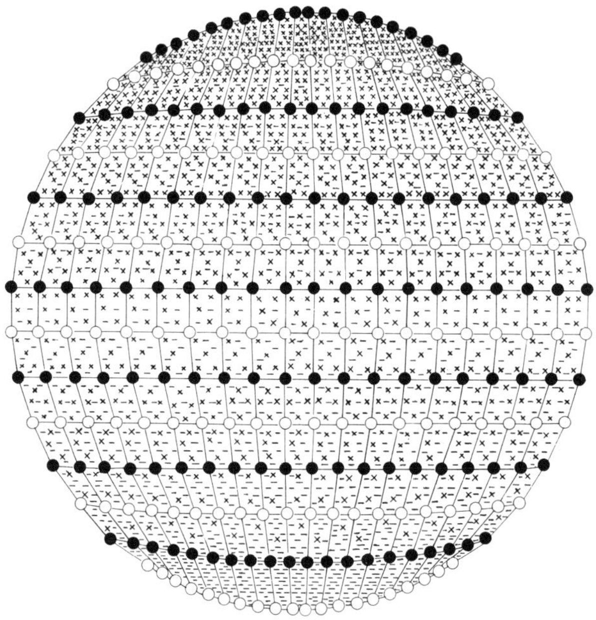

FIG. 12. Diagram of ovum, illustrating the two factors in polarity. There is a lattice made up of black and white circles oriented toward the two poles, and constituting the polarity of direction (Richtungspolarität). In the interstices of the lattice are particles of two kinds designated by × and —, one of which is more concentrated at the animal pole and the other at the vegetative pole, and indicating the polarity of stratification (Schichtungspolarität), in accordance with the Runnström-Hörstadius theory of the sea urchin egg.

lecular orientation with reference to the egg surface is postulated. If the two poles are 180° apart, and no further orientation occurs, the gross figure is symmetrical about an axis and has an infinite number of planes of symmetry. Practically all eggs are bipolar, and the poles, in many cases at least, have a definite relation to the attachment of the egg in the ovary. This fact of being attached must not, however, be assumed to create the fundamental polarity of the egg but simply to orient its constituent particles.

In order to pass to a bilaterally symmetrical structure with only one plane of symmetry all that has to take place macroscopically is a relative shift of the two poles so that they are no longer 180° apart. However, it may be assumed that the molecules are oriented with respect to their surfaces as well as end to end, thus giving the dorso-ventral and the right-left axes of the egg. The freedom of motion in the three axes may differ in degree and with it the relative fixity of the axial polarities. In certain plant eggs, such as those of Fucus, the brown rockweed, polarity may be readily controlled by external factors such as light, electric current, hydrogen ion concentration and position relative to other eggs.[21] In such cases molecular orientation may be assumed to occur in response to the external stimuli.

During the growth period of the egg a large amount of material is transferred to it and is laid down partly in the form of what is frequently called "ground substance" and partly in the form of inclusions. There seem to be regional differences in this ground cytoplasm either in substance or in arrangement along the main axis of the egg, thus constituting a gradient, which becomes more pronounced after fertilization. The inclusions are such bodies as pigment granules, oil droplets, yolk, mitochondria, etc., and they also are localized along the axis of the cell or with reference to the cell surface, forming gradients of materials. In other words there is a primary polarity (*Richtungspolarität*) due to the orientation of dipole molecules, and a polarity of stratification (*Schichtungspolarität*) which may involve the ground substance as well as the visible inclusions.[22] The normal development of egg fragments of molluscs and annelids after strong centrifugation which separates or removes the inclusions[23] indicates that the vectorial qualities of the organism are represented in the ground substance.

The polarization of the dorso-ventral axis, here considered as due primarily to molecular orientation, is likewise associated with localization of material. While this usually does not become visible until after fertilizaton, as in the appearance of the gray crescent of the amphibian egg, there is evidence from experiments that it may occur earlier, just as the primary egg polarity is set up in the ovary. With this the position of the median plane of the embryo is established, though under certain

21. Early observations by Rosenvinge (1888), who showed that light had an effect on the polarity of the Fucus egg, were followed by studies of Winkler (1900) on Cystosira, Küster (1906), Kniep (1907), Lund (1923). Whitaker (1931, 1935), Whitaker and Lowrance (1936) and others have confirmed the observations and greatly extended our knowledge.

22. These useful terms were introduced by Przibram (1913, p. 35), and correspond to the "Richtungsbau und Stoffbau des Eies" of Driesch (1894, p. 100). The historical foundation of these concepts would require a special essay. So far as they are based upon the data of experimental embryology the reader is referred, among others, to Driesch (1894, 1900, pp. 385, 407; 1908), Boveri (1901), Wilson (1903, 1904, 1928), Lillie (1906, 1909), Conklin (1917, 1931), Hörstadius (1935, 1937).

23. Cf. Conklin (1917), Wilson (1929).

conditions the dorso-ventral axis may be reversed[24] or it may be rotated around the primary egg axis and the median plane of the embryo thereby shifted.[25]

Reversal of the primary polarity of the amphibian egg has been brought about by holding the egg in an inverted position during early development [p. 20]. However, this involves mass shifting of the yolk under gravitational force, as Born (1885) first pointed out. The gray crescent, perhaps the whole of the cortical layer, is not shifted and since the dorsal blastoporic lip and with it the embryonic axes are fixed according to the topographic relation of the gray crescent to the yolk mass,[26] the embryo is inverted. Molecular reorientation, according to the hypothesis here adopted, is probably also involved in the adjustment to the imposed conditions. That these adjustments are not readily made is shown by the large number of abnormalities encountered in the experiments.

Hörstadius (1935, p. 404) has succeeded in reversing the polarity of the main axis of the sea urchin egg by the implantation of vegetative material (micromeres) at the pole of the animal half egg. Here too the results are variable, indicating a delicate balance between the polar forces of the material of the egg and those of the graft.

Asymmetry in the third dimension (right-left), resulting in a completely asymmetric organism, as the final expression of the molecular asymmetry of the egg protoplasm, does not usually show itself until relatively late in development, as for instance, in the vertebrate embryo when the heart begins to twist, although indirect evidence of it has been found at much earlier stages. In eggs with spiral cleavage (molluscs, annelids), however, the asymmetry is visible in the egg just before segmentation, being marked by the direction of the first cleavage spindle. So far as

24. According to Hörstadius and Wolsky (1936, p. 76) sea urchin embryos divided in half along the primary egg axis preserve their original axial relations except that dorsal halves show a reversal of the dorso-ventral axis. Pease (1941, 1942a, 1942b) has recently shown that the dorso-ventral axis of the eggs of the sand dollar, Dendraster, may be fixed in relation to chemical gradients in the surrounding medium.

25. The study of these relations in the amphibian embryo goes back to the beginning of experimental embryology, at which time the relations of gravity and the sperm path to the embryonic axes were much under discussion. Newport (1854) and later Roux (1885a, 1887) claimed to have shown by applying sperm locally to the egg that the point of entry of the sperm determines the plane of symmetry of the embryo. Subsequent work (Jenkinson, 1909a, 1909b; Bánki, 1927; Weigmann, 1927) has shown that this relation is not so simple, as Roux himself recognized, and that other factors also affect the position of the median plane. The fact that after pricking with a needle the frog's egg develops parthenogenetically into a normal embryo, the median plane of which bears no fixed relation to the meridian of pricking, shows that there is a preformed plane of symmetry (Brachet, 1911) and the same is borne out by the normal development of polyspermic eggs (Brachet, 1910). Whatever influence the sperm path may have is then merely to shift something that is already laid down.

26. This follows from the work of Weigmann (1927), Penners and Schleip (1928a, b), Motomura (1935), Dalcq and Pasteels (1937).

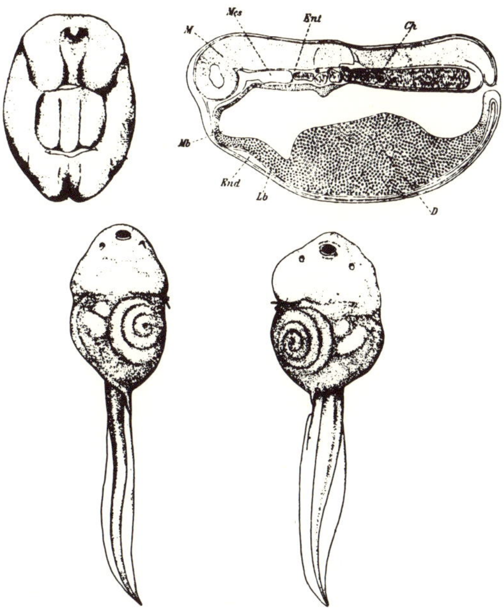

FIG. V–9. *Situs inversus viscerum* in the frog, *R. esculenta,* after rotation of a portion of the neural plate. After Spemann from Meyer, 1913.

known it is not reversible (p. 171). Conklin (1903) advanced the idea that a dextral egg might be transformed into a sinistral one by reversal of the polarity of the egg, but later (1917, p. 364), in the absence of experimental evidence for this view, ascribed the polarity of the egg (Crepidula) to the structure of the spongioplasmic framework of the cell, with which view the hypothesis here advocated is in agreement. Moreover, Hämmerling's (1927) observations on frog embryos which had undergone their early development upside down, i.e. with the polarity of the egg axis inverted, show that their asymmetry is not affected.

On the other hand Spemann (1906a and b) and Pressler (1911) have obtained inverse symmetry (*situs inversus viscerum*) as a result of the rotation of a portion of the archenteric roof of the anuran embryo in the neurula stage [Fig. V–9]. Again in twins and double monsters of the newt, as found by Spemann and Falkenberg (1919) after mechanical constriction of the blastula, *situs inversus* often occurs in the right hand member of the pair [Fig. V–10]. In these cases it might be assumed with Spemann and Falkenberg that the gross mechanical factors operating in normal development or in the case of the left hand twin act in the same sense as the molecular asymmetric factors, while after rotation of the archenteric roof or in the right hand

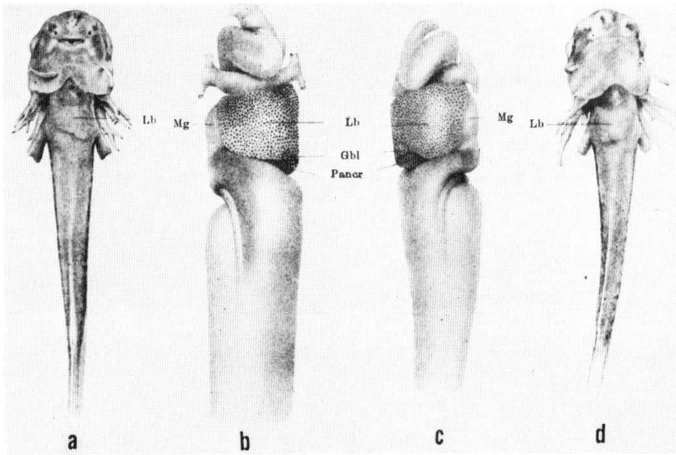

FIG. V–10. *Triton cristatus* twins resulting from constriction through the first furrow of the two or four cell egg.

a and b. Ventral view of right member with reversed liver. b. Reconstruction of gut and heart. *Situs inversus viscerum et cordis.*

c and d. Normal left member. After Spemann and Falkenberg, 1919.

twin they act in the opposite sense and may overcome the latter. However, observations on double embryos (trout and frog) occurring sporadically cast doubt on so simple an explanation, since inversion may occur in either component. Nevertheless, the evidence is against the assumption that the inverse situs is due to the inversion of molecular asymmetry. Similarly in the reversal of the asymmetry of the limb or the ear by reorientation of the rudiment, reinforcement or counteraction of molecular forces may occur. If such is the case, the reversal should take place more readily or with greater velocity in material from one side than in that from the other.[27] This should be put to an experimental test.

During maturation and fertilization there is a solation of the main portion of the egg contents with resultant streaming and rearrangement of materials before cleavage sets in, observed in many different eggs but particularly striking in the ascidian, as described by Conklin (1905). By means of vital dyes it has been shown by Spek (1934a, 1934b) in certain marine worms (Nereis and Chaetopterus) and by Ries and Gersch (1936) in the mollusc, Aplysia, that basic and acid substances which are mixed in the unfertilized egg are shifted toward the poles, the former toward the

27. Driesch (1906, p. 786) raised this question in connection with the development of the two halves of a single sea urchin egg, between which he at first thought there was difference in velocity. Later (1908, p. 138) experiments did not bear out this view.

animal and the latter toward the vegetal, possibly through electrophoresis. These movements must involve disturbances in molecular orientation, but in the new state of equilibrium which supervenes the molecular configuration and surface forces would again tend to bring about an ordered orientation.[28]

Even in the undivided cell further localized activites may be assumed to take place through the interaction between the primary structural material in the ground substance and the material constituting the gradients. The velocity of these reactions would vary from place to place since the concentration of the reacting substances is not uniform. Such reactions would be more effectually localized after cells had divided and cell walls were formed between the different regions of the cytoplasm. In order to account for these localized differentiations it is suggested that the chemical composition prevailing in a particular region activates certain side chains in the molecules of that region more than others and that these reactions therefore differ quantitatively in the several regions. Thus new substances are formed locally, making each region behave differently, although its fundamental protoplasmic composition still remains the same. Differentiations of this kind, through the activation of one type of side chain or another, lead to new situations which start up new side chain activities regionally. These activities are also centered and in themselves show a gradient inversely proportional to a power of the distance from the center. As this type of change goes on, the cells become saturated with the products of the reaction, so that differentiation in other directions is inhibited, thus leading to the condition which is now frequently called "determined." Yet the characters of the specific molecules are left intact, since living protoplasm has the way of working through agents such as enzymes which it may form without changing itself essentially. In some such manner as this the submicroscopic repeat pattern characteristic of the egg may be conceived as giving rise to a system with localized differentiation.

Involved in these reactions are substances arising from the nucleus. Other reactions may be set up or intensified by conditions in the external medium, as for instance when the endodermal region of the sea urchin embryo is extended through the action of lithium salts (endodermization) or when the opposite effect (ectodermization) is produced by sodium thiocyanide.[29]

28. Moore and Miller (1937) report that the clear cytoplasmic zone in unfertilized centrifuged eggs of Strongylocentrotus purpuratus is optically isotropic but that within three minutes after fertilization and afterward a definite measurable birefringence is present which is assumed to be due to a regular grouping (orientation) of the molecules.

29. For the effect of lithium see Herbst (1892, 1893, 1896), MacArthur (1924), Runnström (1928, 1935), Lindahl (1933), Hörstadius (1936a, 1936b). For the effect of sodium thiocyanide see Runnström (1928) and Lindahl (1933).

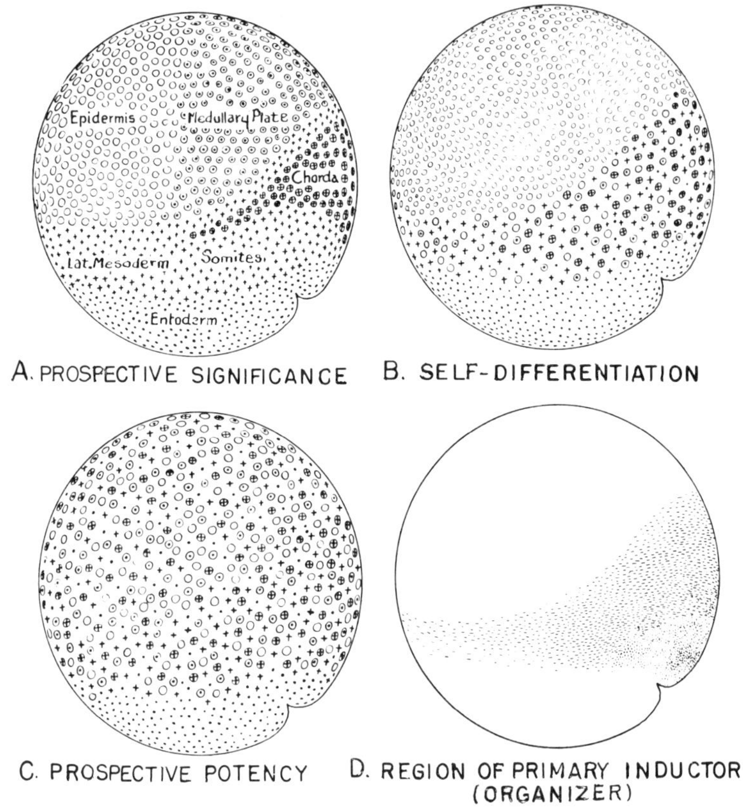

FIG. 13. A, profile view of the early gastrula showing the prospective areas as revealed by the vital staining method. B, similar view showing power of self-differentiation of small pieces when isolated. C, showing prospective potency (powers of differentiation under various tried conditions). D, showing the primary inductor area; stippled area, head organizer; area with dashes, trunk-tail organizer. After Holtfreter (1936).

Some idea of the conditions in the amphibian egg may be gained from the following considerations. Projected back upon the early stages of the egg with the aid of a vital staining technique (Vogt, 1925, 1929), the principal structures of the embryo appear in the form of a definite mosaic (Figs. 13A and D). While this mosaic holds for normal development, an entirely different picture appears when it is seen what the cells of each region may actually do under various other conditions (Figs. 13B and C). This is shown by experiments of Kusche (1929), Bautzmann (1929), Holtfreter (1931, 1936) and others. Many regions are found to have much more diverse powers of differentiation. The mosaic blocks are mixed and the areal boundaries blurred or obliterated. Not until the organizer material of the border zone is rolled in during gastrulation do the main areas reach a stage of differentiation in which they are no longer interchangeable.

The greater part of the animal hemisphere of the egg is at first undifferentiated within itself, though distinguishable from the opposite hemisphere through the primary polarity of the egg. Its further differentiation lags and in fact cannot be detected until after the border-zone material has been turned in under it. Reacting with the axial part of the latter, which consists of the pharyngeal roof, notochord, and the medial part of the mesoderm, the overlying area undergoes a marked change and forms the neural plate. Further differentiations may be assumed to take place in the ectoderm by this same process of activation through agents in the underlying layer of mesoderm or endoderm, or later through the nerve cord itself after it has been folded off from the remainder of the ectoderm.

This mode of development combines unfolding and interaction, thus according partly with each of the old theories of preformation and epigenesis.[30] It may be described as epigenetic in that the various characters of the future organism do not appear at first as such, but are realized through reaction between the several constituents of the egg and also between them and the environment. It is preformational in that there is something in the molecular configuration which determines the physical arrangement and the special chemical reactions that produce in the end the adult organism of the proper species.

Let us now return to the development of the ear, in the consideration of which stress has been laid on the consequences of reorientation of the rudiment by grafting. The importance of this rests upon the fact that in rotation experiments only a simple vectorial change is imposed, from the results of which a simple conclusion may be drawn. As described above, the ear rudiment is at first isotropic about an axis perpendicular to the surface ectoderm, following which at a certain stage evidence of definite antero-posterior polarization is found. The whole ectodermal region is, in fact, so plastic during the stages of gastrulation and early neurulation that neither local nor vectorial differences in its qualities can be readily detected. Just as the neural folds are closing, something occurs which orients in an antero-posterior direction some constituent of the material destined to form the ear. At about the same time the direction of the ciliary beat of the ectodermal cells and the direction of outgrowth

30. Both Roux and Driesch, who are commonly thought of respectively as the chief exponents of the two theories of development, admit that both kinds of process, evolution and epigenesis, occur. Cf. Driesch (*Analytische Theorie*, 1894, pp. 28–29) and Roux (1885b, p. 20).

of the lateral line[31] become fixed, all of which shows that a fundamental change in protoplasmic structure then takes place in the ectodermal layer.

The orientation in question may be due to several factors combined. In the first place the ectoderm of that region is markedly stretched during this period, along with the longitudinal stretching of the central nervous system (Fig. 3). This alone might have an orienting effect similar to the remarkable molecular rearrangement which accompanies the stretching of wool, rubber, etc. Since the mechanical stretching might effect a parallel arrangement without necessarily orienting all of the individual elements in the same sense, it is necessary to assume a further orienting factor to account for the latter. Contact with the central nervous system or the mesoderm may play this role.[32] When twinned ears develop in which the two components are oppositely oriented, as is most common in experiments performed in the critical stages, interaction between the elements of the two halves must take place, such as occurs when twin crystals are formed. This phenomenon of twinning is perhaps the best evidence we have for the paracrystalline nature of living material.

The regular molecular arrangement here postulated might conceivably yield to X-ray analysis, for although the technical difficulties of obtaining X-ray diffraction pictures of living material are great, they have been overcome in part, especially in connection with fibrous structures such as tendon,[33] muscle,[34] and nerve.[35] Muscle shows a definite molecular spacing, indicating orientation in a longitudinal direction, which largely disappears on contraction. Its principal constituent, myosin, when drawn out into a thread, gives a diffraction picture characteristic of crystalline fibers with spacing similar to that of keratin.[36]

31. Twitty (1928); Stone (1931); Tung and Tung (1940).
32. When the hind-brain region of the neural plate is removed on one side and replaced by a similar piece with its anteroposterior axis reversed, disturbances occur in the development of the ear on that side. They are more frequent and generally more far-reaching when the mesoderm underlying the neural plate is included in the graft. The results are, however, irregular, and simple reversal of the asymmetry or twinning of the ear is relatively rare. They are interesting in connection with the work of Tung and Tung (1940), who find that in the anuran embryo reversal of a piece of the mesoderm in the early gastrula stage results in the reversal of the direction of the effective stroke of the cilia in the overlying ectoderm.
33. Gerngross and Katz (1926); Astbury (1936b).
34. Boehm and Schotzky (1930); Astbury (1937).
35. Boehm (1933); Schmitt, Bear and Clark (1935). Bear, Schmitt and Young (1937) have studied the giant nerve fiber of the squid by means of polarized light and find that birefringence of the axoplasm is due to both form and intrinsic anisotropy. They conclude "that the optical properties (of the axoplasm) are determined by a relatively small well oriented fraction of its protein complex," and that "this view is not inconsistent with X-ray diffraction and chemical information obtained from similar material." (p. 518).
36. Boehm and Weber (1932); Astbury and Dickinson (1940).

In the hope that it might be possible to get evidence of oriented structures either in the ear rudiment or in the medullary plate, X-ray diffraction photographs from both dried and fresh specimens of these tissues have been made, but so far with only negative results. Since various conditions in the embryonic tissue used, such as presence of large amounts of water, yolk and other inclusions, would render it difficult to demonstrate regularly oriented structure by this method, the results are not to be taken as disproving the theory.[37]

In summing up the views here expressed, I am conscious of their imperfection in contrast with the extremely delicate and complex transformations they attempt to describe. It may all seem but an echo of Wagner, when he says to Mephistopheles as the Homunculus is about to appear in the phial:

> Was man an der Natur Geheimnisvolles pries,
> Das wagen wir verständig zu probieren,
> Und was sie sonst organisieren liess,
> Das lassen wir krystallisieren.

Experimental embryology has, however, too long remained in isolation, and the problems peculiar to it have not been brought into sufficiently close relation to other biological fields and the underlying physical sciences. Our widening knowledge of the physicochemical properties of protoplasm and the vista opening up through advances in organic chemistry, the newer biophysical methods and X-ray analysis, amply justify the attempt to bring these fields into closer union. In endeavoring to reach a physicochemical description of life one is baffled rather by the bewildering array of possibilities than by such dearth of material as would warrant recourse to so-called vitalistic forces. At least I am unwilling to accept the defeatism of the vitalist, so long as means of investigation by experiment are available.

In the foregoing exposition the specific form of the organism is taken as an expression of the molecular configuration of its protoplasm, in which the protein constituents play the chief part. In the growing egg cell these tend, through the action of molecular forces, to arrange themselves in a definite way, similar to the arrangement of atoms or atom groups in liquid crystals, associating with themselves water, electrolytes and organic constituents. The large amount of water that must enter into the meshes of the crystalline network, i.e. into the actual lattice cells, gives them very great lability, but the intermolecular forces are sufficiently constant to hold the

37. See discussion in Harrison, Astbury and Rudall (1940), p. 352.

molecules in their characteristic configuration and even in some cases to restore such configuration, when mechanically disturbed. The changes that we call differentiation, which is the transformation of a universal repeat pattern of molecular order into a macroscopic pattern of localized organs and tissues, presumably take place through the local activation of particular side chains of the protein molecules. One local change leads to another until the end stage of what is often called segregation seems complete. But it is not a true segregation, for some of the original protein molecules must be assumed to remain unchanged, preserving in each living cell the protoplasmic characteristics of the species. The differentiated structures are products of the original protoplasm rather than the result of its transformation *in toto.*

Such a theory combines various concepts that have come into embryology in recent years: that of the hierarchy of organizers as conceived by Spemann, the idea of morphogenetic fields, and the axial gradient theory of Child. Many others previously have used the crystal analogy, and Przibram (1926) has presented this whole subject in a comprehensive way. In these discussions the crystal has usually been taken as a model, but Needham (1936) goes much further, when, near the end of his stimulating book on "Order and Life" he says, "Liquid crystals, it is to be noted, are not important for biology and embryology, because they manifest certain properties which can be regarded as analogous to those which living systems manifest (models), but because living systems actually *are* liquid crystals, or, it would be more correct to say, the paracrystalline state undoubtedly exists in living cells." This may be an advanced position to take at the present time, but then progress is made only by taking advanced positions.

References

Astbury, W. T. 1933. *Fundamentals of Fibre Structure.* Oxford Univ. Press. London, Humphrey Milford, 187 pp.

———. 1936a. Recent advances in the x-ray study of protein fibres. *J. Text. Inst., 27:* 281–92.

———. 1936b. The x-ray interpretation of protein structure. *Chem. Weekblad, 33:* 778–84.

———. 1937. Protein structure from the viewpoint of x-ray analysis. *C. R. du Lab. Carlsberg,* sér. chim., *22:* 45–53.

———. 1940. The hydrogen bond in protein structure. *Trans. Faraday Soc., 36:* 871–80.

Astbury, W. T., and S. Dickinson. 1940. X-ray studies of the molecular structure of myosin. *Proc. Roy. Soc. (B), 129:* 307–32.

Astbury, W. T., and R. Lomax. 1935. An X-ray study of the hydration and denaturation of proteins. *J. Chem. Soc.,* 846–51.

Astbury, W. T., and W. A. Sisson. 1935. X-ray studies of the structure of hair, wool and related fibres. III. The configuration of the keratin molecule and its orientation in the biological cell. *Proc. Roy. Soc. (A), 150:* 533–51.

Astbury, W. T., and A. Street. 1931. X-ray studies of the structure of hair, wool and related fibres. I. General. *Phil. Trans. Roy. Soc. (A), 230:* 75–101.

Astbury, W. T., and H. J. Woods. 1930. The x-ray interpretation of the structure and elastic properties of hair keratin. *Nature, 126:* 913–14.

———. 1933. X-ray studies of the structure of hair, wool and related fibres. II. The molecular structure and elastic properties of hair keratin. *Phil. Trans. Roy. Soc. (A), 232:* 333–94.

Avery, O. T., and M. Heidelberger. 1923. Immunological relationships of cell constituents of pneumococcus. *J. Exp. Med., 38:* 323–27.

Baltzer, F. 1930. Ueber die Entwicklung des Tritonmerogons Triton taeniatus (♀) × cristatus ♂. *Rev. Suisse Zool., 37:* 325–32.

Baltzer, F., and V. De Roche. 1936. Ueber die Entwicklungsfähigkeit haploider Triton alpestris-Keime und über die Aufhebung der Entwicklungshemmung bei Geweben letaler bastardmerogonischer Kombination durch Transplantation in einen normalen Wirt. *Rev. Suisse Zool., 43:* 495–505.

Bánki, O. 1927. Die Lagebeziehung der Spermiumeintrittsstelle zur Medianebene und zur ersten Furche nach Versuchen mit örtlicher Vitalfärbung am Axolotlei. *Anat. Anz., Ergänz., 63:* 198–209.

Bautzmann, H. 1929. Ueber bedeutungsfremde Selbstdifferenzierung aus Teilstücken des Amphibienkeimes. *Naturwiss., 17:* 818–27.

Bear, R. S., F. O. Schmitt, and J. Z. Young. 1937. The ultrastructure of nerve axoplasm. *Proc. Roy. Soc. (B), 123:* 505–19.

Bernal, J. D. 1940. Structural Units in Cellular Physiology, in *The Cell and Protoplasm.*, pp. 199–205. Pub. Amer. Assoc. Adv. Sci. no. 14. Science Press, 205 pp.

Boehm, G. 1933. Das Röntgendiagramm der Nerven. *Koll.-Zeitsch., 62:* 22–26.

Boehm, G., and K. F. Schotzky. 1930. Das Röntgendiagramm der Muskelkontraktion. *Naturwiss., 18:* 282.

Boehm, G., and H. H. Weber. 1932. Das Röntgendiagramm von gedehnten Myosinfäden. *Koll.-Zeitsch., 61:* 269–70.

Born, G. 1885. Biologische Untersuchungen. I. Ueber den Einfluss der Schwere auf das Froschei. *Arch. mikr. Anat., 24:* 475–545.

Boveri, Th. 1887. Ueber Differenzierung der Zellkerne während der Furchung des Eies von Ascaris megalocephala. *Anat. Anz., 2:* 688–93.

———. 1892. Ueber die Entstehung des Gegensatzes zwischen den Geschlechtszellen und den somatischen Zellen bei Ascaris megalocephala, nebst Bemerkungen zur Entwicklungsgeschichte der Nematoden. *Gesell. Morph. Physiol. i. München, 8:* 114–25.

———. 1899. Die Entwickelung von Ascaris megalocephala mit besonderer Rücksicht auf die Kernverhältnisse. Festsch. z. 70 Geburtstag C. v. Kupffer: pp. 383–430. Jena, Gustav Fischer.

———. 1901. Ueber die Polarität des Seeigel-Eies. *Verh. Phys.-Med. Gesell., Würzburg, 34,* 145–76.

———. 1910. Die Potenzen der Ascaris-Blastomeren bei abgeänderter Furchung. Zugleich ein Beitrag zur Frage qualitativ-ungleicher Chromosomen-Teilung. Festsch. z. 60. Geburtstag R. Hertwig, *3:* 133–214. Jena, Gustav Fischer.

Boycott, A. E., C. Diver, S. L. Garstang, and F. M. Turner. 1930. The inheritance of sinistrality in Limnaea peregra (Mollusca, Pulmonata). *Phil. Trans. Roy. Soc. (B), 219:* 51–131.

Brachet, A. 1910. La polyspermie expérimentale comme moyen d'analyse de la fécondation. *Roux' Arch., 30:* 261–303.

——. 1911. Études sur les localisations germinales et leur potentialité réelle dans l'oeuf parthénogénétique de Rana fusca. *Arch. d. Biol., 26:* 337–63.

Brown, M. G., V. Hamburger, and F. O. Schmitt. 1941. Density studies on amphibian embryos with special reference to the mechanism of organizer action. *J. Exp. Zool., 88:* 353–72.

Butler, Samuel. 1878. *Life and Habit.* London, Trübner, 307 pp.

Carpenter, E. 1937. The head pattern in Amblystoma studied by vital staining and transplantation methods. *J. Exp. Zool., 75:* 103–26.

Cohn, E. J. 1935. The chemistry of the proteins and amino acids. *Ann. Rev. Biochem., 4:* 93–148.

Conklin, E. G. 1897. The embryology of Crepidula, a contribution to the cell lineage and early development of some marine gastropods. *J. Morph., 13:* 1–226.

——. 1903. The cause of inverse symmetry. *Anat. Anz., 23:* 577–88.

——. 1905. The organization and cell-lineage of the Ascidian egg. *J. Acad. Nat. Sci., Philadelphia, 13:* 1–119.

——. 1917. Effects of centrifugal force on the structure and development of the eggs of Crepidula. *J. Exp. Zool., 22:* 311–419.

——. 1931. The development of centrifuged eggs of ascidians. *J. Exp. Zool., 60:* 1–119.

Crampton, H. E. 1894. Reversal of cleavage in a sinistral gasteropod. *Ann. N.Y. Acad. Sci., 8:* 167–70.

Dalcq, A. 1933. La déterminatir ı de la vésicule auditive chez le Discoglosse. *Arch. d'Anat. Micr., 29:* 389–420.

Dalcq, A., and J. Pasteels. 1937. Une conception nouvelle des bases physiologiques de la morphogénèse. *Arch. d. Biol., 48:* 669–710.

Driesch, H. 1894. *Analytische Theorie der organischen Entwicklung.* Leipzig, Engelmann, 185 pp.

——. 1899. Die Lokalisation morphogenetischer Vorgänge. Ein Beweis vitalistischen Geschehens. *Roux' Arch., 8:* 35–111.

——. 1900. Die isolierten Blastomeren des Echinidenkeims. Eine Nachprüfung und Erweiterung früherer Untersuchungen. *Roux' Arch., 10:* 361–410.

——. 1905. Die Entwickelungsphysiologie von 1902–1905. *Ergeb. Anat. Entw., 14:* 603–807.

——. 1906. Studien zur Entwicklungsphysiologie der Bilateralität. *Roux' Arch., 21:* 755–91.

——. 1908. Zur Theorie der organischen Symmetrie. *Roux' Arch., 26:* 130–45.

——. 1929. *The Science and Philosophy of the Organism,* 2d Ed. London, Black, 344 pp.

East, E. M. 1934. The nucleus-plasma problem. Am. Naturalist, *63:* 289–303, 402–39.

Edsall, J. T., and A. L. Von Muralt. 1930. Studies in the physical chemistry of muscle globulin. IV. The anisotropy of myosin and double refraction of flow. *J. Biol. Chem., 89:* 351–86.

Frank, F. C. 1936. Energy of formation of "cyclol" molecules. *Nature, 138:* 242.

Freundlich, H. 1940. Colloid chemistry of development and growth. Second Symposium on Development and Growth. Growth, suppl., pp. 33–54.

Frey-Wyssling, A. 1938. *Submikrosckopische Morphologie des Protoplasmas und seiner Derivate.* Protoplasma-Monographien, Bd. 15. Berlin, Borntraeger, 317 pp.

Gerngross, O., and J. R. Katz. 1926. Ueber die Herstellung sehr stark gedehnter Gelatinepräparate und deren Röntgendiagramm. Gelatine und Kollagen. *Koll.-Zeitsch., 39:* 181.

Glaser, O. C. 1914. On the mechanism of morphological differentiation in the nervous system. *Anat. Rec., 8:* 525–51.

Hadorn, E. 1930. Ueber die Organentwicklung in bastardmerogonischen Transplantaten bei Triton. *Rev. Suisse Zool., 37:* 333–41.

———. 1932. Ueber Organentwicklung und histologische Differenzierung in transplantierten merogonischen Bastardgeweben (Triton palmatus (♀) × Triton cristatus ♂). *Roux' Arch., 125:* 495–565.

———. 1934. Ueber die Entwicklungsleistungen bastardmerogonischer Gewebe von Triton palmatus (♀) × Triton cristatus (♂) im Ganzkeim und als Explantat in vitro. *Roux' Arch., 131:* 238–84.

Hall, E. K. 1939. On the duration of the polarization process in the ear primordium of embryos of Amblystoma punctatum (Linn.). *J. Exp. Zool., 82:* 173–92.

Hämmerling, J. 1927. Die Umkehrung der Polarität des ungefurchten Eies von Rana fusca und ihre Folgeerscheinungen. *Roux' Arch., 110:* 395–416.

———. 1932. Entwicklung und Formbildungsvermögen von Acetabularia mediterranea. *Biol. Zentralbl., 52:* 42–61.

———. 1934. Ueber formbildende Substanzen bei Acetabularia mediterranea, ihre räumliche und zeitliche Verteilung und ihre Herkunft. *Roux' Arch., 131:* 1–81.

Hardy, W. B. 1934. *To Remind, A Biological Essay.* Baltimore, Williams and Wilkins, 45 pp.

Harrison, R. G. 1917. Transplantation of limbs. *Proc. Nat. Acad. Sci., 3:* 245–51.

———. 1921. On relations of symmetry in transplanted limbs. *J. Exp. Zool., 32:* 1–136.

———. 1935. Factors concerned in the development of the ear in Amblystoma punctatum. *Anat. Rec., 64, suppl. 1:* 38–39.

———. 1936. Relations of symmetry in the developing ear of Amblystoma punctatum. *Proc. Nat. Acad. Sci., 22:* 238–47.

———. 1938. Further investigation of the factors concerned in the development of the ear. *Anat. Rec.,* 70, suppl. *3:* 35.

Harrison, R. G., W. T. Astbury, and K. M. Rudall. 1940. An attempt at an x-ray analysis of embryonic processes. *J. Exp. Zool., 85:* 339–63.

Harvey, E. N., and D. A. Marsland. 1932. The tension at the surface of Amoeba dubia with direct observations on the movement of cytoplasmic particles at high centrifugal speeds. *J. Cell. Comp. Physiol., 2:* 75–97.

Heidelberger, M., and O. T. Avery. 1923. The soluble specific substance of pneumococcus. *J. Exp. Med., 38:* 315–21.

———. 1924. The soluble specific substance of pneumococcus. Second paper. *J. Exp. Med., 40:* 253–68.

Heidelberger, M., W. F. Goebel, and O. T. Avery. 1925. The soluble specific substance of pneumococcus. Third paper. *J. Exp. Med., 42:* 503–21.

Herbst, C. 1892. Experimentelle Untersuchungen über den Einfluss der veränderten chemischen Zusammensetzung des umgebenden Mediums auf die Entwicklung der Tiere. I. Versuche an Seeigeleiern. *Zeit. wiss. Zool., 55:* 446–518.

Herbst, C. 1893. II. Weiteres über die morphologische Wirkung der Lithiumsalze und ihre theoretische Bedeutung. *Mitth. Zool. Sta. Neapel, 11:* 136–220.

——. 1896. III. Ueber das Ineinandergreifen von normaler Gastrulation und Lithiumentwickelung (Züchtung von Lithiumlarven mit Entodarm und Mund). IV. Die formative Wirkung des Lithiums auf befruchtete Eier von Asterias glacialis. V. Ueber die Unterdrückung von Entwickelungsprocessen (Wirkung von Kalium rhodanatum und Natrium butyricum). VI. Ueber den Einfluss einiger anderer organischer Salze. Schlussbemerkungen. *Roux' Arch., 2:* 455–516.

Holtfreter, J. 1931. Ueber die Aufzucht isolierter Teile des Amphibienkeimes. II. Züchtung von Keimen und Keimteilen in Salzlösung. *Roux' Arch., 124:* 404–66.

——. 1933. Der Einfluss von Wirtsalter und verschiedenen Organbezirken auf die Differenzierung von angelagertem Gastrulaektoderm. *Roux' Arch., 127:* 619–775.

——. 1936. Regionale Induktionen in xenoplastisch zusammengesetzten Explantaten. *Roux' Arch., 134:* 466–550.

Hörstadius, S. 1935. Ueber die Determination im Verlaufe der Eiachse bei Seeigeln. *Pub. Staz. Zool. Napoli, 14:* 251–479.

——. 1936a. Ueber die zeitliche Determination im Keim von Paracentrotus lividus Lk. *Roux' Arch., 135:* 1–39.

——. 1936b. Weitere Studien über die Determination im Verlaufe der Eiachse bei Seeigeln. *Roux' Arch., 135:* 40–68.

——. 1937. Investigations as to the localization of the micromere-, the skeleton-, and the entoderm-forming material in the unfertilized egg of Arbacia punctulata. *Biol. Bull., 73:* 295–316.

Hörstadius, S., and A. Wolsky. 1936. Studien über die Determination der Bilateralsymmetrie des jungen Seeigelkeimes. *Roux' Arch., 135:* 69–113.

Jäger, G. 1876. *Zoologische Briefe.* Vienna, Braunmüller.*

——. 1878. *Lehrbuch der allgemeinen Zoologie.* II. Abteilung. Leipzig, Günther, 384 pp.

——. 1879. Zur Pangenesis. *Kosmos, 4:* 377–85.

Jenkinson, J. W. 1909a. On the relation between the symmetry of the egg, the symmetry of segmentation, and the symmetry of the embryo in the frog. *Biometrika, 7:* 148–209.

——. 1909b. *Experimental Embryology.* Oxford, Clarendon Press, 341 pp.

Jordan Lloyd, D. 1932. Colloidal structure and its biological significance. *Biol. Rev., 7:* 254–73.

Jordan Lloyd, D., and A. Shore. 1938. *Chemistry of the Proteins.* Philadelphia, Blakiston, 532 pp.

Kaan, H. W. 1926. Experiments on the development of the ear of Amblystoma punctatum. *J. Exp. Zool., 46:* 13–61.

Kniep, H. 1907. Beiträge zur Keimungs-Physiologie und Biologie von Fucus. *Jahrb. wissen. Bot., 44:* 635–724.

Kofoid, C. A. 1895. On the early development of Limax. *Bull. Mus. Comp. Zool., 27:* 33–118.

Koltzoff, N. K. 1928. Physikalisch-chemische Grundlage der Morphologie. *Biol. Zentralbl., 48:* 345–69.

Kusche, W. 1929. Interplantation umschriebener Zellbezirke aus der Blastula und der Gastrula von Amphibien. I. Versuche an Urodelen. *Roux' Arch., 120:* 192–271.

* Not seen in the original.

Küster, E. 1906. Normale und abnorme Keimungen bei Fucus. *Ber. deut. bot. Ges., 24:* 522–28.

Landsteiner, K. 1936. *The Specificity of Serological Reactions.* Springfield, Ill., Thomas, 178 pp.

Lehmann, F. E. 1929. Die Entwicklung des Anlagenmusters im Ektoderm der Tritongastrula. *Roux' Arch., 117:* 312–83.

———. 1933. Das Prinzip der kombinativen Einheitsleistung in der Biologie, im besonderen in der experimentellen Entwicklungsgeschichte, und seine Beziehung zur Gestalttheorie. *Biol. Zentralbl., 53:* 471–96.

Lillie, F. R. 1906. Observations and experiments concerning the elementary phenomena of embryonic development in Chaetopterus. *J. Exp. Zool., 3:* 153–268.

———. 1909. Polarity and bilaterality of the annelid egg. Experiments with centrifugal force. *Biol. Bull., 16:* 54–79.

Lindahl, E. 1933. Ueber "animalisierte" und "vegetativisierte" Seeigellarven. *Roux' Arch., 128:* 661–64.

Lund, E. J. 1923. Electrical control of organic polarity in the egg of Fucus. *Bot. Gazette, 76:* 288–301.

Luther, W. 1934. Untersuchungen über die Umkehrbarkeit der Polarität zwischen Aussen- und Innenseite des Ektoderms von Amphibienkeimen. *Roux' Arch., 131:* 532–39.

Lyon, E. P. 1907. Results of centrifugalizing eggs. I. The specific gravity of eggs and the changes in specific gravity occurring during development. II. Effects of centrifugalizing unfertilized eggs on their development. *Roux' Arch., 23:* 151–73.

MacArthur, J. W. 1924. An experimental study and a physiological interpretation of exogastrulation and related modifications in echinoderm embryos. *Biol. Bull., 46:* 60–87.

Mirsky, A. E., and L. Pauling. 1936. On the structure of native, denatured, and coagulated proteins. *Proc. Nat. Acad. Sci., 22:* 439–47.

Moore, A. R. 1933. On the cytoplasmic framework of the plasmodium, Physarum polycephalum. *Sci. Reports Tôhoku Imp. Univ.,* 4th ser., *8:* 189–92.

———. 1935. On the significance of cytoplasmic structure in plasmodium. *J. Cell. Comp. Physiol., 7:* 113–29.

Moore, A. R., and W. A. Miller. 1937. Occurrence of birefringence in the fertilized egg of the sea urchin. *Proc. Soc. Exp. Biol. Med., 36:* 835–36.

Morgan, T. H. 1927. *Experimental Embryology.* New York, Columbia Univ. Press, 766 pp.

Motomura, I. 1935. Determination of the embryonic axis in the eggs of amphibia and echinoderms. *Sci. Reports Tôhoku Imp. Univ.,* 4th ser., *10:* 211–45.

Moyer, L. S. 1942. *Proteins and Protoplasmic Structure.* A Symposium on the Structure of Protoplasm. A Monograph of the American Society of Plant Physiologists. Ames, Iowa State College Press, 283 pp.

Nägeli, C. 1858–1884. *Die Micellartheorie.* Ostwald's Klassiker der exakten Wissenschaften, No. 227, Ed. by A. Frey. Leipzig, 1928, Akademische Verlagsgesellschaft M.B.H, 143 pp.

Needham, J. 1936. *Order and Life.* New Haven, Yale Univ. Press, 175 pp.

Newport, G. 1854. Researches on the impregnation of the ovum in the Amphibia; and on the early stages of development of the embryo. *Phil. Trans. Roy. Soc. London, 144:* 229–44.

Norris, H. W. 1892. Studies on the development of the ear of Amblystoma. Part I. Development of the auditory vesicle. *J. Morph., 7:* 23–34.

Nussbaum, M. 1880. Zur Differenzierung des Geschlechts im Tierreich. *Arch. mikr. Anat., 18:* 1–121.

———. 1893. Beiträge zur Lehre von der Fortpflanzung und Vererbung. *Arch. mikr. Anat., 41:* 119–45.

Ogawa, C. 1921. Experiments on the orientation of the ear vesicle in amphibian larvae. *J. Exp. Zool., 34:* 17–43.

Owen, R. 1849. *On Parthogenesis, or the Successive Production of Procreating Individuals from a Single Ovum.* A discourse introductory to the Hunterian lectures on generation and development for the year 1849. London, Van Voorst, 76 pp.

Pease, D. C. 1941. Echinoderm bilateral determination in chemical concentration gradients. I. The effects of cyanide, ferricyanide, iodoacetate, picrate, dinitrophenol, urethane, iodine, malonate, etc. *J. Exp. Zool., 86:* 381–404.

———. 1942a. Echinoderm bilateral determination in chemical concentration gradients. II. The effects of azide, pilocarpine, pyocyanine, diamine, cysteine, glutathione, and lithium. *J. Exp. Zool., 89:* 329–45.

———. 1942b. Echinoderm bilateral determination in chemical concentration gradients. III. The effects of carbon monoxide and other gases. *J. Exp. Zool., 89:* 347–56.

Penners, A., and W. Schleip. 1928a. Die Entwicklung der Schultze'schen Doppelbildungen aus dem Ei von Rana fusca. Teil 1–4. *Zeit. wiss. Zool., 130:* 305–454.

———. 1928b. Die Entwicklung der Schultze'schen Doppelbildungen aus dem Ei von Rana fusca. Teil 5 und 6. *Zeit. wiss. Zool., 131:* 1–156.

Peters, R. A. 1929. Co-ordinative bio-chemistry of the cell and tissues. Lecture I. "Cell surfaces" with reference to the part played in the integration of cell processes by adsorption. Harben Lectures, *J. State Med., 37:* 683–709.

Pressler, K. 1911. Beobachtungen und Versuche über den normalen und inversen Situs viscerum et cordis bei Anurenlarven. *Roux' Arch., 32:* 1–35.

Przibram, H. 1906. Kristall-Analogien zur Entwicklungsmechanik der Organismen. *Roux' Arch., 22:* 207–87.

———. 1913. *Experimental-Zoologie.* 4. Vitalität. Leipzig and Vienna, Deuticke, 179 pp.

———. 1926. Die anorganischen Grenzgebiete der Biologie (insbesondere der Kristallvergleich). *Sammlung Borntraeger, 10:* 240 pp. Berlin, Borntraeger.

Ries, E., and M. Gersch. 1936. Die Zelldifferenzierung und Zellspezialisierung während der Embryonalentwicklung von Aplysia limacina L. Zugleich ein Beitrag zu Problemen der vitalen Färbung. *Pub. Staz. Zool. Napoli, 15:* 223–73.

Rinne, F. 1931. *Grenzfragen des Lebens.* Leipzig, Quelle and Meyer, 128 pp.

Rosenvinge, K. 1888. Om ydre faktorers inflydelse paa Organdannelsen. Kopenhagen.†

Roux, W. 1885a. Beiträge zur Entwickelungsmechanik des Embryo. III. Ueber die Bestimmung der Hauptrichtungen des Froschembryo im Ei und über die erste Teilung des Froscheies. *Breslauer ärzt. Zeit.,* 1–54. *Ges. Abh.,* no. 20, *2:* 277–343.

———. 1885b. "Einleitung" zu den "Beiträgen zur Entwickelungsmechanik des Embryo." *Zeit. Biol., 21:* 412–28. *Ges. Abh.,* no. *13:* 1–23.

———. 1887. Beiträge zur Entwickelungsmechanik des Embryo. IV. Die Bestimmung der Medianebene des Froschembryo durch die Copulationsrichtung des Eikernes und des Spermakernes. *Arch. mikr. Anat., 29:* 157–211. *Ges. Abh.,* no. 21, *2:* 344–418.

Runnström, J. 1928. Plasmabau und Determination bei dem Ei von Paracentrotus lividus Lk. *Roux' Arch., 113:* 556–81.

† Not seen in the original.

———. 1935. An analysis of the action of lithium on sea urchin development. *Biol. Bull., 68:* 378–84.
Scarth, G. W. 1927. The structural organization of plant protoplasm in the light of micrurgy. *Protoplasma, 2:* 189–205.
Schmidt, W. J. 1939. Polarisationsoptische Erforschung des submikroskopischen Baues tierischer Zellen und Gewebe. Der experimentelle Weg und einige Beispiele. *Verh. deut. Zool. Ges., Zool. Anz., 12,* suppl., pp. 303–389.
Schmitt, F. O. 1941. Some protein patterns in cells. Third Symposium on Development and Growth, *Growth, 5,* suppl., 1–20.
Schmitt, F. O., R. S. Bear, and G. L. Clark. 1935. X-ray diffraction studies on nerve. *Radiology, 25:* 131–51.
Seifriz, W. E. 1936. *Protoplasm.* New York and London, McGraw-Hill, 584 pp.
Spek, J. 1934a. Ueber die bipolare Differenzierung der Eizellen von Nereis limbata und Chaetopterus pergamentaceus. *Protoplasma, 21:* 394–405.
———. 1934b. Die Reaktion der Protoplasmakomponenten des Asterias-Eies. *Protoplasma, 21:* 561–76.
Spemann, H. 1906a. Ueber eine neue Methode der embryonalen Transplantation. *Verh. deut. Zool. Ges., pp.* 195–202.
———. 1906b. Ueber embryonale Transplantation. *Naturwiss. Rundschau, 21:* 1–7.
———. 1910. Die Entwicklung des invertierten Hörgrübschens zum Labyrinth. Ein kritischer Beitrag zur Strukturlehre der Organanlagen. *Roux' Arch., 30:* 437–58.
Spemann, H., and H. Falkenberg. 1919. Ueber asymmetrische Entwicklung und Situs inversus viscerum bei Zwillingen und Doppelbildungen. *Roux' Arch., 45:* 371–422.
Sponsler, O. L. 1940. Molecular Structure in Protoplasm in *The Cell and Protoplasm,* pp. 166–87. Pub. Am. Assoc. Adv. Sci. no. 14. Science Press, 205 pp.
Sponsler, O. L., and J. D. Bath. 1942. Molecular structure in protoplasm. A Symposium on the Structure of Protoplasm. Monograph of the American Society of Plant Physiologists. Ames, The Iowa State College Press, pp. 41–79.
Stone, L. S. 1931. Induction of the ear by the medulla and its relation to experiments on the lateralis system in Amphibia. *Science, 74:* 577.
Streeter, G. L. 1907. Some factors in the development of the amphibian ear vesicle and further experiments on equilibration. *J. Exp. Zool., 4:* 431–45.
———. 1914. Experimental evidence concerning the determination of posture of the membranous labyrinth in amphibian embryos. *J. Exp. Zool., 16:* 149–76.
Svedberg, T. 1934. Molecular weight analysis in centrifugal fields. *Science, 79:* 327–32.
Szepsenwol, M. J. 1933. Recherches sur les centres organisateurs des vésicules auditives chez des embryons de poulets omphalocéphales obtenus expérimentalement. *Arch. d'Anat. Micr., 29:* 5–94.
Tung, T. C., and Y. F. Yeh-Tung. 1940. Experimental studies on the determination of polarity of ciliary action of anuran embryos. *Arch. de Biol., 51:* 203–18.
Twitty, V. C. 1928. Experimental studies on the ciliary action of amphibian embryos. *J. Exp. Zool., 50:* 319–44.
Vogt, W. 1925. Gestaltungsanalyse am Amphibienkeim mit örtlicher Vitalfärbung mit Agar als Farbträger. *Roux' Arch., 106:* 542–610.
———. 1929. Gestaltungsanalyse am Amphibienkeim mit örtlicher Vitalfärbung. II. Teil. Gastrulation und Mesodermbildung bei Urodelen und Anuren. *Roux' Arch., 120:* 384–706.

Weigmann, R. 1927. Ueber die Bestimmung der Medianebene im Froschei. *Zeit. wiss. Zool., 129:* 48–102.

Weismann, A. 1892a. Aufsätze über Vererbung und verwandte biologische Fragen. Ueber die Vererbung (1883), pp. 73–121. Die Continuität des Keimplasmas als Grundlage einer Theorie der Vererbung. (1885), pp. 191–207. Jena, Gustav Fischer.

———. 1892b. *Das Keimplasma. Eine Theorie der Vererbung.* Jena, Fischer, 628 pp.

Whitaker, D. M. 1931. Some observations on the eggs of Fucus and upon their mutual influence in the determination of the developmental axis. *Biol. Bull., 61:* 294–308.

———. 1935. Induction of polarity in Fucus furcatus by a localized concentration of hydrogen ions. *Proc. Soc. Exp. Biol. Med., 33:* 472–73.

Whitaker, D. M., and E. W. Lowrance. 1936. On the period of susceptibility in the egg of Fucus furcatus when polarity is induced by brief exposure to directed white light. *J. Cell. Comp. Physiol., 7:* 417–24.

Wilson, E. B. 1903. Experiments on cleavage and localization in the Nemertine-egg. *Roux' Arch., 16:* 411–60.

———. 1904. Experimental studies on germinal localization. I. The germ-regions in the egg of Dentalium. *J. Exp. Zool., 1:* 1–72.

———. 1928. *The Cell in Development and Heredity.* New York, Macmillan, 1232 pp.

———. 1929. The development of egg-fragments in annelids. *Roux' Arch., 117:* 179–210.

Winkler, H. 1900. Ueber den Einfluss äusserer Faktoren auf die Teilung der Eier von Cystosira barbata. *Ber. deut. bot. Ges., 18:* 297.

Wrinch, D. M. 1936. The pattern of proteins. *Nature, 137:* 411–12.

———. 1937a. On the pattern of proteins. *Proc. Roy. Soc. (A), 160:* 59–86.

———. 1937b. The cyclol hypothesis and the "globular" proteins. *Proc. Roy. Soc. (A), 161:* 505–24.

Wrinch, D. M., and D. Jordan Lloyd. 1936. The hydrogen bond and the structure of proteins. *Nature, 138:* 758–59.

Yntema, C. L. 1933. Experiments on the determination of the ear ectoderm in the embryo of Amblystoma punctatum. *J. Exp. Zool., 65:* 317–57.

———. 1939. Self-differentiation of heterotopic ear ectoderm in the embryo of Amblystoma punctatum. *J. Exp. Zool., 80:* 1–17.

Zeleny, C. 1904. Experiments on the localization of developmental factors in the nemertine egg. *J. Exp. Zool., 1:* 293–329.

6

Heteroplastic Grafting in Embryology

In the course of his first experiments in embryonic transplantation, Born (1894, 1897) found that parts of embryos of different species, even those of such diversity as the frog and toad, could be united to form single individuals able to continue their normal development. Shortly afterward the method was used by Harrison (1898) to trace the relative movement of developing parts by means of differences in pigmentation, and incidentally it was found that composite individuals of diverse specific origin might even pass metamorphosis successfully. Gradually this method of heteroplastic grafting has come to be of general use in the investigation of problems of morphogenesis and growth, which although of great variety, nevertheless, have sufficient in common to make possible a connected story.

Heteroplastic grafting may be defined as the union of parts of organisms of different species[1] into a single individual, or the combination of individuals of different species in double or multiple organisms living parabiotically. It is thus a realization of the fanciful notion of composite monsters which goes back to antiquity; the Chimera of Greek mythology was a beast with the forepart of a lion, the middle of a goat and the hind part of a dragon. The corporeal chimeras of modern science are less spectacular, being confined to plants and cold blooded animals. Oddly enough the word has been introduced into biology by the botanists, so that it is now customary to speak of *animal chimeras,*[2] by analogy with what Winkler (1907) called chimeras in plants.

Lecture delivered December 14, 1933. Reprinted from *The Harvey Lectures* (Baltimore, Williams and Wilkins, 1935), pp. 116–57.

1. Geinitz (1925) has proposed the use of the word *xenoplastic* to denote grafting between organisms of different genera or those still further removed systematically, restricting the meaning of *heteroplastic* to grafting beween two species of the same genus.

2. The term was introduced into embryology by Spemann (1921). See also the interesting account of animal chimeras by Taube (1922).

Although in plants the scion usually maintains its specific characters and reproduces its kind both vegetatively and sexually, from time to time, as a result of grafting, forms have arisen which seem to be intermediate between the two species of the combination, and which have hence been termed graft hybrids. First produced experimentally by Winkler (1907, 1909, 1910) by grafting tomato and nightshade, these so-called hybrids were shown to be combinations of cells of the two constituent species, arranged in some orderly way and arising from buds formed at the boundary between scion and stock. Their hybrid nature is, therefore, illusory and the term chimera more appropriate. Two classes of these composite plants are now generally recognized. In the first, termed sectorial, a whole sector of the bud, which may involve stem, branches, leaves, flowers and fruit in the resulting plant, is composed of tissue of one species, the remainder being of the other. In the other class, the periclinal chimeras, the epidermis—one, two or three layers—is derived from one species, and the deeper tissues from the other. The mosaic character of the first group is obvious, but the second look superficially like true blends, and their real nature was first recognized by Bauer (1909, 1910) only after careful microscopic study of forms produced experimentally in geraniums and in members of the nightshade family.[3] The mystery of the two most celebrated "graft hybrids," *Cytisus adami* and *Crataegomespilus asnieresii,* was thus solved, for the microscopic study of all parts of those plants showed them to be periclinal chimeras.[4] Both types of chimeras may be produced by grafting in animals, but since they do not reproduce vegetatively they cannot be perpetuated.

Heteroplastic grafting has been used in embryology mainly in the analysis of processes of growth and differentiation. The applicability of the method rests upon the fact that there are related species that differ from one another in pigmentation, rate of growth and development, ultimate size, relative time of appearance of organs, or even in the presence or absence of organs, while at the same time their tissues show a mutual tolerance when combined in one organism.

The apparent inconsistency between the latter circumstance and the experience of those who have used warm blooded animals for grafting is due to differences between lower and higher vertebrates with regard to tolerance for foreign tissues, in which

3. The subject of graft hybrids and chimeras is fully discussed in Baur's "Vererbungslehre" (1930), pp. 287–304, and by Krenke (1933), pp. 601–841.

4. *Cytisus adami* is composed of the deeper tissues of the common golden chain, *Laburnum vulgare,* covered by epidermis of *Cytisus purpureus,* a procumbent form with purple flowers. *Crataegomespilus* is a similar combination of the deeper tissues of the hawthorn, *Crataegus monogyna,* and epidermis of the medlar, *Mespilus germanica.*

respect there is a graded series. The reactions which occur differ in degree, depending upon the combination made and varying from complete tolerance, through leucocytic infiltration and gradual resorption, to rapid disintegration and sloughing of the offending graft. This subject has been thoroughly studied by L. Loeb,[5] who has found that in the mammals there is an appreciable intolerance even between tissues from individuals of the same species and variety, which is less, however, in the case of close blood relationship. There is, thus, not only a species differential but also an individual differential that has to be considered in all such transplantation experiments. In general it may be said that these differentials become more pronounced as we ascend the animal scale or as we pass from younger to older stages of development in a given species. According to Loeb, the first indication of the individual differential in the vertebrate series appears in adult Amphibia.[6] However, in the tailed forms the species differential frequently does not show itself in any appreciable degree, and is not a constant phenomenon. In embryos and larvae there is a widespread tolerance between the tissues of different species, genera and families, and even grafts from anuran to urodele embryos succeed in many combinations. [This type of heteroplastic combination by induction has been investigated experimentally. In general inductors are not specific as shown by many experiments with the organizer. But the induced organ has entirely the character of the species from which it is developed. The most striking case of this is the induction of mouth parts in an anuran embryo graft by the urodele inductor, the plug of endoderm marking the future mouth opening (Spemann and Schotté 1932). This indicates that the character of the induced organ is that of its own species and also that there is a kind of location stimulus, which might be taken to indicate the reality of homologies. (Lecture VI notes.)]

Methods, Materials and Basic Results[7]

The older studies in embryonic transplantation made use of heteroplastic grafting to trace the relative movements of parts by means of differences in pigmentation.[8]

5. Besides contributing extensively to the experimental side Loeb (1921 and 1930) has published two excellent reviews of this subject.

6. Cf. Schultz (1917), Erdmann (1927) and Hadley (1929).

7. I am much indebted to Miss Lisbeth Krause for the drawings and to Miss Sally Wilens for the photographs and the statistical work here presented.

8. Cf. Harrison (1898, 1903), Spemann (1918), Mangold (1923), Mangold and Seidel (1927).

This has been partially superseded by the method of vital staining. The newer studies, dealing with correlative growth and differentiation, which are the main theme of this lecture, were begun about eleven years ago, having been originally suggested by the marked difference in the time of appearance of the fore limb in two species of salamander, *Amblystoma punctatum* (Linn.) and *A. tigrinum* Green. At the beginning of the larval period, when the yolk has been resorbed and the young animals are ready to feed, the fore legs of *A. punctatum* are almost fully developed, having arm, fore arm and hand with two long and two rudimentary digits (Fig. 2). The limbs are capable of independent movements and are used in crawling. In contrast with this, the fore limbs of *A. tigrinum* and the axolotl at the corresponding stage are small nodules of mesenchyme without visible differentiation (Fig. 1). After this latent period, however, they begin to develop much faster and, corresponding to the more rapid growth of *A. tigrinum,* they soon surpass the punctatum limbs in size.

The question naturally arose, what would occur if the limb buds were interchanged before beginning their development. Would they adjust themselves to the new environment and develop at the tempo of the host, or would they maintain their own rate of growth and reach the ultimate size characteristic of their own species? The second alternative proved to be true (Harrison 1924).

The punctatum limb bud on the tigrinum embryo develops rapidly during the embryonic period and forges ahead of the normal control limb of the opposite side (Fig. 3). Its movements are somewhat delayed, but it soon becomes functional and remains so throughout its existence. In the larval period, the normal limb increases rapidly in size and the grafted limb is soon overtaken; at the time of metamorphosis the linear dimensions of the latter average a little over half those of the former (Fig. 17).

The reciprocal experiment gives corresponding results. The tigrinum limb bud on the punctatum embryo grows slowly at first (Fig. 4), but after it once starts, its growth is rapid, and even before it is fully differentiated, it may exceed the normal control limb of the opposite side in length and thickness. At metamorphosis it has become nearly twice as long as its mate (Figs. 13 and 14). The limb is innervated precociously, compared with the donor control, and it functions normally even though extremely disproportionate.

Other combinations have also been used for such experiments. With regard to growth rate and ultimate body size the five species of Amblystoma that have been studied fall into two groups, between which there is a marked difference in both respects. The smaller and more slowly growing species are *A. punctatum* (*maculatum*), *A. texanum* (*microstomum*) and *A. jeffersonianum*. The second group is

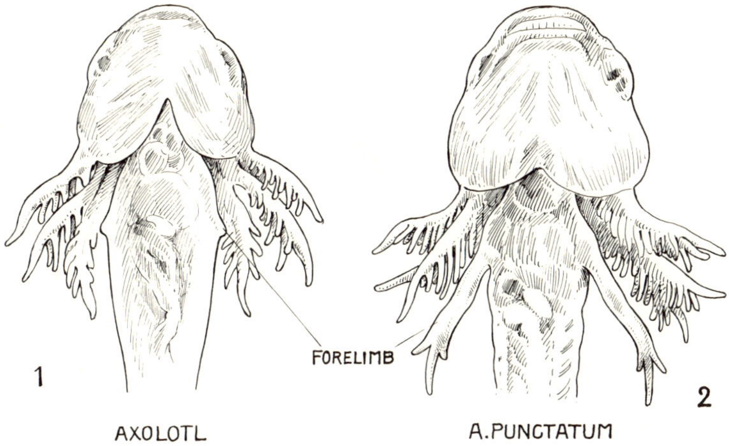

FIG. 1. Axolotl at the beginning of the larval period, yolk entirely resorbed; fore limb a small nodule.

FIG. 2. *Amblystoma punctatum* at the same period; fore limb well differentiated and functional. × 10.

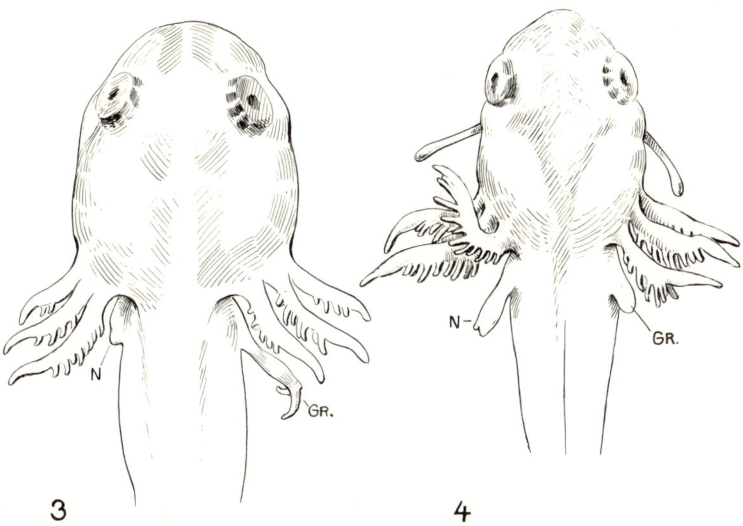

FIG. 3. Larva of *A. tigrinum* (Exp. NE 10) with right limb of *A. punctatum* (GR); the normal limb (N) still remains a small nodule 34 days after operation (larva underfed).

FIG. 4. *A. punctatum* (Exp. NE 13) with right limb (GR) of *A. tigrinum* less advanced than the normal (N) 14 days after operation. × 10.

made up of *A. tigrinum* and *A. mexicanum* (axolotl), both of which grow rapidly and attain great size. *A. opacum* and *Triturus torosus* also belong in the first group.

Fore limb transplantations have been carried out successfully in all twenty combinations of the five Amblystoma species listed above. The measurements are insufficient for plotting relative growth rates by Huxley's method (p. 221), but in all combinations comparable measurements of various parts of the fore limb are available for determining ratios between the species near the close of the larval period. In general these ratios form a consistent whole. Combinations between members of opposite groups yield limbs very disproportionate in size (Figs. 14, 17, 18, and 19), the ratio of the length of the tigrinum to the punctatum appendage being about 1.85. In combinations of members of the same group the limbs are nearly of the same size (Figs. 15, 16, and 33).

Of the three more slowly growing species, *A. jeffersonianum* holds a position in advance of the other two. The ratios of its limb measurements to those of *A. texanum* near the end of larval life are about 1.15 and to those of *A. punctatum* 1.19. Consonant with this, texanum limbs, growing on punctatum larvae, have ratios slightly above unity. Reciprocal experiments give values approximating the arithmetical reciprocals of these figures, as is true for the other combinations. The grafted limb resembles in form that of the donor; in pigmentation it simulates that of the host except when the white axolotl is used. In all cases the transplanted appendages function normally, and in none of the combinations is there lack of tissue tolerance between graft and host.

Similar results are obtained from grafting other organs, and they are particularly clear in the case of the eye, which, on account of its relatively simple configuration, is easily measured in the living animal (Fig. 20). Experiments on this organ have been reported by Koppányi (1923), Harrison (1929a), Stone (1930) and Twitty (1932). In Europe several species of Triton have been used by Rotmann (1929, 1931, 1933) in studies on the gills and the limbs. Combinations of Triton and the axolotl, which show marked disproportion in size, form the basis of some of Copenhaver's (1933) experiments on the limbs and the heart.[9]

The results of R. K. and L. M. Burns (1929) with heteroplastic parabiotic twins are also of special interest, since they afford a means of measuring relative total growth under comparable conditions. Joined together in the tail bud stage the two embryos are approximately equal in length when feeding begins. Owing to the much

9. von Ubisch (1932, 1933) has made an interesting beginning with heteroplastic grafting of sea urchin embryos, a much more difficult material to handle.

greater activity of the tigrinum component and its more rapid growth, after a time it alone captures food and the punctatum partner becomes parasitic. It is adequately nourished, however, thanks to the free interchange of blood, but by the time its growth slows down, which is relatively early, the ratio of the length of the two components is about the same as the ratio between grafted eyes or limbs and their normal controls (Fig. 36).

From the standpoint of a rigorous methodology, the importance of the reciprocal experiment in this work can scarcely be overestimated. In such cases the result obtained in one of the combinations should be the reverse of that shown by the other. Indeed, when concerned with relative growth, the ratio between the two parts in one combination should be close to the arithmetical reciprocal of the ratio in the other combination at a corresponding physiological age (p. 225); any significant deviation from this ratio must be taken to indicate the activity of some factor other than inherent growth capacity.

Again, in composite organs, the value of transplanting each constituent separately, as well as the organ as a whole, is very great, for in this way it is possible to ascertain the proportional amount of the contribution made by each constituent part to the growth of the whole, and thus to check quantitatively the relation between the growth of the whole and that of its constituent parts. This will become more apparent from experiments on the eye, the limb and the gills, described below.

The growth rate of parts relative to that of the whole organism is a very important factor in the genesis of animal form, as was recognized by His (1874) and by D'Arcy Thompson (1917) in "Growth and Form."[10] Since then the subject has been treated comprehensively by Huxley (1932) who has proposed a convenient method for treating relative growth rates quantitatively. It is based on the assumption, for which there is considerable evidence, that the ratio of the growth rate of the part to that of the whole, over the period in question, remains constant and is independent of the absolute rates. This has provoked much theoretical discussion, into which it is not necessary to enter now, for the method has certainly proved itself to be of practical value. In the comparison of the growth of like organs on the opposite side of the body in the same individual or in different individuals under properly controlled treatment, the method is at any rate free from most of the objections that have been urged against it in general.

When a part is growing at a different rate from the whole, its growth is said to be *heterogonic,* to use a term proposed by Pézard; when growing at the same rate, it is

10. pp. 54–57 and 88–103.

termed *isogonic.* Any change in body proportions in a developing organism involves heterogony [heterauxesis], for the concept need not be limited to relative growth rates of part and whole, but may be extended to include relative growth rates of two parts, of the magnitude of any two dimensions, and even of two separate individuals of different species. In the last case, however, there are complications that render the use of the method very difficult (p. 223). Thus not only is the normal growth of eye or limb relative to that of the whole larva, heterogonic, but also the growth of a grafted eye relative to that of the normal control. Both eyes thus grow heterogonically, but the degree of heterogony with respect to body size is different in the two cases.

When one part is growing at a different rate from another, but at a relative rate which is constant for the period concerned, the relation between the two may be expressed by Huxley's equation[11]

$$y = bx^{k}$$

where y is the magnitude of one part at a given instant, x that of the other at the same instant, and k the constant ratio between the growth rates of the two. The constant, b, or fractional coefficient merely gives the magnitude of y when $x = 1$. The constant, k, is, however, very important, and may be termed the differential growth ratio or the growth coefficient.

The value of k is most readily found from a series of measurements of the two parts by plotting them on a double logarithmic grid. The points will lie on a straight line if k is constant, and the value of k is given by the slope of the line. If the relative growth rate varies, the line will not be straight. In many cases, as Huxley (op. cit.) has shown, there is a definite break in the line at a given point, indicating a sudden change in the differential growth ratio.

Measurements of the grafted tigrinum eye in comparison with the normal eye of the punctatum host, in my experiments, show a constant relative growth rate, $k = 1.36$. In the reciprocal experiment, the relative growth rate of the grafted punctatum eye is likewise constant, $k = 0.73$, the reciprocal of which is 1.37 (Table 1). Burns' data for parabiotic twins (heteroplastic pairs) give a fairly satisfactory straight line with a value $k = 1.36$, for the relative total length of *A.* tigrinum to *A. punctatum.*

These values of k, which have been obtained in three different kinds of experiments, show such close correspondence that the relative growth rate of the two species during the larval period, under comparable conditions, might seem to be well

11. Cf. Huxley (1932), pp. 4–11.

established. However, the data for the eye given by Twitty and Schwind (1931) and several sets of unpublished data which Dr. Twitty has kindly placed at my disposal show greater variation in the value of *k*. Nevertheless, the average value of seven different determinations is 1.37. The factors underlying this variation will have to be investigated further, but meanwhile the value may be accepted provisionally as the differential growth ratio of the two species and should be considered as the norm when comparisons are made. Any marked deviation from this value would indicate either that a secondary factor in the internal medium is affecting the growth rates differentially, or that the grafted organ in question is being influenced in its growth by associated structures in the host. The former possibility may be readily eliminated in practice by keeping all of the animals in an experiment under properly controlled conditions with respect to food, temperature, etc.

If the data are insufficient for computing *k,* as is the case when the measurements do not extend over a long enough period of time, then the ratios reached by the end of larval life may be used for comparison—a less satisfactory method but one involving much less labor in recording the data.

Robb (1929) has made the interesting suggestion that a physicochemical analogy for differential growth rates may be found in partition phenomena, of which the simplest type is the distribution of a substance between two immiscible solvents in contact with one another. In this case, the *partition coefficient* is the ratio of the concentrations of the substance in the two solvents after equilibrium is established. It is constant for a given temperature, no matter how much of the substance is present. In chemical reactions between one substance and a mixture of two others, mutually inert, there is also a constant ratio of distribution, or partition coefficient, as in the case of hydrochloric acid reacting with any two of the alkaloids, quinine, codeine, and brucine, provided the amount of acid present is insufficient to convert the total amount of the aikaloids into their respective chlorides. In the organism, the growing cells that react with the internal medium may be supposed to do so in proportion to their affinity for the substances they take from it. These affinities may be different in the various parts of an organism and are almost certainly so in grafts from other species.

Up to this point we have had to do with the comparison of two parts growing in the same organism. When it is desired to compare a grafted organ with its control growing on the donor, the difficulties are enormously increased, for the problem of what constitutes identical conditions in two organisms of different specific constitution is involved. This was not fully appreciated at first, and in the earlier reports of my work a certain amount of confusion was introduced which shows a tendency to be perpetuated. It is, therefore, desirable to digress a moment in an attempt to clarify it.

The comparison of growth rates in two different species of organisms is complicated by an indefinite number of variables, some of which may be quite hidden. Between the two species under discussion, for example, there is an obvious difference that undoubtedly does affect growth rate, i.e. a difference in voracity, which involves a whole complex of neuro-muscular, digestive and assimilative mechanisms, including probably basal metabolism and endocrine function.

Consequent to the tacit assumption, made at the time of the first experiments, that both species of animals were receiving equivalent feeding, a growth-regulating factor (R) was postulated to account for the observation that punctatum grafts on tigrinum larvae were retarded, and reciprocal grafts accelerated, in their growth, as compared with the donor controls. This factor was considered to be of hormonal nature, circulating in the blood of the host, and to act with greater intensity in *A. punctatum* than in *A. tigrinum;*[12] the other constitutional factor, the growth potential of the tissues (G) is obviously greater in the latter species. The observed results were explained by the joint action of the two factors, one in the organ and one in the internal medium, of which there were four combinations.

However, in the experiments of Twitty and Schwind (1931), in which nutrition was better controlled by feeding the animals all they would take, a procedure commonly referred to as "maximal feeding," it was found that the more voracious tigrinum larvae grew relatively much more rapidly and that under those conditions graft and donor control grew at essentially the same rate. In other words, the grafted limb or eye retains its intrinsic growth rate, even when transplanted to another species of animal. There was thus no need to postulate in the present case a special growth-regulating factor other than the "nutrient level." That other substances affecting growth rate differentially do exist is, however, certain. In this connection one need only recall the effect of sex hormones on the growth of secondary sexual characters, or that of the thyroid in accelerating the relative growth of limbs in the frog larva.[13]

The grounds for the magical action of "maximal feeding" in producing the result

12. Experiments to test this were made with the hypophysis, which, it was thought, might supply the hormone, but the results were negative. The punctatum larva with both limb and hypophysis from *A. tigrinum* showed the same disparity in the size of the limbs as cases in which the limb alone was grafted, and similarly a case in which the hypophysis was entirely removed. However, this experiment may not be significant, for the hypophysis itself behaves as other grafts, growing much too large in one combination and too small in the other, so that the difference in size might offset the presumed difference in potency.

13. Cf. Huxley (1932), pp. 176–189.

it does are not clear. In fact it is difficult to define exactly what maximal feeding is. The relative growth rates of *A. tigrinum* and *A. punctatum* can be determined for a given set of conditions, but unless such conditions as temperature, kind of food as well as quantity, and various other factors are rigorously controlled, the results will vary. Unpublished work by Hutchinson and Hewitt[14] shows the importance of considering the kind of food used, even with "maximal feeding." A diet of beef and liver tends to favor tigrinum over punctatum, while small crustacea (Daphnia) alone or white worms (Enchytraeus) alone have the reverse effect, but this may be due in part to the quantity of food ingested and not entirely to the quality. The strain of Amblystoma used may also affect the results, as first suggested by Detwiler (1930). With these facts in view, it is futile to generalize from comparisons of growth rates in animals of two species, or from comparisons of grafts and their donor controls. The best that can be said is that the results hold for the conditions under which the particular experiments were made.

In the above comparisons a further complication arises in the question of relative age. As Twitty has pointed out, physiological age does not necessarily coincide with chronological age, even in two individuals of the same species, if kept under different conditions. If, in two species with similar life and growth cycles, the absolute length of a cycle is different, measurements plotted on an absolute time scale would show a distortion of the relative growth curve. It is possible, however, to reduce the two chronologies to a common scale;[15] certain morphological criteria, such as completion of resorption of yolk, metamorphosis, etc., can be used in fixing points on the two life cycles or growth curves from which the common scale can be computed.

When the two organs compared are by grafting made to be parts of the same organism, most of the above difficulties are obviated, for their internal milieu as well as their external environment may then be assumed to be identical, whatever the nutrient level or the hormonal conditions of the host may be. Questions of feeding and external conditions then recede in importance, since only if the tissues of the two species are differentially affected will the relative growth rate change. The possibility that this may occur must be kept in mind, but it is unlikely that within the range of normal conditions there will be great variation. Twitty and Elliott (1934) have actually found that low temperature retards the regulative growth of eyes grafted

14. I am much indebted to Dr. Cranford Hutchinson and Dr. Dorothy Hewitt Hutchinson for permission to cite their work in advance of publication.
15. Cf. Waddington (1933) and Needham (1934, p. 106). The latter paper sets forth in a very interesting way the application of Huxley's method to the study of chemical changes in the organism (chemical heterogony).

from young to old larvae. On the other hand they have advanced evidence, from experiments on eyes similarly transplanted, to show that the nutrient level, provided it is sufficient to maintain some growth, has no effect on growth regulation, the relative growth of grafts being the same in both slowly and rapidly growing hosts. In these experiments, however, the factor of relative age is introduced, and evidence is still lacking to show whether variations in nutrient level and temperature actually do affect the differential growth ratio, when the relative age of the two components remains the same.

The Relations of Parts in the Development of Composite Organs

The Eye

There are two main constituents of the eye, the optic cup, an outgrowth of the fore brain, and the lens, which develops from the overlying ectoderm. Are these components independent in their growth or are they mutually regulating?

In some species of amphibians the early development of the lens is entirely dependent upon influences emanating from the optic cup; in others it is already self-differentiating before contact with the optic cup is established. This relation has been investigated very fully[16] and need not be considered further here, since we are concerned more particularly with growth regulation. The latter may be studied by grafting heteroplastically each of the two main constituents separately (Harrison 1929a).

To transplant the optic vesicle by itself, one may proceed in either one of two ways. The organ may be placed directly under a loose flap of ectoderm, which is then healed back in place. Since this is difficult to do with precision, the better method is first to graft the ectoderm heteroplastically from one embryo to the other and later, when it has healed, to take the eye with its new covering back to an embryo of the first species.

The number of cases in this series is small. It is clear, however, that the punctatum optic vesicle in a tigrinum embryo lags considerably behind the normal host eye, though less than when eye vesicle and ectoderm are grafted together, and that in the reciprocal experiment the transplanted vesicle, while growing more rapidly than the normal, has a lower relative growth rate than grafts of the whole eye (Table 1).

16. See the excellent review by Mangold (1931).

TABLE 1. Differential Growth Rates of the Eye

Material Transplanted	Combination	k	$\frac{1}{k}$ in Reciprocal Combination	k(P/T) × k(T/P)	i	$\frac{1}{i}$ in Reciprocal Combination	i(P/T) × i(T/P)
Whole eye	*P/T*	0.73	0.74	0.99	0.64	0.56	1.15
	T/P	1.36	1.37		1.80	1.56	
Vesicle	*P/T*	0.87	0.90	0.97	0.77	0.76	1.01
	T/P	1.11*	1.15		1.31	1.30	
Lens	*P/T*	0.87	0.82	1.06	0.81	0.79	1.03
	T/P	1.22	1.15		1.27	1.23	

For P/T $k_v \times k_l = 0.87 \times 0.87 = 0.76$
$k_o = 0.73$
For T/P $k_v \times k_l = 1.11 \times 1.22 = 1.35$
$k_o = 1.36$

For P/T $i_v \times i_l = 0.77 \times 0.81 = 0.62$
$i_o = 0.64$
For T/P $i_v \times i_l = 1.31 \times 1.27 = 1.66$
$i_o = 1.80$

P/T = punctatum graft on tigrinum.
T/P = tigrinum graft on punctatum.
k = relative growth rate of grafted eye (diameter) and normal control on opposite side of the animal.
i = the ratio of diameter of grafted and normal eyes at close of experiment.
The subscripts v, l and o refer to the constants for optic vesicle, lens and whole eye, respectively.
* In this case the relative growth curve shows a decided bend. The value given for k is that which would have produced the end result, had it remained constant. For this reason k is here no more significant than the ratio, i.

After grafting lens ectoderm alone the results are similar. The tigrinum eye with punctatum lens is appreciably retarded in its growth in comparison with the normal eye of the opposite side, while the punctatum eye with tigrinum lens is actually stimulated thereby to more rapid growth than the eye with its own lens. However, the structures here compared, the two eye bulbs, are both derived from the host, so that the measurements reveal not the relative growth rates of graft and host but the influence of the grafted foreign lens on the eye with which it is associated.

The values of the relative growth rates (k), as well as the final ratios or indices (i) of eye diameter, are given in Table 1. They are based upon relatively few determinations (Harrison 1929a) and should not be considered as final but as subject to revision when more adequate data are available. They are offered here as an example of how the method of heteroplastic grafting may be used for the quantitative study of development. The values of k, given in the third column of the table, have been determined from the experimental data by Huxley's method. The reciprocals in the fourth column are calculated from the values in the third, as a means of checking the latter. As a further check, the fifth column gives the products of the values of k in reciprocal experiments. These should be close to unity where the specific qualities of the host do not influence the relative growth rate. Still another check is offered by comparing the relative growth rate, when the whole eye is transplanted, with the product of the rates for grafts of the two constituents separately. The close correspondence shown in these cases (see middle section of table) indicates that the combined effect of the two constituents on the growth of the eye bulb is a simple multiplicative one.

In the last three columns of the table the final ratios or eye indices (i) are given in a similar way. The comparisons, using these constants, are not so reliable, since the relative age of the hosts of the two species is involved. The discrepancy shown in the value for whole eye grafts on *A. tigrinum* probably indicates that the final measurements were taken too early.

The corresponding relative growth rates of the lens cannot be computed from the data available, and the final lens indices, given previously (Harrison 1929a), are less reliable than those for the optic cup. Nevertheless the data at hand show that the growth rate of the lens is affected by that of the optic cup. In other words, there is a mutual adjustment between the two components. The effect of the lens on the bulb is fully as great as, if not greater than, the effect of the bulb on the lens, and the mutual adjustment seems to be more perfect, i.e. the lens-eye ratio approaches more nearly the normal in those combinations in which initially the lens is too large for the eye cup. The study of a few early stages of punctatum eyes with tigrinum lenses shows that the lens grows unhindered at first and may almost fill out the whole cup (Fig. 23); also that the effect on the size of the eye bulb scarcely begins to be shown until after complete resorption of the yolk.

The eyes that are out of scale with the rest of the organism have a far-reaching effect on the associated organs, extending to the central nervous system and to the extrinsic muscles. Although extirpation of the eye does not do away with the muscles entirely, they are smaller and lose their typical arrangement in part. In both the small and the large grafted eyes, as Twitty (1932) has shown, the size of the muscles is very neatly adjusted to that of the eye (Figs. 24 and 25). This is not due to hypertrophy or atrophy of the individual muscle fibers, but to their actual increase or decrease in number.

The Fore Limb

Aside from nerves and vessels, which are not included in the grafted material but grow in from the host, the two main constituents of the limb are the ectoderm and the mesoderm. The former gives rise only to epidermis and skin glands, while the latter forms muscles and skeleton.

Although dissenting opinion has been expressed, the weight of evidence points to the view that the mesoderm is the dominant element in determining the character of the limb. The results of heteroplastic grafting of the two constituents separately fully bear out this view. Such experiments have been carried out independently by Rotmann (1929, 1931, 1933) with two species of Triton, and by Harrison (1931) with Amblystoma.

The experiments show clearly that the form of the limb, as well as its growth rate and ultimate size, is fixed in large measure by the mesoderm; however, some influence of the ectoderm can be detected. The limbs of the two species differ markedly in form; the digits in *A. punctatum* are nearly cylindrical, while in *A. tigrinum* they are broad at the base, each tapering to a blunt point, and are much more distinctly webbed. Some of the observations seem to indicate that in the punctatum limb with tigrinum ectoderm the web is slightly more definite than in the normal punctatum limb, and that in the reciprocal combination the reverse is true. This, however, does not always hold, and the difference in any case is scarcely measurable.

The effect of the ectodermal covering upon the early development of the limb bud is more definite. It will be recalled that the tigrinum limb, though ultimately much larger, is retarded at first, having a very long latent period. When, in this species, the mesodermal bud is covered by punctatum ectoderm, the latent period is shortened a little and the early phases of growth and differentiation are accelerated (Fig. 5). Forty cases out of 47, or 85 per cent, including a few experiments with *Triturus*

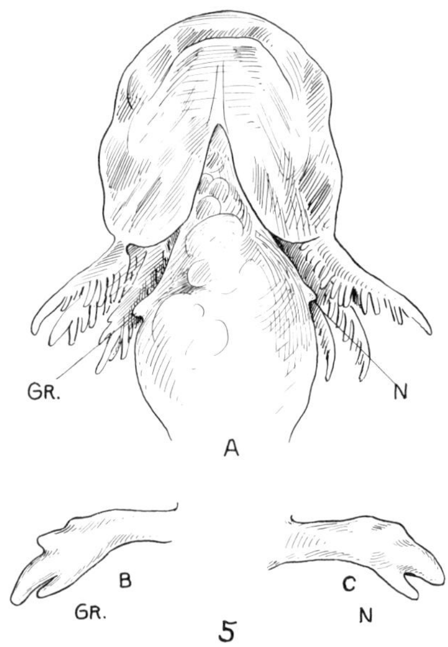

FIG. 5. A. Larva of *A. tigrinum* (Exp. NEE 20), ventral view, 15 days after operation; the right limb (GR), covered by ectoderm of *A. punctatum,* is more advanced than the normal left (N). B. The right grafted limb, 27 days after operation, is still advanced in its differentiation beyond C, the normal. × 10.

torosus and the axolotl, showed this effect. In the reciprocal combination the effect in the opposite sense is neither so marked nor so constant.

The records are too incomplete for plotting relative growth curves, but the final measurements indicate definitely a slight effect of the ectodermal graft on the size of the limb (Figs. 26 and 29). All of the measurements that have been studied statistically show small but consistently significant differences from the normal, ranging from 2 to 9 per cent, whichever species furnishes the ectoderm. Punctatum limbs with tigrinum ectoderm are relatively larger, while in the reciprocal combination the limbs are smaller.

The results of Rotmann's (1933) experiments, in which ectoderm of *Triton taeniatus* was grafted over the mesodermal bud of *T. cristatus,* are to some extent different; no initial effect was observed, but in late larval life a deficiency in dimensions of the grafted appendages, estimated at from 10 to 15 per cent, was found. No data are given for the reciprocal experiment.

Grafting the mesoderm alone (Figs. 28 and 31), as might be expected from the preceding, gives results very similar to those of grafting both layers together (Figs.

27 and 30). This experiment is best done in two stages, since it is almost impossible to handle the mesoderm by itself. Ectoderm of *A. punctatum* is first grafted over the limb region of a tigrinum embryo. After it is well healed, usually the next day, the limb bud with its transplanted covering is grafted back to an embryo of the former species. The limbs developing from such grafts are similar to those from whole buds, both in form and in dimensions, as is also the case in the reciprocal combination. In view of the effect of the ectoderm, when grafted alone, we should expect them to be slightly reduced in size, as compared with whole buds, but the evidence on this point is inconclusive.

Schwind (1931) has obtained mosaic limbs, or sectorial chimeras, by grafting portions of the limb bud heteroplastically. In such cases the two components of the rudiment maintain their specific characters; hands may, for instance, have one or more digits from each species. The outlying parts of the girdle, such as suprascapula and coracoid, which are developed from the host, are, however, influenced in their growth by the more centrally lying tissue of the graft; the girdle adapts itself to the size of the free limb (Schwind 1932; Harrison unpublished).

The Gills

The respiratory organs of lower vertebrates are complicated with regard to their embryonic origin. They are essentially a modification of the pharyngeal wall, involving all three of the germ layers and containing a fourth element, the mesectoderm, which grows down from the ganglion crest and forms the visceral cartilages, including jaws, hyoid and branchials. This one of the many roles of the ganglion crest was demonstrated by Landacre (1921) in normal development and by Stone (1922) experimentally. Heteroplastic grafts give independent evidence for this view and afford, besides, a means of studying the interaction of the constituents in regulating growth rates.

The branchial system of Amblystoma consists of a series of inverted arches between which, on each side, there are openings or gill slits leading from the pharynx to the exterior. Three of these bear plume-like external gills, which are the principal respiratory organs during larval life, disappearing at metamorphosis. The gill slits are developed out of pouches of the pharynx, which grow laterally until they reach the ectoderm, where they break through to the exterior. The mesoderm of the body wall, which in this region is not split into splanchnopleure and somatopleure, is thus cut into a series of arches, into which the mesectoderm grows, comporting itself in a very peculiar way. It first extends ventrally between ectoderm and mesoderm, then wraps itself around the mesodermal core, gradually withdrawing toward the inner side of the latter, where it takes up its permanent position between mesoderm and endoderm and differentiates into cartilage.

The mesoderm thus, curiously enough, does not give rise to the skeleton of the gills except to a small cartilage in the ventral mid line known as the second basibranchial (Stone). As far as is known, it does give rise to the muscles and probably to some connective tissue. The ectoderm forms the outer epithelium of the gills, but, according to Greil (1906), a layer of endoderm from the visceral pouches grows out under it. This circumstance may have something to do with the importance of the inner layer in the development of the gills (cf. Severinghaus 1930).

In the embryo, about the time of closure of the neural folds and a little later, the layers may be readily separated from one another and transplanted in a great variety of ways, as first done by Ekman (1913). In heteroplastic experiments (Harrison 1929) the practice has been to graft the material in normal orientation, so that the only variable introduced is that of species.

The ideal mode of experimentation for analyzing the intrinsic growth factors would be to graft each constituent separately and the various combinations of two, three and four of them in reciprocal pairs. Of the 30 possible combinations, 12 have been carried out, including all four layers together, ectoderm alone, mesoderm alone, ganglion crest (mesectoderm) alone, ectoderm and mesoderm combined, and ectoderm, mesoderm and endoderm combined. Some of these have been done only in small numbers.

The whole complex is most readily grafted after the ganglion crest has grown down, at which time an oval piece, which includes the whole thickness of the pharyngeal wall, may be interchanged between the embryos of the two species (Fig. 10, D and E). The main difficulty is in healing along the ventral border. When this is delayed, the circulation through the branchial vessels is rarely established, and since the circulating blood is an important factor in the expansion of the gills, such cases are useless for the study of their growth.

In the punctatum embryos with tigrinum pharyngeal wall the earliest development of the gills is essentially normal. The precocity that one might expect from the more rapid development of the tigrinum embryo is frequently counteracted by the delay due to the operation. The most striking feature in the grafted embryo is the marked lateral bulge in the posterior part of the head, which sets in shortly before the tigrinum control begins to feed and is due mainly to the precocious development of the visceral skeleton (Fig. 6). The gills themselves also become longer than the normal punctatum gills, provided circulation is established[17] (Figs. 8 and 37, B). In the reciprocal experiment the posterior part of the head is sharply indented and the gills shorter (Figs. 7, 9, and 37, A). The measurements taken are insufficient for plotting a curve of relative growth, but the length index at maturity for tigrinum gills on punctatum larvae averages 1.50, and for punctatum gills on tigrinum 0.68. These figures show less difference between grafted and host organs than in the case of either eye or limb—a result attributable to functional regulation, to which process the gills are known to be highly susceptible.

An early experiment by Spemann (1921), followed by a more complete study by Rotmann (1931), indicates that in Triton the ectoderm has some influence upon the time of appearance of the external gills and upon their size during early development. My own experiments have been rather inconclusive on these points, but in my first report (Harrison 1929) it was stated that the ectoderm had no effect on the growth rate. Later experiments indicate a slight influence of this layer on the rate in early stages, confirming the observations of Spemann and Rotmann.

Measurements of later stages show that the ectoderm does have quite an appreciable effect on the subsequent growth of the gills (Fig. 40, A and B). Including only the second and third gills, which are less subject to traumatic influences than the first, there is an index of 1.13 for punctatum gills with tigrinum ectoderm and one of 0.81 for the reciprocals. Neither the form of the gills nor their pigmentation is noticeably altered by the strange ectodermal covering.

Experiments with the branchial mesoderm alone or in combination with ectoderm (Fig. 39) have not given constant results, owing probably to increased trauma. With more numerous experiments at hand, I am doubtful of upholding the statement made in 1929 that this layer has a slight influence on the size of the gills. Rotmann's results also point to a negative answer to this question.

Far more conclusive are the experiments with the ganglion crest, which is shown to have a profound effect upon the development of the branchial system, particularly the visceral skeleton. Transplantations of this constituent are best done in two stages. When the neural folds are just closing, the whole of the hind brain, together with covering ectoderm, is interchanged between the two embryos (Fig. 10, A and B). If left to develop for about two days, or until the tail bud is well defined, the ganglion crest grows down from the graft into the pharyngeal wall (Fig. 10, C, shaded area). The second operation, which consists in transplanting the pharyngeal wall, with the

17. This is in agreement with Severinghaus (1930), who found further that the grafted gills in punctatum larvae were even larger than those of the donor, in spite of heavy feeding.

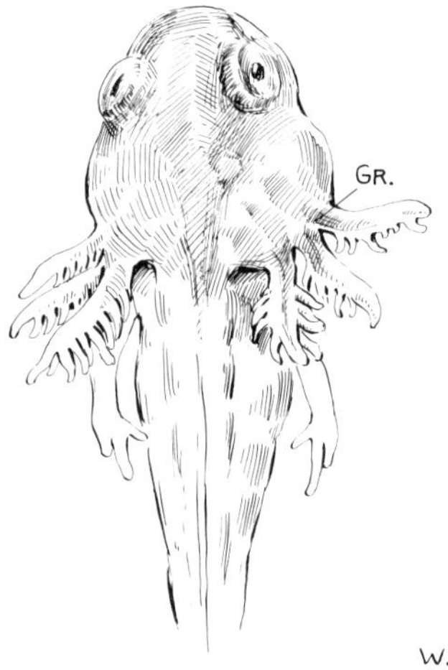

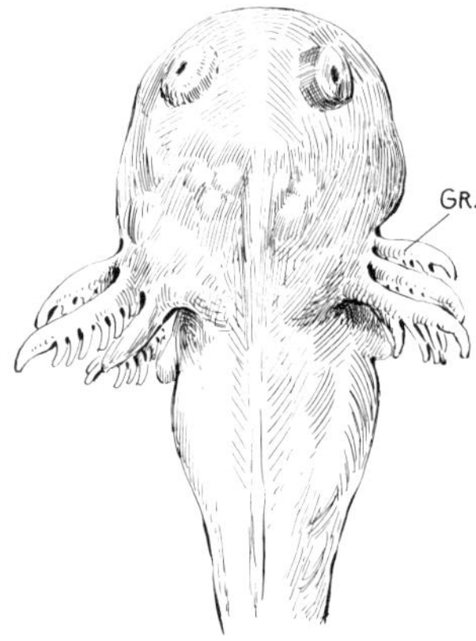

W.Br. 57
23 DAYS.

6

7

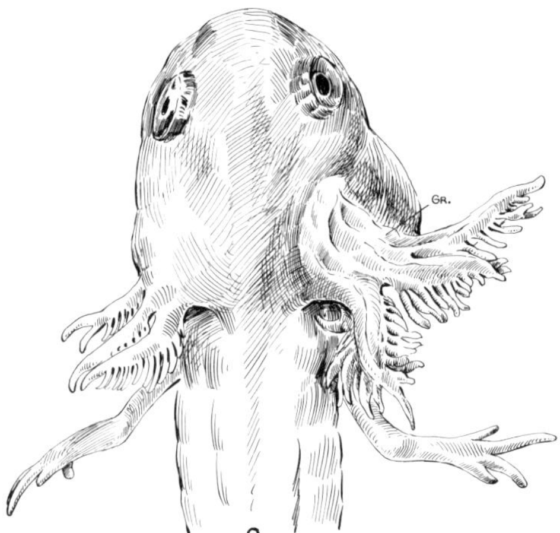

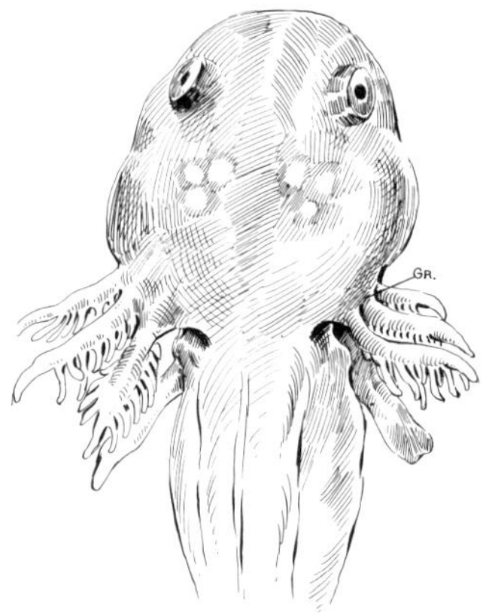

8

W.Br. 57
36 DAYS.

9

FIG. 6. *A. punctatum* with gill complex (GR) transplanted from a white axolotl.
FIG. 7. Reciprocal experiment. × 7.5.
FIG. 8. Same larva as in Fig. 6, 13 days later.
FIG. 9. Same larva as in Fig. 7, 13 days later. × 6.7.

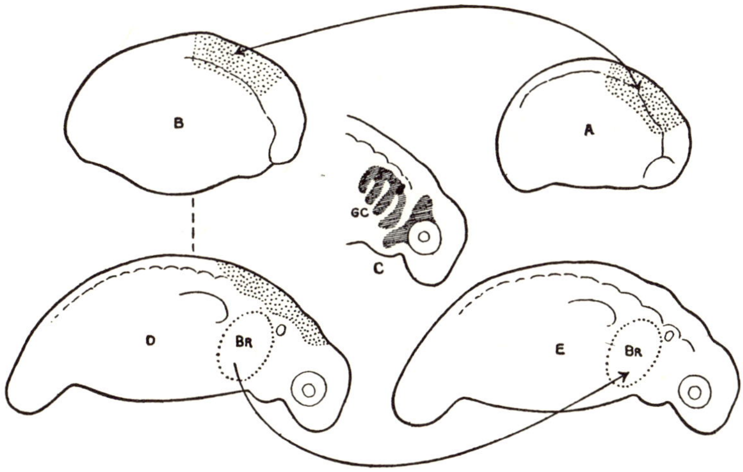

FIG. 10. Diagram to show the method of grafting the ganglion crest (mesectoderm) of the gill region. In A and B the dotted areas represent the parts interchanged at the first operation. In D and E, showing embryos two days later, the areas enclosed in the dotted oval are interchanged at the second operation. In C, the shaded area (GC) shows the distribution of the ganglion crest in the head.

foreign ganglion crest, to its proper position in an embryo of the host species, is then carried out (Fig. 10, D and E, *Br*).

Embryos so fashioned are superficially like those in which the whole pharyngeal wall is interchanged. In punctatum larvae with tigrinum ganglion crest, the posterior part of the head bulges (Fig. 38, A, left side), the operculum is relatively too small and the gill plates are exposed from below. In the reciprocal combination, the posterior quadrant of the head is indented (Fig. 38, B, left side), and the operculum closes down over the gill plates. The gill arches in the first combination are very large, being both thicker and longer than the normal; they are bowed laterally to such a degree that the animals have difficulty in swallowing on account of the wide slits (Fig. 35). Early stages show the precocious development of tigrinum cartilages in comparison with the punctatum tissues surrounding them. The punctatum gill arches in the tigrinum embryo are correspondingly small and lag in the early stages of their differentiation.

Another instructive experiment is to graft the composite pharyngeal wall to an embryo of the donor species, whereby an individual is obtained in which ectoderm, mesoderm and endoderm are foreign to the host, and only the ganglion crest or mesectoderm is of the same species. The punctatum larvae with the tigrinum grafts show a moderate bulge in the posterior quadrant of the head, but much less than when the whole pharyngeal wall is taken and considerably less than when the ganglion crest alone is from tigrinum. In the reciprocal cases, the indentation of the posterior part of the head is much less than when the whole wall is taken, and a little less than in experiments with the ganglion crest alone. One is led to the conclusion that the greater part of the disproportion in the former case is due to structures developed from the mesectoderm, and that a lesser though appreciable remainder is due to mesodermal elements. The latter is borne out by the experiments in which mesoderm by itself or with ectoderm was interchanged. In these cases the branchial arches themselves are of normal size, though they may be bowed differently from the normal owing to differences in the dorsal part of the head just described.

The gills themselves, as far as the measurements available show, are intermediate in size between those of the two species used, when the ganglion crest alone is transplanted. They are apparently not so disproportionate as when the whole pharyngeal wall is taken, but the observations are too few and too variable to warrant a definite statement regarding the extent to which this layer influences their growth. The complementary experiment, in which all but the ganglion crest is taken, supports the conclusion that this layer is an important factor in fixing the size of the gills.

Prevention of the ganglion crest from growing into the pharyngeal walls results in the absence of the branchial cartilages, as previously shown by Stone (1926, 1929) by means of a different method. In such cases the gills themselves are extremely defective (Fig. 32).

The Heart

Heteroplastic transplantation of the embryonic heart in whole or in part, which has been carried out by Copenhaver (1927, 1930, 1933), affords a means of studying other problems than growth, although its bearing on the latter topic is not without special interest.

The heart rudiment is best grafted when very young, before it begins to beat. The tigrinum heart after transplantation has at first a rapid rate of growth, and after a time it may attain a mass double that of the control in the normal punctatum individual. Later, however, the relative size of the graft decreases until, at metamorphosis, it is normal for the host. The failure of the grafted heart to reach and maintain its specific size may be ascribed to the regulative action of function. In the reciprocal combination the organ at first grows less rapidly in proportion to the whole embryo, soon becoming much too small. In the first cases reported the embryos became oedematous and did not survive. Dr. Copenhaver informs me, however, that he has now obtained thirteen cases which lived until the time of complete yolk resorption or later, one of them surviving metamorphosis. The regulation of the heart by hypertrophy begins quite early and is completed in late larval life, by which time it reaches full size in proportion to the body of the host. In other words it grows, under functional regulation, far beyond the size it would have attained under natural conditions. The heart is thus an organ in which intrinsic growth capacity ultimately gives way entirely to functional regulation in determining the size to which it grows.

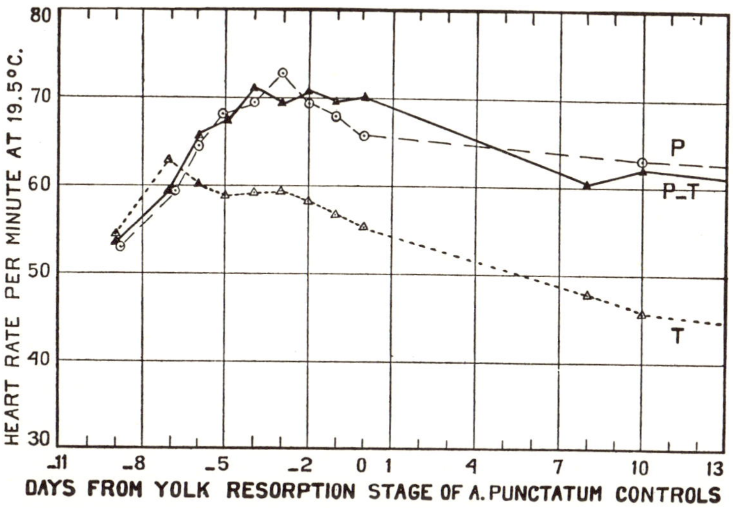

FIG. 11. Graph to show the pulse rate in the grafted heart and its controls; P, punctatum heart; T, tigrinum heart; P–T, punctatum heart grafted to tigrinum. Through the courtesy of Dr. W. M. Copenhaver.

Copenhaver (1930) has also made experiments in which only the anterior or the posterior part of the heart rudiment was transplanted. In these cases, too, no matter how combined, the tigrinum component grows more rapidly than the punctatum, so that at first there is not only a discrepancy in size between graft and host but also between the two parts of the composite heart. Later there is a mutual regulation of the two parts, and the size differences become less and less until the whole organ is regulated to normal proportions with respect to the size of the larva. These experiments succeed in both reciprocal combinations. The punctatum grafts on tigrinum, though for a time too small for the embryo, nevertheless succeed in establishing themselves. The moderate oedema that ensues is later cleared. The functional regulation between

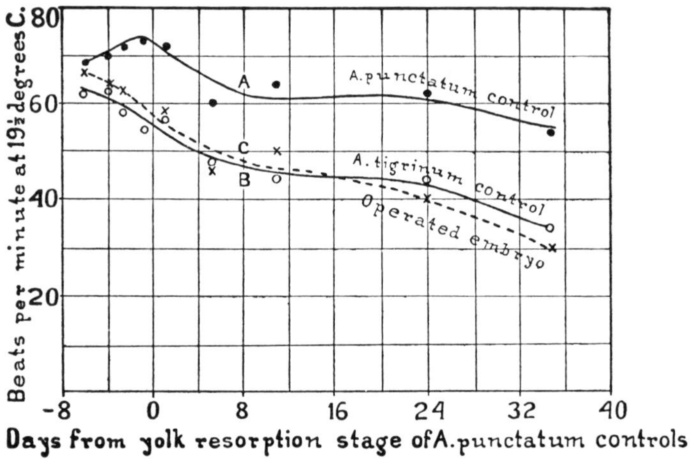

FIG. 12. Pulsation rate of the heart in *A. punctatum* with the posterior part of the heart (sinus and atrium) from *A. tigrinum* (Exp. PS 30); also the rate in the two controls. After Copenhaver.

body size and heart size is more quickly completed when only part of the heart is disproportionate than when the whole is.

One of the most interesting features in these cases is the rate of pulsation of the transplanted hearts. The normal rate varies, of course, with temperature and with the stage of development, but at corresponding stages there are very large differences between the rates in the two species used. After the yolk is entirely resorbed the rate for *A. punctatum* at 19.5° C. is in the neighborhood of 60 per minute, while that for *A. tigrinum* is around 40. The whole tigrinum heart when grafted to a punctatum embryo beats for some time at about the same rate as the normal tigrinum control, and while it does rise above the latter in later larval life, it never approaches the rate of the punctatum control. The pulse curve of the punctatum heart in a tigrinum individual follows even more closely the curve for the donor species (Fig. 11). Since the operation is done before nerves have developed, the innervation of the grafted heart must be entirely from the host, as is also the blood supply. That its pulse rate remains approximately that of the donor control shows that it is determined mainly by its inherent qualities and is influenced by the nervous system and the blood stream only to a minor degree.

The experiments with half hearts also show clearly that the posterior part sets the pace. The pulse has been recorded in all four types of cases, and in three of them the rate of the species furnishing the posterior half is found to obtain (Fig. 12). In only one combination, the tigrinum embryo with the posterior part of the heart from punctatum, is the pulse rate intermediate between the normal rates of the respective species, and even here, during the first month of larval life, the rate approximates that of *A. punctatum*.

The Ear

Richardson's (1932) experiments with the internal ear show that the grafted rudiment tends to maintain its intrinsic growth rate, but is influenced by another factor acting in the opposite sense. In the same interspecific combinations as those previously considered, individuals are obtained with ears that are either too large or too small. The ratio of graft to normal is, however, very low in punctatum larvae with tigrinum ears, and very high in the reciprocal combinations. The volume index of the grafted tigrinum ears averages only 1.9, while in the case of eyes and limbs it is approximately 1.8^3 or 5.8. According to Miss Richardson the secondary factor modifying the intrinsic growth rate of the graft is probably a mechanical one; the ear vesicle is a delicate thin-walled structure, requiring space in which to expand, and the tigrinum ear, located on the smaller punctatum head, is more restricted than in its normal environment. In the reciprocal cross, the punctatum ear finds itself in surroundings that permit of expansion beyond the normal.

The cartilaginous ear capsule, which is formed around the membranous labyrinth, is also affected by the size of the latter. The large tigrinum ear in the punctatum larva acquires a larger capsule than the normal ear of the host, but it is not so large in proportion to the size of the labyrinth. Correspondingly, in the reciprocal experiment, the cartilaginous capsule of the grafted ear is actually smaller than the normal but again not proportionately so in respect to the mass of the labyrinth.

The Spinal Cord

Perhaps the most extreme cases of size regulation in a grafted organ is to be found in the spinal cord, where, according to Detwiler (1931, 1932 a and b), the graft soon accommodates itself to the dimensions of the nervous system of the host. The tigrinum cord at a given age is relatively thinner than the punctatum, and for a considerable period its excess growth is mainly in length. When the brachial region of the spinal cord of a punctatum embryo, including the third, fourth and fifth trunk segments, is replaced by a similar tigrinum graft, the latter at first elongates at the same rate as the normal tigrinum cord. At the same time it increases so rapidly in thickness as to surpass controls of both species.

For twenty days or more the grafted cord is both too long and too thick, and it contains more cells per cross section than the corresponding regions of normal larvae. Then a process of size regulation sets in, and at about thirty days the effect of the early excess growth has largely disappeared. The grafted piece ultimately becomes adjusted precisely to the size of the host. Punctatum grafts on tigrinum are also regulated in a corresponding way, but in this combination, the process is one of increased growth on the part of the graft.

While this regulation may be assumed to be in part due to the mechanical conditions surrounding the grafted segments of the spinal cord, which prevent its expansion much beyond the normal, other factors must also be called into play. It is known, for instance, from other experiments that the number of cells in many parts of the central nervous system is profoundly influenced by the stimuli which reach them through sense organs or through their central connections. Such factors would tend to regulate the grafts to proper proportions with respect to the rest of the central nervous system and the peripheral regions connected with them.

Wieman (1926), who made the first heteroplastic transplantations of the spinal cord, found that the grafts had no effect on the rate and manner of growth of the limbs innervated by them.

Effects upon the Nervous System of Grafted Organs That Are out of Scale

Within the past fifteen years a large mass of evidence has accumulated to show the effects of peripheral loading upon the growth and differentiation of the nervous system. Heteroplastic grafting affords a means of instituting unusual loading under conditions that deviate little from the normal. The eye and other sense organs, as well as the limbs, have been used for this purpose.

In the retina the number of rods and cones is proportional to the square of the radius of the eye, so that in the large grafted tigrinum eye there may be nearly four times as many of these elements as in the punctatum eye of the opposite side (Harrison 1929a). The number of cells in the ganglionic layer is not proportionately so large, since, as the eye grows, the differentiation of these cells lags, but the absolute number in the grafted eye is nevertheless much larger, varying from about 50 to over 100 per cent in excess of the normal. The number of optic nerve fibers is correspondingly large. If the proper connections with the brain are established, the unequal chiasma is very plain (Fig. 21), the larger nerve crossing to the opposite side of the midbrain, where cellular hyperplasia is induced (Fig. 22). Twitty (1932) finds that the amount of hyperplasia varies from 12 to 22 per cent, as measured by the number of cells, while according to sectional area of gray matter it varies from 9 to 16.5 per cent, indicating that cell size does not quite keep pace with cell multiplication.

Corresponding effects in the opposite sense occur in the reciprocal combination, in which the transplanted punctatum eye remains small. In cases where the optic nerve from the grafted eye does not connect with the brain, the hypoplasia is more pronounced than when the small eye is properly connected.

The nasal pit of *A. tigrinum* likewise grows large when grafted to the punctatum embryo, but according to figures given by Burr (1930) the percentage increase over the normal host control is relatively low. For example, in one animal, preserved two months after operation, the grafted olfactory organ is only one-third larger than normal by weight. This would indicate, in comparison with the indices for eyes and limbs, that the surrounding parts had an inhibiting effect on the growth of the olfactory epithelium. The larger nasal organ has a larger olfactory nerve, and the hemisphere of the brain with which it is connected is increased in volume by about 9 per cent. No important structural changes were noted in the fore brain beyond a slight increase in the number of olfactory glomeruli.

The effect of transplanted tigrinum ear vesicles on the acoustic area of the hind brain (Richardson 1932) is to increase the number of cells, but the amount of hyperplasia varies considerably. Hypoplasia following the substitution of a punctatum ear vesicle in a tigrinum embryo is detectable though less striking.

The disproportionate growth of grafted limbs also affects the nervous system, but in a different way. The intrinsic nerve trunks are approximately proportional to the size of the limb, and in the sensory nerves at least there is an actual change in the number of neurones involved, for the spinal ganglia have fewer cells when the grafted limb is small and a larger number when it is large (Schwind 1931). These responses are, however, not so constant as those of the mid brain after transplantation of the eye. The motor neurones in the spinal cord do not show similar adjustments, the number remaining approximately normal—a confirmation of Detwiler's conclusion that the number of motor neurones is unaffected by the peripheral field.[18] The exi-

18. This is a controversial question and there are probably differences between urodele amphibians on the one hand and anurans and birds on the other.

gency of a disproportionate peripheral area to be innervated is probably met by the motor nerves through regulation of the frequency of axonal branching.

Pigmentation

The pigmentation of the grafted organs, particularly the limbs, raises questions of a nature altogether different from those previously considered, though none the less interesting. Each species has certain characteristics of pigmentation, and the axolotl occurs in two varieties, one dark and one albinotic. The pigment cells are very diversified, but three main types are recognized: cells with dark granular pigment, or melanophores; cells with a diffuse yellow lipoid pigment, or xanthophores; cells containing guanin, a granular pigment of metallic lustre, silver or golden, known as guanophores. All types, particularly the melanophores, show differences in color, form and distribution. Each spieces of Amblystoma thus has not only a characteristic ground shade but also a specific color pattern.

The transplanted limbs, in respect to their pigmentation, resemble the limbs of the host species in all combinations except those in which the white axolotl takes part. The resemblances show themselves in early larval life but become more striking after metamorphosis (Figs. 18, 19, and 47 to 50). The pigment pattern, on

---

Plate 1

FIG. 13. *A. punctatum* with large right limb from *A. tigrinum,* seventy days after operation (Exp. NE 13).

FIG. 14. Same individual in metamorphosis two days later; ventral view of preserved specimen. × 1.2.

FIG. 15. *A. jeffersonianum* with right fore limb from *A. texanum,* about eleven months after operation. (Exp. NE 239). × 2/3.

FIG. 16. *A. texanum* with right fore limb from *A. jeffersonianum,* reciprocal of animal shown in Fig. 15 (Exp. NE 239). × 2/3.

FIG. 17. *A. tigrinum* with small right limb from *A. punctatum,* seventy-nine days after operation (Exp. NE 10). × 1.2.

FIG. 18. *A. jeffersonianum* with large right fore limb from *A. mexicanum* about seven months after operation (Exp. NE 190). × 2/3.

FIG. 19. Axolotl (*A. mexicanum*) with small right limb from *A. jeffersonianum,* reciprocal of animal shown in Fig. 18 (Exp. NE 190); age, not recorded, about 15 months. × 2/3.

FIG. 20. Larva of *A. punctatum* with large right eye from *A. tigrinum,* four months after operation (Exp. Oc 20). × 1.2.

13
14
15
16
17
18
19
20

Plate 2

FIG. 21. Cross-section through the diencephalon of a larva of *A. punctatum* (Exp. Oc 23) 41.5 mm. long, showing the difference between the two optic nerves; the large one on the right (left of figure) comes from the larger transplanted axolotl eye. × 38.

FIG. 22. Cross-section through the mid brain of a similar specimen (Exp. Oc 10), showing the enlargement of the left side (right of figure), which receives the larger optic nerve. × 38.

FIG. 23. Cross-section through the head of an embryo of *A. punctatum* with grafted lens ectoderm (Exp. LE 15). The lens developed from the graft (left side of figure) is larger than the normal. × 38.

FIG. 24. Ventral view of a dissection, showing brain, eyes and eye muscles, of a recently metamorphosed *A. punctatum* with grafted tigrinum eye (gr.). The muscles of the grafted bulb are distinctly enlarged. 1, retractor bulbi; 2, external rectus; 3, inferior rectus; 4, inferior oblique; N, optic nerve. After Twitty.

FIG. 25. Ventral view of a dissection, showing the eye muscles of an 85 mm. larva of *A. tigrinum* with a grafted punctatum eye (gr.). The muscles of the grafted eye are reduced in size. × ca. 5. After Twitty.

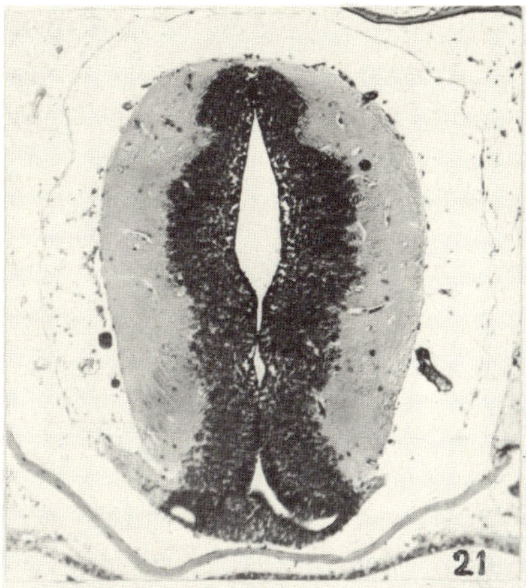
21

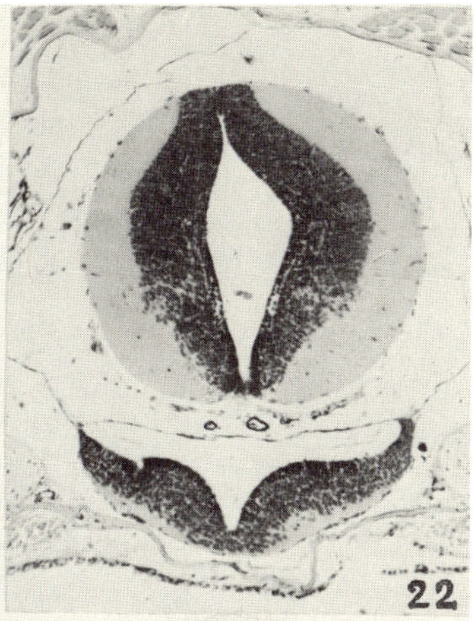
22

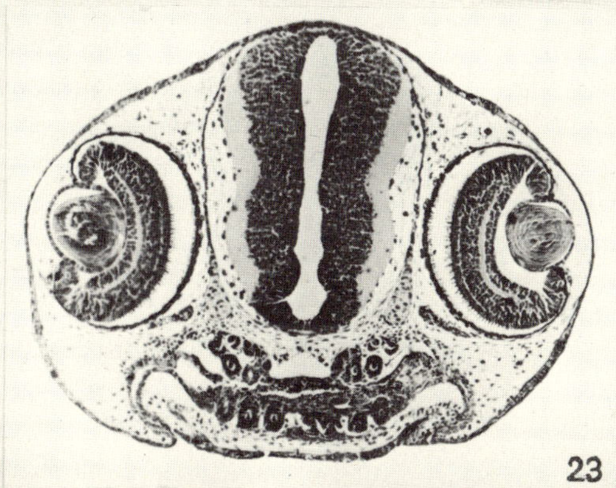
23

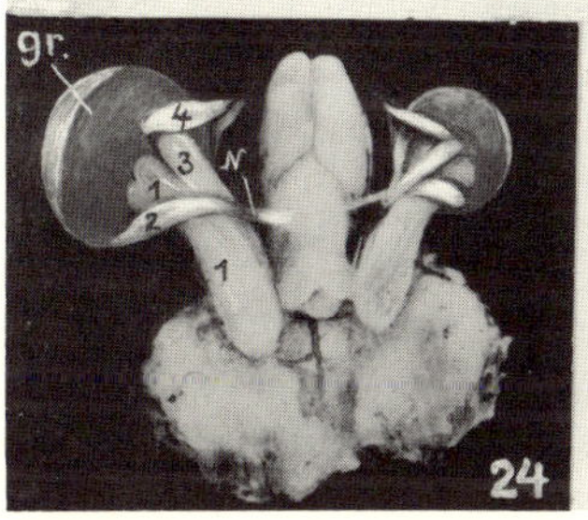
gr.
4
3
1
2
N
1
24

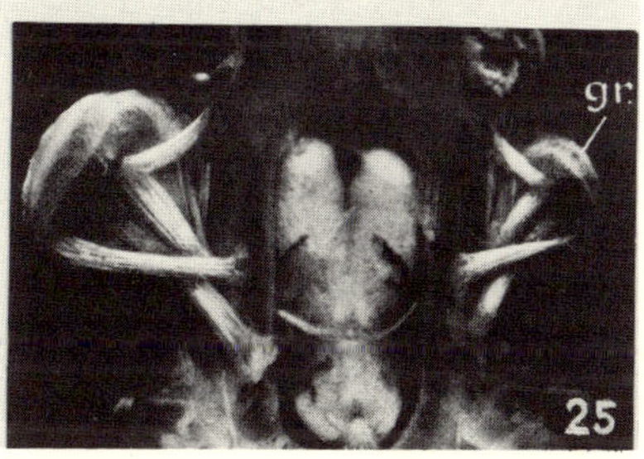
gr.
25

close examination, is found to be not a perfect imitation of that of the host but a very close approximation. This is seen clearly in the reciprocal combinations of the axolotl and *A. jeffersonianum* (Figs. 41 to 44). In the latter species there is a purplish brown ground color, which is flecked with small bluish white spots. After metamorphosis both ground color and spots are divided into small areas, each surrounding a skin gland (Fig. 41). Since the latter are larger in the axolotl (Fig. 42), the texture of the skin appears coarser, though the general pattern is very similar in both limbs.

These facts can be satisfactorily explained only through a full knowledge of the source of the pigment cells. Recently DuShane (1934) has made an important advance in this field, showing, in corroboration of a considerable mass of casual observations by others, that pigment cells, at least in large part, are derived from the ganglion crest. Thence they migrate ventrally in the early motile stages of the embryo.

The round or oval yellow spots in *A. punctatum,* not found in any of the other species, are an important feature in the study of pigment. They occur on the head, on the back, where they form a double row, and also on the limbs (Fig. 48), where,

---

Plate 3

FIGS. 26 to 31. Set of six specimens to show the relative effect of grafting the whole limb bud and each of its constituents separately in reciprocal experiments between *A. punctatum* (figs. 29 to 31) and *A. tigrinum* (figs. 26 to 28); operations on the right fore leg; all material of each species from the same bunch of eggs. × 1.

FIG. 26. Ectoderm alone P/T (Exp. NEE 75); the grafted limb is almost of the same size as the normal; 87 days after operation.

FIG. 27. Whole limb bud P/T (Exp. NE 155); 89 days.

FIG. 28. Mesoderm alone P/T (Exp. NEM 27); 88 days.

FIG. 29. Ectoderm alone T/P (Exp. NEE 74); 87 days.

FIG. 30. Whole limb bud T/P (Exp. NE 155); 89 days.

FIG. 31. Mesoderm alone T/P (Exp. NEM 28); 88 days.

FIG. 32. *A. punctatum,* in which the ganglion crest was excluded from the branchial region on the left side (Exp. EBrME 6). Note the almost total absence of gills on that side; 85 days after operation. × 1.3.

FIG. 33. *A. tigrinum* with right fore limb from white axolotl (Exp. NE 186), seven months after operation. × half.

FIG. 34. *A. jeffersonianum* with right fore limb of white axolotl (Exp. NE 243), about ten months after operation. × 1.

FIG. 35. Larva of *A. punctatum* with mesectoderm (ganglion crest) of *A. tigrinum* (Exp. BrME 7), 88 days after operation; ventral view of cleared specimen stained in methylene blue to show cartilage; note the very large branchial arches on the right (left of figure). × 4.7.

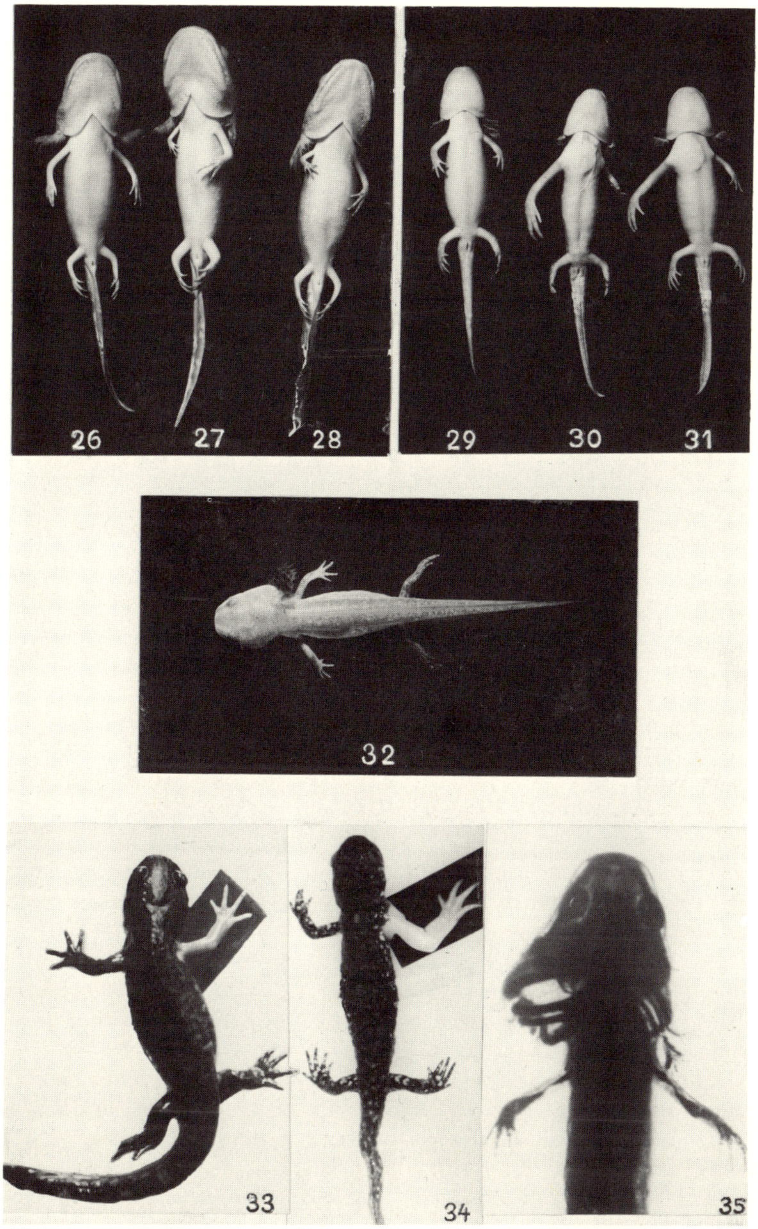
26
27
28
29
30
31
32
33
34
35

Plate 4

FIG. 36. Series of parabiotic pairs, *A. tigrinum* on the left and *A. punctatum* on the right in each case, showing the relative growth of the two species. Original photographs by Dr. R. K. Burns.

FIG. 37. Reciprocal pair (Exp. WBr 62), showing the effect of interchange of the whole pharyngeal wall (right side) between *A. tigrinum* (A) and *A. punctatum* (B). In A, note the small grafted gills and the indented head; in B, the very large gills arising from the projection on the side of the head. × 1.

FIG. 38. A Larva of *A. punctatum* (Exp. BrME 12) with ganglion crest from *A. tigrinum.* The left side of the head is bulged out owing to the abnormally large branchial arches. B. Larva of *A. tigrinum* with punctatum ganglion crest (Exp. BrME 2). × 1.

FIG. 39. A. Larva of *A. tigrinum* with gill ectoderm and mesoderm on right side from *A. punctatum* (Exp. Br 111); gills irregular in length. B. *A. punctatum* with gill ectoderm and mesoderm from *A. tigrinum* (Exp. Br 139); right half of head a little broader. × 1.

FIG. 40. Reciprocal pair (Exp. BrE 181) in which gill ectoderm was interchanged between *A. tigrinum* (A) and *A. punctatum* (B). × 1.

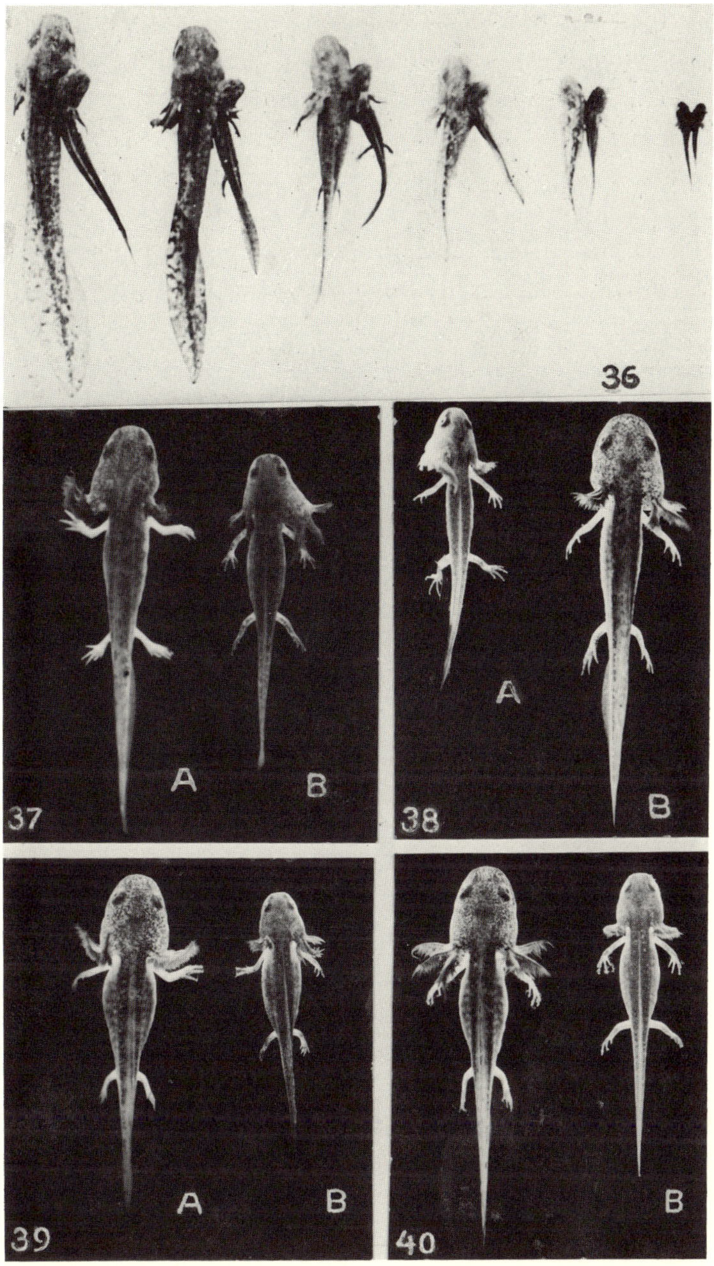
36
A
B
37
A
B
38
A
B
39
B
40

however, they are smaller and their distribution inconstant; sometimes a limb may lack them entirely (Fig. 45). On limbs of all of the other four species, when grafted to punctatum hosts, such spots have been found (Figs. 46 and 47) and they seem to occur with greater regularity in the transplanted limbs than in normal specimens. No yellow spots have ever been found on punctatum limbs growing on any of the other species (Fig. 50).

The grafts to or from the white axolotl give an entirely different picture (Figs. 33, 34, 51, and 54) as Ruud (1929) has shown by embryonic grafts between the white and dark varieties and Schaxel (1922) by transplanting the blastema of regenerating limbs. A limb bud from a pigmented animal grafted to a white axolotl develops pigment and remains throughout life very dark (Fig. 51 compared with 52). Close examination reveals that the pigmentation is due to melanophores, which do not have the form of those of the donor species (Fig. 53) but are characteristic of the

---

Plate 5
Photographs of colored drawings by Lisbeth Krause.

FIGS. 41 to 44. Details of skin pattern on dorsal side of hands in animals shown in figures 18 and 19 (Exp. NE 190), in which the right fore limb was interchanged between *A. jeffersonianum* and axolotl. × 6.3.

FIG. 41. Normal jeffersonianum; note the large whitish patches broken up into small fields by the skin glands.

FIG. 42. Axolotl limb on jeffersonianum; note the larger size of the glands.

FIG. 43. Jeffersonianum limb on axolotl.

FIG. 44. Normal axolotl limb; animal still in larval condition and the skin glands do not show.

FIG. 45. Normal left limb of metamorphosed *A. punctatum* (Exp. NE 227), 62 mm. long; in this case yellow spots are lacking. × 3.8.

FIG. 46. Grafted axolotl limb in the same case (Exp. NE 227), much larger than the normal and with four round yellow spots characteristic of *A. punctatum*. × 3.8.

FIGS. 47 to 50. Normal and transplanted limbs in two comparable cases of grafting right fore limb between *A. punctatum* and *A. jeffersonianum*. The two horizontal pairs of figures represent limbs growing on the same animal; the two vertical pairs, limbs derived from the same species. × 4.

FIG. 47. Jeffersonianum limb growing on punctatum, showing one of the yellow spots on the back of the hand near the ulnar border (Exp. NE 253).

FIG. 48. Normal punctatum limb of same animal, with yellow spot in different position above elbow.

FIG. 49. Normal jeffersonianum limb showing characteristic irregular white mottling.

FIG. 50. Grafted punctatum limb on the same animal, showing similar markings (Exp. NE 241).

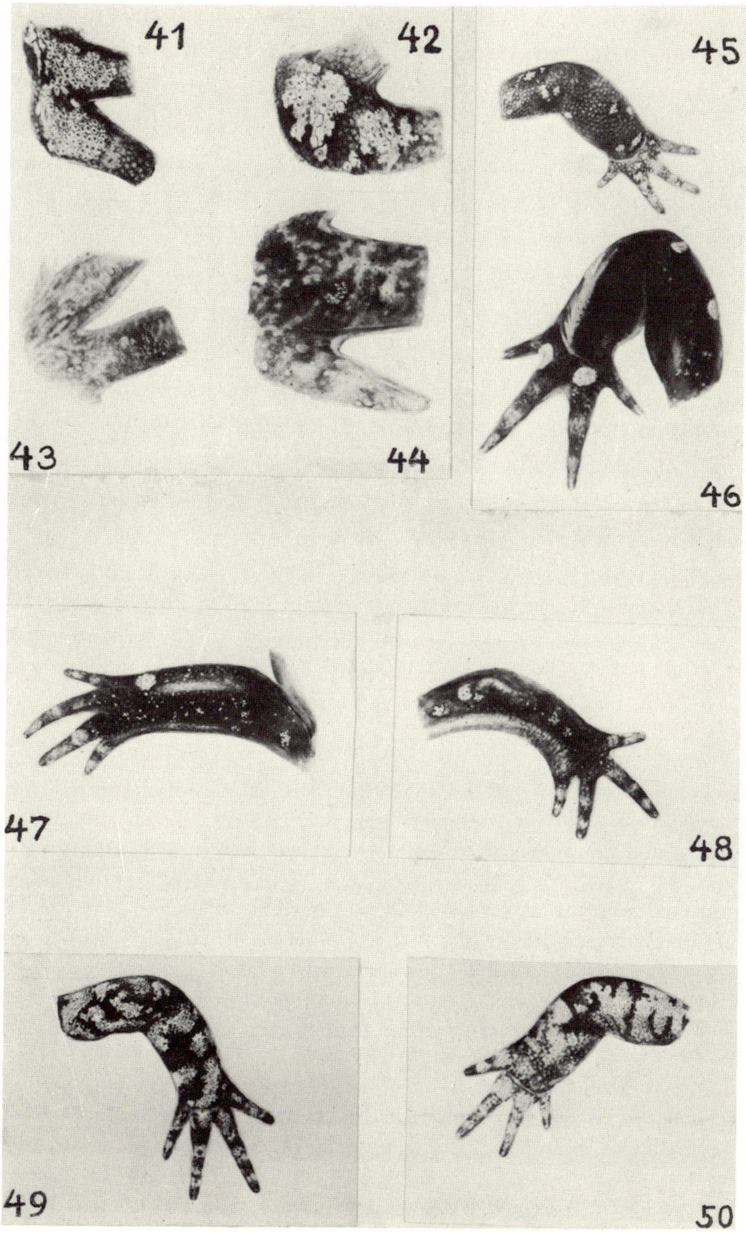
41
42
45
43
44
46
47
48
49
50

black axolotl. The negative of this picture is given in grafts of white axolotl limbs on pigmented embryos of any of the species; five such combinations, including the one with *Tr. torosus,* have been made. In several such experiments very striking sectorial chimeras have been obtained (Fig. 56), in which a streak of host tissue, running from shoulder to finger tips, is present in each grafted limb. In the one figured, which is the case of a white axolotl graft on a punctatum embryo, there is one of the characteristic yellow spots on the host in the pigmented area. It stops at the white area on one side.

Periclinal chimeras with white axolotl tissue shed some light on the respective roles of ectoderm, mesoderm and ganglion crest in the production of pigment. Ectoderm from a pigmented embryo, grafted to the limb bud of the white form, brings about the formation of a moderate number of melanophores and xanthophores (Figs. 57 and 58). This is interpreted, in accordance with DuShane, as due to the activation of pigment production in the potential melanophores of the host by a substance found in the grafted epithelium, but lacking in the epithelium of the white axolotl. Similar results are obtained, when the mesoderm of *A. punctatum* is grafted alone to a white axolotl embryo. However, when a periclinal chimeric limb develops upon a pigmented host, as when limb mesoderm of *A. punctatum* is covered by ec-

---

Plate 6
Photographs of colored drawings by Lisbeth Krause.

FIGS. 51 to 54. Normal (left) and grafted (right) limbs in a reciprocal pair of *A. tigrinum* and white axolotl (Exp. NE 189). The horizontal pairs each represent limbs growing on the same animal and the vertical pairs limbs derived from the same animal. × 4.

FIG. 51. Tigrinum limb growing on white axolotl and showing characteristic black axolotl pigmentation.

FIG. 52. Normal white axolotl limb without pigment.

FIG. 53. Normal tigrinum limb.

FIG. 54. White axolotl limb, without pigment, growing on *A. tigrinum.*

FIG. 55. Limb (periclinal chimera) growing on *A. punctatum* (Exp. FEEM 2) ; ectoderm from *A. jeffersonianum,* mesoderm from white axolotl; the limb, while containing considerable pigment, is much lighter than the normal. × 3.8.

FIG. 56. Limb (sectorial chimera) growing on *A. punctatum* (Exp. NE 237) grafted bud from white axolotl, but a band of host tissue runs on the post-axial border to the tips of three of the fingers; in the pigmented area note the round yellow spot, characteristic of the host species. × 3.

FIG. 57. Limb (periclinal chimera) formed by grafting ectoderm from *A. punctatum* to a white axolotl (Exp. NEE 132), animal 50 mm. long, three months after operation; numerous melanophores and xanthophores present. × 4.

FIG. 58. Normal limb on opposite side of host. × 4.

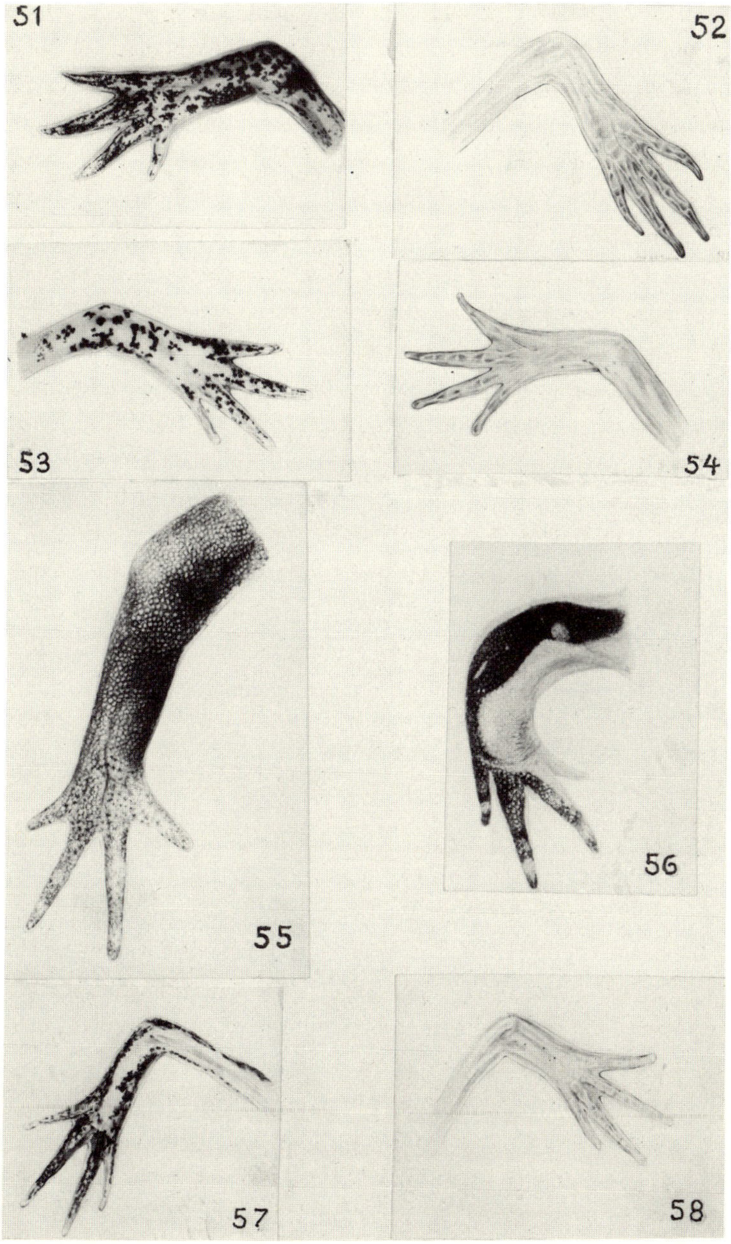
51
52
53
54
55
56
57
58

toderm of the white axolotl, the resulting limb, while not so dark as the normal, nevertheless has much more pigment than in the reciprocal case. With mesoderm alone grafted from the white axolotl to a pigmented embryo, the approach toward a normally pigmented limb is also not inconsiderable (Fig. 55). When dark limb mesoderm alone is grafted to a white animal, some pigment cells are developed, both melanophores and xanthophores, but they are only a little more numerous than when ectoderm alone is taken.

The facts in these experiments are explained, in harmony with DuShane's results, by the circumstance that the ganglion crest gives rise to the cells which develop pigment and that a second factor is necessary to activate the process. The latter may be furnished both by mesoderm and ectoderm of other parts of the body and the effect of the two layers is summative. The difference between the pigmentation of similar chimeric limbs when growing on white and on pigmented animals remains to be accounted for.

The above explanation of the host pigmentation of grafted limbs differs from the one given before knowledge of the derivation of pigment cells from the ganglion crest was available (Harrison, 1931). The identity of both *milieu interieur* and nervous control, which was then held solely accountable for the results, contributes undoubtedly to bring about the similarity between the two limbs of the same individual, even though they be of different specific origin, but the common origin, in the ganglion crest, of the cells which produce the pigment is certainly the more important factor.

Conclusion

I shall make no attempt to summarize, for too many diverse questions have been considered. The one thread that runs through the whole is the method of experimentation. By making reciprocal transplantations in as many combinations of species as were available, and by transplanting both whole organs and their constituent parts separately, a series of quantitative results have been obtained that check with one another to a remarkable degree, when equated in any significant way. It is realized now that the measurements should have been more complete and that they might have been taken so as to be much more useful, but in spite of these defects in application, it is believed that the method and results here presented open the way to further advance in our knowledge of development and growth.

References

Baur, E. 1909. Pfropfbastarde, Periclinalchimären und Hyperchimären. *Ber. deutsch. bot. Ges.*, *27:* 603–05.

———. 1910. Pfropfbastarde. *Biol. Zentralbl.*, *30:* 497–514.

———. 1930. *Einführung in die Vererbungslehre,* 11th ed. Berlin, Borntraeger, 478 pp.

Born, G. 1894. Die künstliche Vereinigung lebender Theilstücke von Amphibien-Larven. *Jahresber. Schles. Ges. vaterländische Cultur.* Medicinische Section. 11 pp.

———. 1897. Ueber Verwachsungsversuche mit Amphibienlarven. *Roux' Arch.*, *4:* 349–465, 517–623.

Burns, R. K., and L. M. Burns. 1929. The growth of the whole organism and of the limbs in two species of Amblystoma united in parabiosis. *J. Exp. Zool.*, *53:* 455–77.

Burr, H. S. 1930. Hyperplasia in the brain of Amblystoma. *J. Exp. Zool.*, *55:* 171–90.

Copenhaver, W. M. 1927. Results of heteroplastic transplantations of the heart rudiment in Amblystoma embryos. *Proc. Nat. Acad. Sci.*, *13:* 484–88.

———. 1930. Results of heteroplastic transplantation of anterior and posterior parts of the heart rudiment in Amblystoma embryos. *J. Exp. Zool.*, *55:* 293–318.

———. 1933. Transplantations of heart and limb rudiments between Amblystoma and Triton embryos. *J. Exp. Zool.*, *65:* 131–57.

Detwiler, S. R. 1930. Some observations upon the growth, innervation, and function of heteroplastic limbs. *J. Exp. Zool.*, *57:* 183–203.

———. 1931. Heteroplastic transplantations of embryonic spinal-cord segments in Amblystoma. *J. Exp. Zool.*, *60:* 141–71.

———. 1932a. Further experiments upon accelerated growth in heteroplastic spinal-cord grafts. *J. Comp. Neur.*, *56:* 465–502.

———. 1932b. Growth acceleration and regulation in heteroplastic spinal-cord grafts. *J. Exp. Zool.*, *61:* 245–77.

DuShane, G. P. 1934. The origin of pigment cells in Amphibia. *Science*, *80:* 620–21.

Ekman, G. 1913. Experimentelle Untersuchungen über die Entwicklung der Kiemenregion (Kiemenfäden und Kiemenspalten) einiger anuren Amphibien. *Morph. Jahrb.*, *47:* 419–592.

Erdmann, R. 1927. Verwandtschaftsbezeihungen der Anurenfamilien, geprüft durch Implantationsversuche gezüchteter Haut. *Roux' Arch.*, *112:* 739–806.

Geinitz, B. 1925. Embryonale Transplantation zwischen Urodelen und Anuren. *Roux' Arch.*, *106:* 357–408.

Greil, A. 1906. Ueber die Homologie der Anamnierkiemen. *Anat. Anz.*, *28:* 257–72.

Hadley, C. E. 1929. The compatibility of the skin of Rana pipiens and Rana clamitans as tested by transplantation. *J. Exp. Zool.*, *54:* 127–55.

Harrison, R. G. 1898. The growth and regeneration of the tail of the frog larva. Studied with the aid of Born's method of grafting. *Roux' Arch.*, *7:* 430–85.

Harrison, R. G. 1903. Experimentelle Untersuchungen über die Entwicklung der Sinnesorgane der Seitenlinie bei den Amphibien. *Arch. mikr. Anat.* *63:* 35–149.

———. 1924. Some unexpected results of the heteroplastic transplantation of limbs. *Proc. Nat. Acad. Sci.*, *10:* 69–74.

———. 1929. Heteroplastic transplantation in amphibian embryos. *X*[e] *Congrès Intern. de Zoologie. Budapest.* 642–50.

Harrison, R. G. 1929a. Correlation in the development and growth of the eye studied by means of heteroplastic transplantation. *Roux' Arch. 120:* 1–55.

———. 1931. Experiments on the development and growth of limbs in the Amphibia. *Science, 74:* 575–76.

His, W. 1874. *Unsere Körperform.* Vogel, Leipzig, 224 pp.

Huxley, J. S. 1932. *Problems of Relative Growth.* New York, Dial, 276 pp.

Koppányi, T. 1923. Die Replantation von Augen. IV. Ueber das Wachstum der replantierten Augen. *Roux' Arch., 99:* 60–63.

Krenke, N. P. 1933. *Wundkompensation, Transplantation und Chimären bei Pflanzen.* Monographien aus dem Gesamtgebiet der Physiologie der Pflanzen und der Tiere, vol. 29. Berlin, Springer, 934 pp.

Landacre, F. L. 1921. The fate of the neural crest in the head of the urodeles. *J. Comp. Neurol., 33:* 1–43.

Loeb, L. 1921. Transplantation and individuality. *Biol. Bull., 40:* 143–80.

———. 1930. Transplantation and individuality. *Physiol. Rev., 10:* 547–616.

Mangold, O. 1923. Transplantationsversuche zur Frage der Spezifität und der Bildung der Keimblätter. *Roux' Arch., 100:* 198–301.

———. 1931. Das Determinationsproblem. 3. Teil. Das Wirbeltierauge in der Entwicklung und Regeneration. *Ergeb. Biol., 7:* 193–403.

Mangold, O., and F. Seidel. 1927. Homoplastische und heteroplastische Verschmelzung ganzer Tritonkeime. *Roux' Arch., 111:* 593–665.

Needham, J. 1934. Chemical heterogony and the ground-plan of animal growth. *Biol. Rev., 9:* 79–109.

Richardson, D. 1932. Some effects of heteroplastic transplantation of the ear vesicle in Amblystoma. *J. Exp. Zool., 63:* 413–40.

Robb, R. C. 1929. On the nature of hereditary size limitation. II. The growth of parts in relation to the whole. *Brit. J. Exp. Biol., 6:* 311–24.

Rotmann, E. 1929. Ueber den Anteil des Mesoderms und Ektoderms an der Formbildung der Amphibienextremität. *Naturwiss., 17:* 878.

———. 1931. Die Rolle des Ektoderms und Mesoderms bei der Formbildung der Kiemen und Extremitäten von Triton. *Roux' Arch., 124:* 747–94.

———. 1933. Die Rolle des Ektoderms und Mesoderms bei der Formbildung der Extremitäten von Triton. II. Operationen im Gastrula- und Schwanzknospenstadium *Roux' Arch., 129:* 85–119.

Ruud, G. 1929. Heteronom-orthotopische Transplantationen von Extremitätenanlagen bei Axolotlembryonen *Roux' Arch., 118:* 308–51.

Schaxel, J. 1922. Ueber die Natur der Formvorgänge in der tierischen Entwicklung. *Roux' Arch., 50:* 498–525.

Schultz, W. 1918. Gleichlauf von Verpflanzung und Kreuzung bei Froschlurchen. Schlüsse auf die Zusammenfügung der Plasmabausteine. *Roux' Arch., 43:* 361–81.

Schwind, J. L. 1931. Heteroplastic experiments on the limb and shoulder girdle of Amblystoma. *J. Exp. Zool., 59:* 265–95.

———. 1932. Further experiments on limbs containing tissue of two species. *J. Exp. Zool., 63:* 365–69.

Severinghaus, A. E. 1930. Gill development in Amblystoma punctatum. *J. Exp. Zool., 56:* 1–29.

Spemann, H. 1918. Ueber die Determination der ersten Organanlagen des Amphibienembryo I–VI. *Roux' Arch.*, *43:* 448–555.

———. 1921. Die Erzeugung tierischer Chimären durch heteroplastiche embryonale Transplantation zwischen Triton cristatus und taeniatus. *Roux' Arch.*, *48:* 533–70.

Stone, L. S. 1922. Experiments on the development of the cranial ganglia and the lateral line sense organs in Amblystoma punctatum. *J. Exp. Zool.*, *35:* 421–96.

———. 1926. Further experiments on the extirpation and transplantation of mesectoderm in Amblystoma punctatum. *J. Exp. Zool.*, *44:* 95–131.

———. 1929. Experiments showing the role of migrating neural crest (mesectoderm) in the formation of head skeleton and loose connective tissue in Rana palustris. *Roux' Arch.*, *118:* 40–77.

———. 1930. Heteroplastic transplantation of eyes between the larvae of two species of Amblystoma. *J. Exp. Zool.*, *55:* 193–261.

Taube, E. 1922. Tierische Chimären. *Naturwiss.* Wochenschrift, N. F., *21:* 457–65.

Thompson, D' A. W. 1917. *Growth and Form.* Cambridge, University Press, 793 pp.

Twitty, V. C. 1932. Influence of the eye on the growth of its associated structures, studied by means of heteroplastic transplantation. *J. Exp. Zool.*, *61:* 333–74.

Twitty, V. C., and H. A. Elliott. 1934. The relative growth of the amphibian eye, studied by means of transplantation. *J. Exp. Zool.*, *68:* 247–91.

Twitty, V. C., and J. L. Schwind. 1931. The growth of eyes and limbs transplanted heteroplastically between two species of Amblystoma. *J. Exp. Zool.*, *59:* 61–86.

Ubisch, L. von. 1931. Ueber Keimblatt-Chimären. *Verh. deut. Zool. Ges.*, pp. 174–84.

———. 1933. Keimblattchimären. *Naturwiss.*, *21:* 325–29.

Waddington, C. H. 1933. Heterogony and the chemical ground-plan of animal growth. *Nature*, *131:* 134 pp.

Wieman, H. L. 1926. The effect of heteroplastic grafts of the spinal cord on the development of the limb of Amblystoma. *J. Exp. Zool.*, *45:* 335–48.

Winkler, H. 1907. Ueber Pfropfbastarde und Pflanzliche Chimären. *Ber. deut. bot. Ges.*, *25:* 568–76.

———. 1909. Weitere Mitteilungen über Pfropfbastarde. *Zeit. Bot.*, *1:* 315–45.

———. 1910. Ueber das Wesen der Pfropfbastarde. *Ber. deut. Bot. Ges.*, *28:* 116–18.

7

Recapitulation and Conclusion

In the lectures we have taken up a good many different subjects. Often it may have been difficult to see the connection between some of them. Now it becomes my duty to point out these relations and, like the embryo itself when disturbed, to build the parts into a whole. This is after all the central problem of embryology—the relations of the parts to each other and to the whole.

Living matter occurs at two integrative levels, that of the individual organism or person and that of the cell. That these two levels merge is shown by the facts that cell and organism are one in the Protozoa and that every organism is at one time or other in its career a single cell, the ovum from which it develops.

This statement is not vitiated but reinforced by the occurrence of certain alternative modes of reproduction in some organisms, such as: the reproduction of ascidians by means of winter buds, consisting of masses of undifferentiated cells; the formation of the fruiting phases of Mycetozoa (especially in the Acrasieae) by the cooperation of great numbers of independent myxamoebae; the reconstitution of sponges and hydroids out of loose cells; the limitations placed on the activities of cells in culture media outside the body, which nevertheless suffice to permit unlimited longevity and reproduction.

We may assert, then, that each and every living being can be encompassed in the organization of a single cell of its species. This alone raises the cell to an object of extreme importance and makes one wonder what is meant by the phrase "inadequacy of the cell theory."

The problem of the embryologist would be relatively simple if the egg could do

From Lecture VI Notes

p. 56 Faust **VI** Conclusion

Meph.

Ein Teil von jener Kraft,
die stets das Böse will und stets das Gute schafft.

Faust

Was ist mit diesem Rätselwort gemeint?

Meph

Ich bin der Geist der stets verneint!
Und das mit Recht; denn Alles, was entsteht
Ist wert daß es zu Grunde geht;
Drum besser wär's daß nichts entstünde.
So ist denn Alles, was ihr Sünde,
Zerstörung, kurz das Böse nennt,
Mein eigentliches Element.

Faust.

Du nennst dich einen Teil und stehst doch
ganz vor mir?

Meph.

Bescheidne Wahrheit sprech' ich dir.
Wenn sich der Mensch, die kleine Narrenwelt,
Gewöhnlich für ein Ganzes hält:
Ich bin ein Teil des Teils, der Anfangs Alles war,
Ein Teil der Finsterniß, die sich das Licht gebar,
Das stolze Licht, das nun der Mutter Nacht
Den alten Rang, den Raum ihr streitig macht.

Johann Wolfgang von Goethe
Faust, 1. Teil. Scene Studierzimmer
(In Jubiläumsausgabe, 13. Bd., p. 54–55, lines 1335–1345)

nothing more than produce a normal organism, for then we might describe the egg as simply containing the organism folded up within it, and development as a mere unfolding, a theory that actually prevailed well through the eighteenth century and was revived again in the late nineteenth.

However, Driesch's discovery in the early 90's that a whole organism would develop out of a single blastomere or part of a blastula made serious difficulties for this simple preformation theory, since in the experimentally modified development, not a single cell plays exactly the same role it would have played had it been left to develop in the whole embryo. Driesch was quick to see in this the central problem of development and if I have been critical of his way out of the difficulties by invoking a non-material entelechy, I must pay him my respect for his astuteness in grasping the problem. He must have felt the same incredulity and wonderment as Faust when he says to Mephistopheles, "Du nennst dich einen Teil und stehst doch ganz vor mir?" Astonishment was heightened by the further discovery that two eggs in the early stages could be fused into a single organism. All of this led Driesch to an extreme generalization that the role of a given cell in development is a function of its position in the whole.

This, however, left out the regional differences in the egg, which later research has clearly brought out, as described in the sea urchin and the Amphibia, and which are found in greater or lesser intensity in all eggs. The regional differences are gradual and relative rather than absolute (Boveri). In other words there is an intensity gradient for each kind of differentiation activity. The center of such an activity, as for instance at the animal or vegetative pole of the egg, does not contain material that is necessary for the activity to assert itself, for large areas may be removed and the remainder of the egg may still produce a normal embryo.

Such facts and a host of others have led many to postulate in the living organism, a microstructure with a repeat pattern which is oriented and polarized. In other words, the egg and early embryo consist of fields—gradients or differentiation centers in which the specific properties drop off in intensity as the distance from the field center increases, but in which any part within limits may represent any other.

In regions in which differentiation is not primary but dependent on an organizer or inductor as is the case in the ectoderm of the vertebrate embryo, two fields are involved, the inducing and the induced—each with its intensity gradients, thresholds of activity and sensitivity and each with time limits on its function. In normal development the elements are attuned in time and space and while their interaction may overcome interferences, if not too severe, it has proved impossible to obtain normal development through the action of unnatural inductors.

The fields or gradients have been held by many to be due to concentration of materials but it is my opinion that they are to be accounted for primarily by the dipole nature of the protein components of the living substance. This leads to their orientation in the cell and secondarily to the accumulation of different substances in graded concentration from one pole or region to the opposite. These substances in some cases may be essential to the particular differentiation of the region and in other cases not. Proper orientation of the finer elements (micellae) is necessary and when disturbed leads to an abnormal outcome unless the elements can reorient themselves.

All of this is a tremendous over-simplification. When one thinks of the complexity of the conditions that have to be encompassed in the living cell, it is hard to see how they can be accommodated. The interacting factors of organization and development within the living system to be accounted for are: the species characters; all the possibilities of differentiation; all of the means of carrying out these possibilities through the action of enzymes which must be very numerous; and the relations between the genes in the nucleus and the cytoplasmic agents which carry out their bidding.

I have said nothing about these things, mainly because of ignorance, but their importance is obvious and embryology will surely profit greatly from any advances in these fields, just as it will from the efforts of biophysicists and biochemists who may be attracted to study its problems.

As prophetic of what our successors may look forward to, let me quote in conclusion a noteworthy passage from von Baer, in the dedication of his great work, "On the Development of Animals," which is in the form of a letter to his friend and colleague, Pander: "Many will win prizes but the palm will be taken by that fortunate one to whom it is reserved, to resolve the formative forces of the animal body into the general forces of the universe." He hastens to add, however, "The tree out of which his cradle will be fashioned has not yet germinated." [From Lecture Notes.]

Supplementary References Cited in Notes from the Silliman Lectures

von Baer, K. E. 1827. *De Ovi Mammalium et Hominis Genesi. Epistolam ad Academiam Imperialem Scientiarum Petropolitanam,* Leipzig, Leopold Voss, 41 pp.

Barron, D. H. 1945. The role of the sensory fibers in the differentiation of the spinal cord in sheep. *J. Exp. Zool., 100:* 431–43.

Bataillon, E. 1910. L'embryogénèse complète provoguée chez les Amphibiens par piqûre de l'oeuf vierge, larves parthénogénétiques de Rana fusca. *Compt. Rend. Acad., Sci., 150:* 996.

Bateson, W. 1894. *Materials for the Study of Variation,* London, Macmillan, 598 pp.

——— 1913. *Problems of Genetics. Silliman Memorial Lectures,* New Haven, Yale University Press and London, Oxford University Press, 258 pp.

Bellamy, A. W. 1919. Differential susceptibility as a basis for modification and control of early development in the frog. *Biol. Bull., 37:* 312–61.

Berrill, N. J. 1935. Cell division and differentiation in asexual and sexual development. *J. Morph., 57:* 353–427.

———, and A. Cohen 1936. Regeneration in Clavelina lepadiformis. *J. Exp. Biol., 13:* 352–62.

Bierens de Haan, J. A. 1913a. Ueber homogene und heterogene Keimverschmelzungen bei Echiniden. *Roux' Arch. f. Ent.-mech., 36:* 473–536.

——— 1913b. Ueber die Entwicklung heterogener Verschmelzungen bei Echiniden. *Roux' Arch. f. Ent.-mech., 37:* 420–32.

Bonner, J. T. 1947. Evidence for the formation of cell aggregates by chemotaxis in the development of the slime mold Dictyostelium discoideum. *J. Exp. Zool., 106:* 1–26.

Born, G. 1885. Biologische Untersuchungen. I. Ueber den Einfluss der Schwere auf das Froschei. *Arch. f. mikr. Anat., 24:* 475–545.

Bragg, W. H. and W. L. Bragg 1934. *The Crystalline State,* Vol. 1, New York, The Macmillan Co., 352 pages.

Brien, P. and E. Brien-Gavage 1927. Contribution à l'étude de la Blastogénèse des Tuniciers.

IV. Recherches sur le Bourgeonnement de Perophora Listeri Weigm. (Origine mésoblastique du septum stolonial). *Rec. Inst. Zool. Torley-Rousseau, 1:* 31–85.

Brien, P. and E. Brien-Gavage 1932. Contribution à l'étude de la régénération naturelle chez les Spongillidae Spongilla lacustris (L); Ephydatia fluviatilis (L). *Arch Zool. Exp. et Generale, 74:* 461–506.

——— 1937. La réorganisation de l'eponge après dissociation par filtration et phénomènes d'involution chez Ephydatia fluviatilis. *Arch. Biol., 48:* 185–276. Liege.

Burr, H. S. 1941. Field properties of the developing frog's egg. *Proc. Nat. Acad. Sciences, 27:* 276–81.

Carrel, A. 1914. Present condition of a strain of connective tissue twenty-eight months old. *J. Exp. Med., 20:* 1–2.

Chabry, L. 1887. *Contribution a l'Embryologie Normale et Tératologique des Ascidies simples.* Paris, Ancienne Librairie Germer Baillière et Cie, 154 pages.

Child, C. M. 1915. Axial gradients in the early development of the starfish. *Amer. J. Physiol., 37:* 203–19.

——— 1928. The physiological gradients. *Protoplasma, 5:* 447–76.

——— 1941. *Patterns and Problems of Development,* Chicago, Univ. of Chicago Press, 811 pages.

Chuang, H. 1938. Spezifische Induktionsleistungen von Leber und Niere im Explantatversuch. *Biol. Zbl, 58:* 472–80.

——— 1939. Induktionsleistungen von frischen und gekochten Organteilen (Niere, Leber) nach ihrer Verpflanzung in Explantate und verschiedene Wirtsregionen von Tritonkeimen. *Roux' Arch. f. Entw.-mech., 139:* 556–638.

——— 1940. Weitere Versuche über die Veränderung der Induktionsleistungen von gekochten Organteilen. *Roux' Arch. f. Entw.,-mech., 140:* 25–38.

Copenhaver, W. M. 1939. Some observations on the growth and function of heteroplastic heart grafts. *J. Exp. Zool., 82:* 239–71.

Dalcq, A. and J. Pasteels 1938. Potentiel morphogénètique regulation et "axial gradients" de Child. Mise au point des "bases physiologiques de la morphogenese." *Bull. Acad. Roy. Méd. Belgique, serie 6, 3:* 261–308.

Deiters, O. F. 1865. Untersuchungen über Gehirn und Rückenmark des Menschen und der Säugetiere, herausgegeben von Max Schultze, Braunschweig. (see Heidenhain).

Driesch, H. 1897. Betrachtungen über die Organisation des Eies und ihre Genese. *Roux' Arch. f. Entw.-mech., 4:* 75–124.

——— 1900. Studien über das Regulationsvermögen der Organismen. 4. Die Verschmelzung der Individualität bei Echinidenkeimen. *Roux' Arch. f. Entw.-mech., 10:* 411–34.

——— 1902a. Ueber ein neues harmonisch-äquipotentielles System und über solche System überhaupt. *Roux' Arch. f. Entw.-mech., 14:* 227–46.

——— 1902b. Studien über das Regulationsvermögen der Organismen. 6. Die Restitutionen der Clavellina lepadiformis. *Roux' Arch. f. Entw.-mech., 14:* 247–87.

——— 1905. Skizzen zur Restitutionslehre. *Roux' Arch. f. Entw.-mech., 20:* 21–29.

Dürer, A. Les quatres livres d'Albert Dürer de la proportion des parties et pourtraicts des corps humaine. Arnheim, folio, 1613 (from Growth and Form, D'Arcy Thompson, p. 1053, 1942).

DuShane, G. P. 1935. An experimental study of the origin of pigment cells in Amphibia. *J. Exp. Zool., 72:* 1–31.

——— 1938. Neural fold derivatives in the Amphibia. Pigment cells, spinal ganglia and Rohon-Beard cells. *J. Exp. Zool., 78:* 485–503.

——— 1939. The role of embryonic ectoderm and mesoderm in pigment production in Amphibia. *J. Exp. Zool., 82:* 193–215.

——— 1943. The embryology of vertebrate pigment cells. Part I. Amphibia. *Quart. Rev. Biol., 18:* 109–27.

——— 1944. The embryology of vertebrate pigment cells. Part II. *Birds. Quart. Rev. Biol., 19:* 98–117.

Ebeling, A. H. 1922. A ten year old strain of fibroblasts. *J. Exp. Med., 35:* 755–59.

Fick, R. 1893. Ueber die Reifung und Befruchtung des Axolotleies. *Zeit. f. wissen. Zool., 56:* 529–614.

Fischer, A. 1927. Gewebezüchtung. *Handbuch der Biologie der Gewebezellen in vitro,* München, Rudolph Müller and Steinicke, 508 pages.

Fischer, F. G. und E. Wehmeier 1933. Zur Kenntnis der Induktionsmittel in der Embryonalentwicklung. *Naturwiss., 21 Jahrg.:* 518.

Föyn, B. 1927a. Studien über Geschlecht und Geschlechtszellen bei Hydroiden. I. Ist *Clava squamata* eine gonochoristische oder hermaphrodite Art? *Roux' Arch. f. Entw.-mech., 109:* 513–34.

Föyn, B. 1927b. Studien über Geschlecht und Geschlechtszellen bei Hydroiden. II. Auspressungsversuche an *Clava squamata* (Müller) mit Mischung von zellen aus Polypen desselben oder verschiedenen Geschlechts. *Roux' Arch. f. Entw.-mech., 110:* 89–148.

Galtsoff, P. S. 1929. Heteroagglutination of dissociated sponge cells. *Biol. Bull., 57:* 250–60.

Held, H. 1907. Kritische Bermerkungen zu der Verteidigung der Neuroblasten und der Neuronentheorie durch R. Cajal. *Anat. Anz., 30:* 369–93.

Harrison, R. G. 1913. The life of tissues outside the organism from the embryological standpoint. *Trans. Congr. Amer. Phys. and Surg., 9:* 63–76.

——— 1918. Experiments on the development of the fore-limb of Amblystoma, a self-differentiating equipotential system. *J. Exp. Zool., 25:* 413–61.

——— 1925. The effect of reversing the medio-lateral or transeverse axis of the fore-limb bud in the salamander embryo. *Roux' Arch. f. Entw.-mech., 106:* 469–502.

Heidenhain, M. 1911. Plasma und Zelle, Vol. 2, p. 706. Jena, G. Fischer.

Herbst, C. 1901. *Formative Reize in der tierischen Ontogenese,* Leipzig, Arthur Georgi, 125 pages.

Herlitzka, A. 1895. Contributo allo studio della capacità evolutiva dei due primi blastomeri nell' uovo di tritone (Triton cristatus). *Roux' Arch. f. Entw.-mech., 2:* 352–69.

——— 1897. Sullo sviluppo di embrioni completi da blastomeri isolati di uova di tritone (Molge cristata). *Roux' Arch. f. Entw.-mech., 4:* 624–58.

Holtfreter, J. 1934a. Ueber die Verbreitung induzierender Substanzen und ihre Leistungen im Triton-Keim. *Roux' Arch. f. Entw.-mech., 132:* 307–83.

——— 1934b. Formative Reize in der Embryonalentwicklung der Amphibien, dargestellt an Explanationsversuchen. *Arch. f. Exp. Zellforschung, 15:* 281–301.

——— 1935. Morphologische Beeinflussung von Urodelenektoderm bei xenoplastischer Transplantation. *Roux' Arch. f. Entw.-mech., 133:* 367–426.

——— 1939. Gewebeaffinität, ein Mittel der embryonalen Formbildung. *Arch. exp. Zellf. 23:* 169–209.

Holtfreter, J. 1946. Structure, motility and locomotion in isolated embryonic amphibian cells. *J. Morph.*, *79:* 27–62.

——— 1947. Observations on the migration, aggregation and phagocytosis of embryonic cells. *J. Morph.*, *80:* 25–55.

Hydén, H. 1943. Protein metabolism in the nerve cell during growth and function. *Acta Physiol. Scandinav.*, *6 (suppl. 17)*: 5–136.

Janda, V. 1924. Die Regeneration der Geschlechtsorgane bei Rhynchelmis limosella Hoffm. *Zool. Anz.*, *59:* 257–69.

Japp, F. R. 1898. Stereochemistry and vitalism. *Chem. News*, *77:* 139, 149.

Jolly, J. 1903. Sur la durée de la vie et de la multiplication des cellules animales en dehors de l'organisme. *Compt. Rend., Soc. Biol.*, *55:* 1266–68.

——— 1904. Recherches expérimentale sur la division indirecte des globules rouges. *Arch. d'Anat. Micr.*, *6:* 455–632.

Keibel, F. 1897–1938. *Normentafeln zur Entwicklungsgeschichte der Wirbelthiere*, (16 volumes in 9). Jena, Gustave Fischer.

Kuhl, W. 1937. Untersuchungen über das Verhalten künstlich getrennter Furchungszellen und Zellaggregate einiger Amphibienarten mit Hilfe des Zeitrafferfilmes (Laufbild-und Teilbild-Analyse). *Roux' Arch. f. Entw.-mech.*, *136:* 593–671.

Lambert, R. A. and F. M. Hanes 1911. Characteristics of Growth of Sarcoma and Carcinoma cultivated *in vitro*. *J. Exp. Med.*, *13:* 495–504.

LeBel, J. A. 1874. Sur les relations qui existent entre les formules atomiques des corps organiques et le pouvoir rotatoire de leurs dissolutions. *Bull. Soc. Chim., Paris* (*Ser.* 2), *22:* 337–47.

Levi, G. 1919. Il ritmo e le modalità della mitosi nelle cellule viventi coltivate "in vitro." *Arch. Ital. d. Anat. e.d. Embri.*, *15:* 243–64.

Lewis, M. R. 1915. Rhythmical contraction of the skeletal muscle tissue observed in tissue culture. *Amer. J. Physiol.*, *38:* 153–61.

———, and W. H. Lewis 1917. The contraction of smooth muscle cells in tissue cultures. *Amer. J. Physiol.*, *44:* 67–74.

Lindahl, P. E. 1936. Zur Kenntnis der physiologischen Grundlagen der Determination im Seeigelkeim. *Acta Zool.*, *17:* 179–365.

Leob, L. 1945. *The Biological Basis of Individuality*, Springfield, Ill., Charles C. Thomas, 711 pages.

Mangold, O. 1920. Fragen der Regulation und Determination an umgeordneten Furchungsstadien und verschmolzenen Keimen von *Triton*. *Roux' Arch. f. Entw.-mech.*, *47:* 249–301.

——— 1929. Experimente zur Analyse der Determination und Induktion der Medullarplatte. *Roux' Arch. f. Entw.-mech.*, *117:* 586–696.

——— und H. Spemann 1927. Ueber Induktion von Medullarplatte durch Medullarplatte im jüngeren Keim, ein Beispiel homoeogenetischer oder assimilatorischer Induktion. *Roux' Arch. f. Entw.-mech.*, *111:* 341–422.

Marsh, G. and H. W. Beams 1945. The orientation of pollen tubes of Vinca in the electric current. *J. Cell. and Comp. Physiol.*, *25:* 195–204.

Marsh, G. and H. W. Beams 1946. *In vitro* control of growing chick nerve fibers by applied electric currents. *J. Cell. and Comp. Physiol.*, *27:* 139–57.

Maximow, A. 1925. Tissue-cultures of young mammalian embryos. *Publ. 361, Carnegie Institution of Washington*, pp. 47–113.

McClendon, J. F. 1909. On the totipotence of the first two blastomeres of the frog's egg. *Amer. Nat., 43:* 384.

——— 1910. The development of isolated blastomeres of the frog's egg. *Amer. J. Anat., 10:* 425–30.

Meyer, R. 1913. Die ursächlichen Beziehungen zwischen dem Situs viscerum und Situs cordis. *Roux' Arch. f. Entw.-mech., 37:* 85–107.

Muir, R. and J. Ritchie 1907. *Manual of bacteriology,* 4th Edition, New York, The MacMillan Co., 605 pages.

Needham, J. 1942. *Biochemistry and Morphogenesis,* Cambridge, University Press, 787 pages.

Newman, H. H. and Patterson, J. T. 1910. The development of the nine-banded armadillo from the primitive streak stage to birth; with special reference to the question of specific polyembryony. *J. Morph., 21:* 359–424.

Nicholas, J. S. 1922. The effect of the rotation of the area surrounding the limb bud. *Anat. Rec., 23:* 30.

——— 1924a. Ventral and dorsal implantations of the limb bud in Amblystoma punctatum. *J. Exp. Zool., 39:* 27–41.

——— 1924b. Regulation of posture in the fore limb of Amblystoma punctatum. *J. Exp. Zool., 40:* 113–59.

Nijhoff 1904. Report in Clinic—quintuplets from the Groningen Clinic. *Zentralblatt f. Gynäkologie, 28:* 90 and 750.

Oppel, A. 1912. Causal-morphologische Zellenstudien. IV. Die Explantation von Säugetiergeweben—ein der Regulation von seiten des Organismus nicht unterworfenes Gestaltungsgeschehen. *Roux' Arch. f. Entw.-mech., 34:* 132–67.

Oyama, J. 1930. Normal plates of the development of the newt, Diemictylus (Triturus) pyrrhogaster (Boie). *Zool. Mag., Tokyo, 42:* 465–473. (In Japanese.)

Pasteels, J. 1937. Sur l'origine de la symétrie bilatérale des amphibiens anoures. *Arch. d'Anat. Micro., 33:* 279–300.

Pasteur, L. 1848a. Mémoire sur la relation qui peut exister entre la forme cristalline et la composition chimique, et sur la cause de la polarisation rotatoire. *Compt. Rend. d. Acad. d. Sci., 26:* 535–38.

——— 1848b. Recherches sur les relations qui peuvent exister entre la forme cristalline, la composition chimique et le sens du pouvoir rotatoire. *Compt. Rend. d. Acad. d. Sci., 27:* 401–11.

——— 1848c. Sur les relations qui peuvent exister entre la form cristalline, la composition chimique et le sens de la polarisation rotatoire. *Annales de Chimie et de Physique, Paris, 3 ser., 24:* 442–59.

——— 1853a. Transformation de l'acide tartrique en acide racémique. *Compt. Rend. d. Acad. d. Sci., 36:* 973.

——— 1853b. Sur les relations qui peuvent exister entre la form cristalline, la composition chimique et le phénomène rotatoire moléculaire. *Anales de Chimie et de Physique, Paris, 3 ser., 38:* 437–83.

Pflüger, E. 1883. Ueber den Einfluss der Schwerkraft auf die Theilung der Zellen und auf die Entwicklung des Embryo. *Pflüger's Arch., 32:* 1–81.

Piatt, J. 1940. Nerve-muscle specificity in Amblystoma, studied by means of heterotopic cord grafts. *J. Exp. Zool., 85:* 211–41.

Piatt, J. 1949. Heterotopic transplantation of brachial cord segments followed by reimplantation into the orthotopic site. *J. Exp. Zool., 111:* 1–25.
Pollister, A. W. and J. A. Moore 1937. Tables for the normal development of Rana sylvatica. *Anat. Rec., 68:* 489–96.
Psalms 139: 14, 16. *Holy Bible, King James Version.*
Ranvier, L. A. 1875. *Traité technique d'histologie.* Paris, F. Savy, 976 pages.
——— 1898. Recherches expérimentales sur le mécanisme de la cicatrisation des plaies de la cornée. *Arch. d'Anat. Micr., 2:* 44–64, 177–88.
Raper, K. B. 1935. Dictyostelium discoideum, a new species of slime mold from decaying forest leaves. *J. Agricul. Research, 50:* 135–47.
——— 1937. Growth and development of Dictyostelium discoideum with different bacterial associates. *J. Agricul. Research, 55:* 289–316.
——— 1939. Influence of culture conditions upon the growth and development of Dictyostelium discoideum. *J. Agricul. Research, 58:* 157–98.
——— 1940a. The communal nature of the fruiting process in the Acrasieae. *Amer. J. Botany, 27:* 436–48.
——— 1940b. Pseudoplasmodium formation and organization in Dictyostelium discoideum. *J. Elisha Mitchell Sci. Soc., 56:* 241–82.
——— 1941. Developmental patterns in simple slime molds. *Growth Suppl., 5:* 41–76.
Reverberi, G. 1936. La segmentazione dei frammenti dell'uovo non fecondata di Ascidie. *Pubbli. d. Staz. Zool. d. Napoli, 15:* 198–216.
——— 1937. Ricerche sperimentali sulla struttura dell'uovo fecondato delle Ascidie. *Pont. Acad. Sci., Commentationes, 1:* 135–72.
Reverberi, G. e M. Pitotti 1939. Differenziazioni fisiologiche nell'uovo delle Ascidie. *Pont. Acad. Sci., Commentationes, 3:* 469–88.
Richards, O. W. and G. A. Riley 1937. The growth of amphibian larvae illustrated by transformed coordinates. *J. Exp. Zool., 77:* 159–67.
Roux, W. 1883. Ueber die Bedeutung der Kerntheilungsfiguren. Eine hypothetische Erörterung, *Ges. Abh. II:* 125–43.
——— 1884. Beiträge zur Entwicklungsmechanik des Embryo. II. Ueber die Entwickelung der Froscheier bei Aufhebung der richtenden Wirkung der Schwere. *Bres. Ärztl. Zeitschrift 1884, Ges. Abh. II:* 256–76.
——— 1895. Ueber den "Cytotropismus" der Furchungszellen des Grasfrosches (Rana fusca). I. *Roux' Arch. f. Entw.-mech., 1:* 43–68, II–V, *Ibid:* 161–202.
——— 1896. Ueber die Selbstordnung (Cytotaxis) sich "berührender" Furchungszellen des Froscheies durch Zellenzusammenfügung, Zellentrennung und Zellengleiten. *Roux' Arch. f. Entw.-mech., 3:* 381–468.
Schultze, M. 1865. Ein heizbarer Objekttish und seine Anwendung bei Untersuchung der Blutes. *Arch. f. mikr. Anat., 1:* 1–42.
Shumway, W. 1940. Stages in the normal development of Rana pipiens. *Anat. Rec., 78:* 139–47.
Smuts, J. C. 1926. *Holism and evolution,* New York, The Macmillan Co., 362 pages.
Speidel, C. C. 1935a. Studies of living nerves. III. Phenomena of nerve irritation and recovery, degeneration and repair. *J. Comp. Neur., 61:* 1–80.
——— 1935b. Studies of living nerves. IV. Growth, regeneration, and myelination of the peripheral nerves in salamanders. *Biol. Bull., 68:* 142–163.

Speidel, C. C. 1936. Studies of living nerves. V. Alcoholic neuritis and recovery. *J. Comp. Neur., 64:* 77–113.

——— 1947. Correlated studies of sense organs and nerves of the lateral line in living tadpoles. I. Regeneration of denervated organs. *J. Comp. Neur., 87:* 29–56.

Spemann, H. 1919. Experimentelle Forschungen zum Determinations- und Individualitätsproblem. *Naturwissen., 32:* 1–33.

——— 1921. Die Erzeugung tierischer Chimären durch heteroplastische embryonale Transplantation zwischen *Triton cristatus* und *taeniatus. Roux' Arch. f. Ent.-mech., 48:* 533–70.

——— 1929. Ueber den Anteil von Organisator u. Wirtskeim am Zustandekommen der Induktion. *Naturwissen., 17:* 287–89.

——— 1938. *Embryonic Development and Induction,* New Haven, Yale Univ. Press, 401 pages.

——— and H. Mangold 1924. Ueber Induktion von Embryonalanlagen durch Implantation artfremder Organisatoren. *Roux' Arch. f. Ent.-mech., 100:* 599–638.

——— and O. Schotté 1932. Über xenoplastische Transplantation als Mittel zur Analyse der embryonalen Induktion. *Naturwissen., 20:* 463–67.

Sperry, R. W. 1941. The effect of crossing nerves to antagonistic muscles in the hind limb of the rat. *J. Comp. Neur., 75:* 1–19.

——— 1943. Functional results of crossing sensory nerves in the rat. *J. Comp. Neur., 78:* 59–90.

——— 1945. Restoration of vision after crossing of optic nerves and after contralateral transplantation of eye. *J. Neurophysiol., 8:* 15–28.

Stone, L. S. 1922. Experiments on the development of the cranial ganglia and the lateral line sense organs in Amblystoma punctatum. *J. Exp. Zool., 35:* 421–496.

——— 1944. Functional polarization in retinal development and its reestablishment in regenerating retinae of grafted salamander eyes. *Proc. Soc. Exp. Biol. and Med., 57:* 13–14.

——— 1947. Return of vision and functional polarization in the retinae of transplanted eyes. *Trans. Opthal. Soc., 67:* 349–368.

Swett, F. H. 1928a. Further experiments on the determination of the mediolateral axis of the forelimb of Amblystoma punctatum (Linn.). *J. Exp. Zool., 50:* 51–70.

——— 1928b. Studies on the shoulder girdle of Amblystoma punctatum (Linn.). I. Determination of its dorsoventral axis. *J. Exp. Zool., 51:* 389–401.

——— 1938. Experiments upon the relationship of surrounding areas to polarization of the dorsoventral limb-axis in Amblystoma punctatum (Linn.). *J. Exp. Zool., 78:* 81–100.

——— 1939a. The speed of prospective-polarity reversal in transplanted limbs of Amblystoma punctatum (Linn.). *J. Exp. Zool., 81:* 127–34.

——— 1939b. Further experiments upon the establishment and the reversal of prospective dorsoventral limb-axis polarity. *J. Exp. Zool., 82:* 305–14.

——— 1940. Establishment of definitive polarity in the dorsoventral axis of the fore-limb girdle in Amblystoma punctatum (Linn.). *J. Exp. Zool., 86:* 69–76.

Taylor, A. C. 1943. Development of the innervation pattern in the limb bud of the frog. *Anat. Rec., 87:* 379–413.

Toivonen, S. 1940. Ueber die Leistungsspezifität der abnormen Induktoren im Implantatversuch bei Triton. *Ann. Acad. Sci. Fennicae, Ser. A, 55:* 7–150.

Toivonen, S. 1948. Implantationsversuche mit in verschiedener Weise vorbehandelten abnormen Induktoren bei Triton. *Ann. Zool. Soc. Zool. Bot. Fennicae 'Vanamo', 13:* 1–19.

Twitty, V. C. 1936. Correlated genetic and embryological experiments on Triturus. I. Hybridization: Development of three species of Triturus and their hybrid combinations. II. Transplantation: The embryological basis of species differences in pigment pattern. *J. Exp. Zool., 74:* 239–302.

——— 1944. Chromatophore migration as a response to mutual influences of the developing pigment cells. *J. Exp. Zool., 95:* 259–90.

——— 1945. The developmental analysis of specific pigment patterns. *J. Exp. Zool., 100:* 141–78.

———, and D. Bodenstein 1939. Correlated genetic and embryological experiments on Triturus. III. Further transplantation experiments on pigment development. IV. The study of pigment cell behavior *in vitro*. *J. Exp. Zool., 81:* 357–98.

———, and M. C. Niu 1948. Causal analysis of chromatophore migration. *J. Exp. Zool., 108:* 405–38.

von Ubisch, L. 1933. Untersuchungen über Formbildung. IV. Ueber Plutei ohne Darm, Fortsätze ohne Skelet und die Grenzen der virtuellen Keimbezirke. *Roux' Arch. f. Ent.-mech., 129:* 45–67.

Van't Hoff, J. H. 1874. Sur les formules de structure dans l'espace. *Arch. Neerland. d. Sci. exact. e. nat., 9:* 445–54.

Vogt, W. 1913. Ueber Zellbewegegungen und Zelldegenerationen bei der Gastrulation von Triton taeniatus. I. Untersuchung lebender Embryonalzellen. *Anat. Hefte, 48:* 1–64.

Voigtländer, G. 1932. Neue Untersuchungen über den "Cytotropismus" der Furchungszellen. *Roux' Arch. f. Entw.-mech., 127:* 151–215.

Waddington, C. H., and D. M. Needham 1935. Studies on the nature of the amphibian organization centre. II. Induction by synthetic polycyclic hydrocarbons. *Proc. Roy. Soc., London, Ser. B., 117:* 310–17.

Waddington, C. H., J. Needham, and J. Brachet 1936. Studies on the nature of the amphibian organization centre. III. The activation of the evocator. *Proc. Royal Soc., London, Ser. B., 120:* 173–98.

———, W. W. Nominski, R. Lembery, and A. Cohen 1936. Studies on the nature of the amphibian organization centre. IV. Further experiments on the chemistry of the evocator. *Ibid.,* 198–207.

Wehmeier, E. 1934. Versuche zur Analyse der Induktionsmittel bei Medullarplatteninduktion von Urodelen. *Roux' Arch. f. Entw.-mech., 132:* 384–423.

Weidenreich, F. 1912 Die Lokalisation des Pigmentes und ihre Bedeutung in Ontogenie und Phylogenie der Wirbeltiere. *Zeit. f. Morph. u. Anthropol., 2:* 59–140.

Weismann, A. 1892. *Aufsätze über Vererbung und verwandet biologische Fragen.* Ueber die Vererbung (1883): pp. 73–121. Die Continuität des Keimplasmas als Grundlage einer Theorie der Vererbung (1885) pp. 191–207. Jena, Gustav Fischer.

Weiss, P. 1923. Die Regeneration der Urodelenextremität als Selbstdifferenzierung des Organrestes. *Naturwissen., 11:* 669–77.

——— 1926. Morphodynamik. Eine Einblick in die Gesetze der organischen Gestaltung an Hand von experimentellen Ergebnissen. *Abh. z. theor. Biol., 23:* 1–43.

——— 1928. Morphodynamische Feldtheorie und Genetik. Verh. 5. Intern. Kongr. Vererbungswiss, Berlin 1927. *Zeit. f. Indukt. Abst. u. Vererb. 2. Suppl.,* pp. 1567–74.

——— 1935. The so-called organizer and the problem of organization in amphibian development. *Physiol. Revs., 15:* 639–74.

——— 1937. Further experimental investigations on the phenomenon of homologous response in transplanted amphibian limbs. IV. Reverse locomotion after the interchange of right and left limbs. *J. Comp. Neur., 67:* 269–315.

——— 1939. *Principles of Development,* New York, Henry Holt and Company, 601 pages.

——— 1941. Nerve pattern: The mechanics of nerve growth. *Growth, 5:* 163–203.

——— 1945. Experiments on cell and axon orientation in vitro: The role of colloidal exudates in tissue organization. *J. Exp. Zool., 100:* 353–86.

———, and H. B. Hiscoe 1948. Experiments on the mechanism of nerve growth. *J. Exp. Zool., 107:* 315–95.

Wilson, H. V. 1911. Development of sponges from dissociated tissue cells. *Bull. Bureau of Fisheries, 30:* 1–30.

——— 1932. Sponges and biology. *Amer. Naturalist, 66:* 159–70.

Wilson, H. V., and J. T. Penney 1930. The regeneration of sponges (Microciona) from dissociated cells. *J. Exp. Zool., 56:* 73–147.

Woerdeman, M. W. 1933. Ueber den Glykogenstoffwechsel tierischer "Organisatoren." *Proc. Sec. Sci., Kon-Akad. Wetensch., Amsterdam, 36:* 423–26.

Woodruff, L. L. 1912. A five-year pedigreed race of Paramecium without conjugation. *Proc. Soc. Exp. Biol. and Med., 9:* 121–23.

Ross G. Harrison's Publications 1893–1948

1893

Ueber die Entwicklung der nicht knorpelig vorgebildeten Skelettheile in den Flossen der Teleostier. *Arch. f. mikr. Anat., 42:* 248–78; 3 plates.

1894

The development of the fins of teleosts. *Johns Hopkins Univ. Circ.,* 111.

The metamerism of the dorsal and the ventral longitudinal muscles of the teleosts. *Johns Hopkins Univ. Circ.,* 111.

Ectodermal or mesodermal origin of the bones of teleosts? *Anat. Anz., 10:* 138–43.

1895

Die Entwicklung der unpaaren und paarigen Flossen der Teleostier. *Arch. f. mikr. Anat., 46:* 500–78; 4 plates.

1898

Grafting experiments on tadpoles, with special reference to the study of the growth and regeneration of the tail. *Science, 7:* 198–99.

The growth and regeneration of the tail of the frog larva. Studied with the aid of Born's method of grafting. *Arch. f. Entw.-Mech., 7:* 430–85. 1 plate. Also in *Johns Hopkins Hospital Bull.*

1901

The histogenesis of the peripheral nervous system in Salmo salar. Read before Amer. Morph. Soc., Dec., 1900; *Biol. Bull. 2:* 352–53.

Ueber die Histogenese des peripheren Nervensystems bei Salmo salar. *Arch. f. mikr. Anat., 57:* 354–444. 3 plates.

On the occurrence of tails in man, with a description of the case reported by Dr. Watson. *Johns Hopkins Hospital Bull. 12:* 96–101; 2 plates.

1903

On the differentiation of muscular tissue when removed from the influence of the nervous system. *Proc. Amer. Assoc. Anatomists,* 1902. *Amer. Journ. Anat., 2:* iv–v.

Experimentelle Untersuchungen über die Entwicklung der Sinnesorgane der Seitenlinie bei den Amphibien. *Arch. f. mikr. Anat., 63:* 35–149; 3 plates.

1904

An experimental study of the relation of the nervous system to the developing musculature in the embryo of the frog. *Amer. J. Anat., 3:* 197–220.

Diskussion zu den Vorträgen von Schultze und v. Koelliker. *Verh. d. Anat. Gesellsch., 80th Versamml., Jena.* p. 52.

Neue Versuche und Beobachtungen über die Entwicklung der peripheren Nerven der Wirbeltiere. Sitz.-ber. d. *Niederrhein. Gesellsch. f. Natur. u. Heilkunde. Bonn. Jahrg.* 1904, pp. 1–7.

Neue Versuche und Beobachtungen über die Entwicklung der peripheren Nerven der Wirbeltiere. (Zusammenfassung.) *Bull. VI. Internat. Zool.-Kongr.,* Bern., p. 6.

1905

Karyokinetic division in the spinal ganglion cells of Triton larvae. *Proc. Amer. Assoc. Anatomists,* 1904. *Amer. J. Anat., 4:* XIII.

1906

Further experiments on the development of peripheral nerves. *Amer. J. Anat., 5:* 121–31.

The development of the nerve elements in vertebrates. (Abstract.) *Brit. Med. J., 2:* 1702.

1907

Experiments in transplanting limbs and their bearing on the problems of the development of nerves. (Abstract.) *Anat. Rec., 1:* 58–9.

Experiments in transplanting limbs and their bearing on the problems of the development of nerves. *J. Exp. Zool., 4:* 239–81.

Experiments in transplanting limbs and their bearing on the problems of the development of nerves. (Abstract.) *Science, 26:* 145–46, No. 657.

Experiments in transplanting limbs and their bearing on the problems of the development of nerves. (Abstract in German.) *Arch. f. Entw.-Mech., 24:* 673–74.

Observations on the living developing nerve fiber. *Anat. Rec., 1:* 116–18, and *Proc. Soc. Exp. Biol. and Med., 4:* 140–43.

The Bateson lectures. *Yale Alumni Weekly, 17:* 198–99.

1908

Regeneration of peripheral nerves. *Anat. Rec., 1:* 209.

Embryonic transplantation and development of the nervous system. *Anat. Rec., 2:* 385–410. (Harvey Lecture.)

1910

The development of peripheral nerve fibers in altered surroundings. *Arch. f. Entw.-Mech.*, *30:* 15–33.

The outgrowth of the nerve fiber as a mode of protoplasmic movement. *J. Exp. Zool.*, *9:* 787–846; 3 plates.

1911

The stereotropism of embryonic cells. *Science, 34:* 279–81.

1912

Experimental biology and medicine. *Physician and Surgeon, 34:* 49–65.

The cultivation of tissues in extraneous media as a method of morphogenetic study. *Anat. Rec.*, *6:* 181–93.

1913

Anatomy: its scope, methods and relations to other biological sciences. *Anat. Rec.*, *7:* 401–10.

The life of tissues outside the organism from the embryological standpoint. *Trans. Congr. Amer. Physicians and Surgeons, 9:* 63–76.

1914

The Osborn Memorial Laboratories. I. The Zoological Laboratory. *Yale Alumni Weekly, 23:* 667–72.

Science and practice. *Science, 40:* 571–81.

The reaction of embryonic cells to solid structures. *J. Exp. Zool. 17:* 521–44.

1915

Experiments on the development of the limbs in amphibia. *Proc. Nat. Acad. Sci., 1:* 539–44.

1916

On the reversal of laterality in the limbs of Amblystoma embryos. *Anat. Rec., 10:* 197–98.

1917

Further experiments on the laterality of transplanted limbs. *Anat. Rec., 11:* 1–2.

Transplantation of limbs. *Proc. Nat. Acad. Sci., 3:* 245–51.

1918

Experiments on the development of the fore limb of Amblystoma, a self-differentiating equipotential system. *J. Exp. Zool., 25:* 413–61.

1920

Experiments on the lens in Amblystoma. *Proc. Soc. Exp. Biol. and Med., 17:* 199–200.

1921

On relations of symmetry in transplanted limbs. *J. Exp. Zool., 32:* 1–136.
The development of the balancer in Amblystoma. *Anat. Rec., 21:* 66.
Experiments on the development of the gills in the amphibian embryo. *Biol. Bull., 41:* 156–70; 1 plate.

1924

Some unexpected results of the heteroplastic transplantation of limbs. *Proc. Nat. Acad. Sci., 10:* 69–74.
Experiments on the development of the internal ear. *Science, 59:* 448.
Neuroblast versus sheath cell in the development of peripheral nerves. *J. Comp. Neur., 37:* 123–205; 5 plates.

1925

The development of the balancer in Amblystoma, studied by the method of transplantation and in relation to the connective-tissue problem. *J. Exp. Zool., 41:* 349–427; 3 plates.
The effect of reversing the medio-lateral or transverse axis of the fore-limb bud in the salamander embryo (Amblystoma punctatum Linn.). *Arch. f. Entw.-Mech., 106:* 469–502.
Heteroplastic transplantations of the eye in Amblystoma. *Anat. Rec., 31:* 299.

1928

On the status and significance of tissue culture. *Arch. f. Zellforsch., 6:* 4–27, and X. *Congr. Internat. Zool.,* pp. 112–35.

1929

Heteroplastic transplantation in amphibian embryos. X. *Congr. Internat. Zool.,* pp. 642–50.
Correlation in the development and growth of the eye studied by means of heteroplastic transplantation. *Arch. f. Entw.-Mech., 120:* 1–55.

1930

With P. Pasquini. Esperimenti d'innesto sul cestello branchiale di "Clavelina lepadiformis" (Müller). *Rend. R. Accad. Naz. d. Lincei, 11:* 139–46, Ser. 6.

1931

Experiments on the development and growth of limbs in the amphibia. *Science, 74:* 575–76.

1933

Some difficulties of the determination problem. *Amer. Nat. 67:* 306–21.

1934

Presentation of the Daniel Giraud Elliot medal and honorarium for 1930 to Dr. George Ellett Coghill. *Science, 79:* 553.

1935

Heteroplastic grafting in embryology. *Harvey Lectures, 1933–34,* pp. 116–57.

On the origin and development of the nervous system studied by the methods of experimental embryology (The Croonian Lecture). *Proc. Roy. Soc. Lond., Ser. B., 118:* 155–96.

Factors concerned in the development of the ear in Amblystoma punctatum. *Anat. Rec., Suppl. No. 1:* 38.

1936

Relations of symmetry in the developing ear of Amblystoma punctatum. *Proc. Nat. Acad. Sci., 22:* 238–47.

Response on behalf of the medallist. *Science, 84:* 565–67.

Relations of symmetry in the developing ear. *The Collecting Net, 11:* 217–26.

1937

Embryology and its relations. *Science, 85:* 369–74.

1938

Further investigation of the factors concerned in the development of the ear. *Anat. Rec., 70, Suppl. No. 3:* 35.

Die Neuralleiste. *Anat. Anz., 85:* 3–30.

1939

Fundamental cancer research. Report of a committee appointed by the surgeon general. *Reprint No. 2008, Public Health Reports, 53:* 2121–30.

1940

With V. T. Astbury and K. M. Rudall. An attempt at an x-ray analysis of embryonic processes. *J. Exp. Zool., 85:* 339–63.

Cellular differentiation and internal environment. *Amer. Assoc. Adv. of Sci., No. 14:* 77–97.

Henry McElderry Knower. *Science, 92:* 419–21.

Sixth Pacific Science Congress. Opening address by the President (Ross G. Harrison), Chairman of the National Research Council. *Proc. Sixth Pacif. Sci. Congr., 1:* 50–3.

1944

National Research Council; and its action in field of medical sciences. *Proc. Mid-Year Meeting Amer. Pharm. Manuf. Assoc.,* Sixth Annual Scientific Award Ceremony, Dec. 11 and 12, pp. 43–50.

1945

Relations of symmetry in the developing embryo. *Trans. Conn. Acad. Arts and Sciences, 36:* 277–330.

Walter Bradford Cannon (1871–1945). *A Memorial Exercise held at the Harvard Medical School,* Nov. 5, pp. 29–37.

Facsimile of first announcement of the *Journal of Experimental Zoology.* Retrospect 1903–1945. *J. Exp. Zool., 100:* vii–xxxi.

1947

Studies in experimental embryology. *Carnegie Inst. Wash. Year Book, 46:* 171–72.

Wound healing and reconstitution of the central nervous system of the amphibian embryo after removal of parts of the neural plate. *J. Exp. Zool., 106:* 27–84.

Lorande Loss Woodruff, 1879–1947. *Science, 106:* 533–34.

Frank Rattray Lillie, 1870–1947. *Amer. Phil. Soc. Year Book,* pp. 264–70.

1948

Studies in experimental embryology. *Carnegie Inst. Wash. Year Book, 47:* 205–06.

With B. H. Willier, H. B. Bigelow, and E. G. Conklin. Addresses at the Lillie Memorial Meeting, Woods Hole, August 11, 1948. *Biol. Bull. 95:* 151–62.

Memorial Tributes

Abercrombie, A., 1961. "Ross Granville Harrison," *Biographical Memoirs of Fellows of The Royal Society, 7:* 110–26.

Autrum, H., 1960. "Ross Granville Harrison," *Bayerische Akademie der Wissenschaften, Jahrbuch,* pp. 165–69.

Dalcq, A. M., 1959. "Notice biographique sur M. le Professeur R. G. Harrison," *Bulletin de l' Académie Royale de Médecine de Belgique, VIe Serie, No. 10, 24:* 768–74.

Grassé P., 1960. "Notice nécrologique sur Ross Granville Harrison," *Comptes Rendus des Scéances de l' Académie des Sciences, Decembre, 251:* 2622–23.

Nicholas, J. S., 1961. "Ross Granville Harrison," *Year Book of the American Philosophical Society,* pp. 114–20.

——— 1961. "Ross Granville Harrison," *National Academy of Sciences, Biographical Memoirs, 35:* 130–62.

Pasquini, P., 1960. "Ross Granville Harrison," *Acta Embryologiae et Morphologiae Experimentalis, 3:* 119–30.

Wegner, R. N., 1961. "Ross Granville Harrison," *Anatomischer Anzeiger, 109:* 458–65.

Weiss, P., 1960. "Ross Granville Harrison," *Memorial minute for the Board of Trustees, Rockefeller Institute Quarterly,* p. 6.

Index

Numbers in italics denote pages with illustrations.

Silliman Volumes in Print

S. Chandrasekhar, *Ellipsoidal Figures of Equilibrium*
Theodosius Dobzhansky, *Mankind Evolving*
Ellen Drake, ed., *Evolution and Environment*
Rene Dubos, *Man Adapting*
Ragnar Granit, *Receptors and Sensory Perception*
John von Neumann, *The Computer and the Brain*
Sir Charles Sherrington, *The Integrative Action of the Nervous System*
Kenneth W. Spence, *Behavior Theory and Conditioning*